CONTENTS

OUTDOOR RECREATION MANAGEMENT

It is now widely recognized that recreation is as important as work. This book analyses leisure and outdoor recreation in terms of both their management and their wider importance to society.

Specifically, *Outdoor Recreation Management*:

- clarifies the links between leisure, recreation, tourism and resource management;
- reviews contemporary outdoor recreation management theories and concepts;
- critically examines approaches to outdoor recreation planning and management in diverse recreational settings;
- considers the future of outdoor recreation and the potential influences of economic, social, political and technological developments.

The wide-ranging analysis considers such issues as provision for people with special needs, the impact of outdoor recreation on the environment and outdoor recreation in both urban and rural contexts.

John J. Pigram is Director, Centre for Water Policy Research, University of New England, Australia.

John M. Jenkins is Senior Lecturer, Department of Leisure and Tourism Studies, University of Newcastle, Australia.

ROUTLEDGE ADVANCES IN TOURISM

Series Editors: Brian Goodall and Gregory Ashworth

OUTDOOR
RECREATION
MANAGEMENT

John J. Pigram and John M. Jenkins

London and New York

First published 1999 by Routledge
11 New Fetter Lane, London EC4P 4EE

Simultaneously published in the USA and Canada
by Routledge
29 West 35th Street, New York, NY 10001

Routledge is an imprint of the Taylor & Francis Group

Typeset in Great Britain by Hodgson Williams Associates,
Tunbridge Wells and Cambridge
Printed and bound in Great Britain by
Biddles Ltd, Guildford and King's Lynn

British Library Cataloguing-in-Publication Data
A catalogue record for this book is available from the British Library

Library of Congress Cataloging-in-Publication Data
Pigram, John J.
Outdoor recreation management / John J. Pigram, John M. Jenkins,
p.   cm.
Includes bibliographical references.
1. Outdoor recreation–Management. 2. Outdoor recreation–Social
aspects. 3. Leisure–Social aspects. I. Jenkins, John M., 1961–
II. Title.
GV191.66.P5 1999   98-48278
790´. 06´9–dc21   CIP

ISBN 0-415-15999-7 (hbk)
ISBN 0-415-16000-6 (pbk)

TABLES

FIGURES

ix

LIST OF FIGURES

PREFACE

Recreation is as important as work; perhaps more important for some. This book is the culmination of our lengthy interest in leisure and outdoor recreation, and their importance to society. While we have taken particular care to put forward a balanced treatment of outdoor recreation management concepts, issues and applications, there are several important considerations for the reader. First, the publication of leisure, recreation and tourism materials has increased markedly over the last two decades. In reviewing much of that literature, we were pragmatic: earlier works were included where they retained relevance, and extensive references to recent publications are provided. Second, the contents and arguments presented are clearly influenced by our geographical backgrounds and a commitment to integrating outdoor recreation management and resource conservation. Third, we are both particularly fascinated by the extent and nature of accessibility and conflict in outdoor recreation, and by the diversity of potential and actual management responses. Moreover, we are not convinced that societies have moved to a leisured existence, and nor are we easily swayed by economic arguments for privatisation of outdoor recreation resources. Nonetheless, we do see greater opportunities for public–private partnerships and local entrepreneurship. Unfortunately, however, many such opportunities have been thwarted by over-conservative planning ideologies and approaches; by inter-agency rivalry and dated organisational cultures; and by political squabbles.

Industrialised societies have entered an era dominated by policy and planning ideologies, which support the value of market forces in determining resource allocation and distribution. This is having a significant impact on outdoor recreation supply and management. Simultaneously, the non-voluntary acquisition of greater amounts of leisure time is a major social problem requiring government intervention. Unemployment, redundancy and early retirement mean an increasing proportion of the population has long periods of unobligated time. Coping without work and with an unstructured existence is a difficult process. Recreation can fill the void, if the conditioning, which makes employment so critical, can be overcome. In short, people from all walks of life should have access to rewarding recreational opportunities.

A number of people helped in the preparation of this book. Much appreciated assistance was given by Julie Hodges, Jan Hayden, Megan Wheeler, Micaela Saint, Michele Coleman, Shelagh Lummis, Kerry Beaumont and Linda Zakman. Rudi Boskovic did most of the cartography. We also would like to thank Casey Mein and Routledge for their patience and support.

1

LEISURE AND RECREATION IN A BUSY WORLD

> Outdoor recreation issues may be relatively neglected in our
> national political discourse, but they are not trivial and never will
> be on our shrunken planet.
>
> (Carroll 1990: xvii)

This book presents a comprehensive, non-specialised introduction to outdoor recreation management, as an area both of academic study and of real-world significance. Its underlying principle is the potential of recreation to contribute to pleasurable, satisfying use of leisure. Outdoor recreation is recognised as an important form of resource use, and much attention is given to how resources can be managed to provide a quality environment for sustained and satisfying recreational use.

Our decision to focus on outdoor recreation management, and to do so in an international context, was influenced by the wide-ranging and often fragmented research in the field, the restricted geographical focus of many texts, and the need to provide an overview of past and present understandings of outdoor recreation research and activity spanning mainly the developed world. Consequently, this book:

- clarifies the links between leisure, recreation, tourism and resource management;
- reviews contemporary outdoor recreation and resource management concepts and issues;
- critically examines approaches to outdoor recreation planning and management in diverse recreational settings; and
- considers the future of outdoor recreation and the potential influences of economic, social, political and technological developments.

Leisure, recreation and tourism are widely recognised as important elements in people's lives, and are receiving increasing academic attention and respectability (e.g. Mercer 1980a; Chubb and Chubb 1981; Patmore 1983; Van Lier and Taylor 1993; Lynch and Veal 1996). They are vital social issues (e.g. see Owen 1984) and rewarding forms of human experience, constituting 'a major aspect of economic development and government responsibility' (Kraus 1984: 3). Outdoor recreation brings joy and pleasure to many people, with the provision of appropriate recreational opportunities 'critical to the satisfaction of an individual's need for cognitive and aesthetic stimulation, one of six needs identified by Maslow (1954) as basic to human well-being' (Faulkner 1978, in Walmsley and Jenkins 1994: 89). Put simply, 'in the framework of our civilisation, tourism and recreation have moved from the relatively unimportant margins to a very salient position' (Mieczkowski 1990: 347).

This chapter places outdoor recreation in its broader societal context. It defines relevant terms, clarifies related concepts, and discusses the significance of leisure and outdoor recreation in industrialised nations. Approaches to the study of outdoor recreation and the focus of the present book are outlined.

Key definitions and concepts: leisure, recreation and tourism

Leisure means different things to different people, and thus there are many definitions or conceptualisations of leisure (e.g. see Pieper 1952; DeGrazia 1962; Parker 1971; Kaplan 1975; Godbey and Parker 1976; Patmore 1983; Lynch and Veal 1996). However, three main aspects are commonly noted. First, leisure equates with the enjoyment and satisfaction derived from free-time activities. Second, leisure represents a spiritual condition or state of mind, with the emphasis on self-expression and subjectively perceived freedom (Neulinger 1982). Third, leisure, in one or more of the above contexts, may be associated with activity.

Aristotle viewed leisure as the state of being *free* from the necessity to labour. Freedom is generally considered the key element of leisure. Thus, many definitions link the notion of leisure with free time – periods which are relatively free of economic, social or physical constraints. In these terms, leisure is a residual component – discretionary time over and beyond that needed for existence (Clawson and Knetsch 1966). There are several problems with this point of view in that it assumes the dominance of a work rather than leisure ethic, and it fails to give due recognition to the difficulty in distinguishing obligated time from free time. In particular:

- leisure can be experienced within the context of primary role obligations – leisure and work can become indistinguishable. Professional athletes, writers or, more generally, people who derive relaxation and revitalisation from their work, blur the divisions between work and leisure. Perhaps some professional athletes do not consider the financial or other tangible rewards from their pursuits as payment for work, but rather as rewards for being skilled and highly competitive at their chosen recreational activity;
- the pursuit of leisure can be influenced significantly by personal associations, values and choices. The leisure of parents or guardians may be constrained or eroded if they feel obliged to commit time to the amusement of their dependants; this 'obligated' time, however, may be one of their few means of escape from the work place. If the differentiating factors, then, are freedom of choice and freedom from necessity to fulfil occupational and family duties

2

and expectations or other obligations (Farina 1980), then the leisure-work dichotomy is tenuous; work can acquire some characteristics of leisure (Jamrozik 1986), and much leisure may take a form which is not the preferred choice of an individual;

• this point of view fails to recognise that leisure is a fundamental and essential component of people's lives – leisure time is needed for psychological and physical well-being, perhaps as much as work. One of the primary needs of people is leisure that affords psychological strength and refreshment (Perez de Cuellar 1987).

The concept of leisure clearly implies more than the antithesis of the necessity to labour or work. Lack of employment does not necessarily equate with leisure. Unemployed people do not always make a conscious choice between work and non-work/leisure. For them, an abundance of time free from work is often dictated by their ability to secure employment. Frustration and anti-social behaviour can occur because of the difficulty of occupying time out of work with meaningful, fulfilling or 're-creative' activities. Enforced 'idleness' or 'free time', as a result of unemployment, underemployment, disability, redundancy or early retirement, is a fact of life in many countries; so much so that the work ethic, which has typified Protestant society for generations, may no longer be relevant to many people. However, imposing a leisure ethic in its place can only be appropriate if this new-found leisure is free of guilt, discomfort and anxiety about survival (Bannon 1976). Such an outcome seems unlikely, and is certainly not a prominent aim in the public policies and programmes of modern industrialised societies.

The industrial system has long held out one rather striking promise to its participants. That is, the eventual opportunity for a great deal more leisure... The notion of a new era of greatly expanded leisure is, in fact, a conventional conversation piece. Nor will it serve much longer to convey an impression of social vision. The tendency of the industrial system is not in this direction... To argue for less work and more leisure, as a natural goal of industrial man, is to misread the character of the industrial system. There is no intrinsic reason why work must be more unpleasant than non-work... To urge more leisure is a feckless exercise so long as the industrial system has the capacity to persuade people that goods are more important. Men will value leisure over work only as they find the uses of leisure more interesting or rewarding than those of work, or as they win emancipation from the management of their wants, or both. Leisure is not wanted *per se* but only as these prerequisites are provided.
(Galbraith 1972: 357–9)

Leisure might take place in time free from work, but it is becoming increasingly commodified. Leisure requires that people have money to purchase 'time', recreational access and supporting resources. As Kando (1975: 15) remarked:

3

the coming of automation, cybernation and affluence would logically seem to produce the leisure society. However, this does not occur because of two developments. First, the society's value system is such that the new status hierarchy places an increasing premium on work; second, the society's economic structure – corporate capitalism – demands costly mass consumption and spectacular mass recreation rather than freedom in leisure.

The view that modern technology will create widespread leisure is easily challenged. If anything, modern technology has led to technocratic consumption and much regimented and institutionalised recreation, and costly mass spectacles (Kando 1975: 16) (e.g. the internationalisation of sporting culture such as the American Football 'Superbowl', the America's Cup, the Commonwealth and Olympic Games, Formula 1 Grand Prix, and the Rugby Union and Soccer World Cups via the media and the marketing of associated clothing and equipment). We have entered a broad phase of big-business 'spectatorism', often at home, 'oriented towards high profile sporting events, live theatre extravaganzas, concerts, festivals or the like' (Mercer 1994a: 20). In this context, recreation serves two functions. To paraphrase Kando (1975: 15), from the standpoint of the individual, it restores a person's energy to work. From society's standpoint, it fulfils a major functional prerequisite, namely that of sustaining the economic system. A similar viewpoint was expressed by Braverman (1975: 278–9):

> the atrophy of community and the sharp division from the natural environment leaves a void when it comes to the 'free' hours… the filling of the time away from the job also becomes dependent on the market, which develops to an enormous degree those passive amusements, entertainments, and spectacles that suit the restricted circumstances of the city and are offered as substitutes for life itself. Since they become the means of filling all the hours of 'free' time, they flow profusely from corporate institutions which have transformed every means of entertainment and 'sport' into a production process for the enlargement of capital. By their very profusion, they cannot help but tend to a standard of mediocrity and vulgarity which debases popular taste, a result which is further guaranteed by the fact that the mass market has a powerful lowest-common-denominator effect because of the search for maximum profit. So enterprising is capital that even where the effort is made by one or another section of the population to find a way to nature, sport, or art through personal activity and amateur or 'underground' innovation, these activities are rapidly incorporated into the market so far as is possible.

At the individual level, perceptions of leisure depend very much on a person's subjective, individual and social/political circumstances, and on their view of the world (e.g. see Parker 1983). The sharp distinction implied between discretionary

4

time and time needed for existence is blurred, as leisure is seen to overlap with other uses of time (see Figure 1.1). Despite these qualifications, it is probably true to say that for most people, leisure remains closely associated with uncommitted time. For the purpose of this book:

> Leisure consists of relatively self-determined activity–experience that falls into one's economically free time roles, that is seen as leisure by participants, that is psychologically pleasant in anticipation and recollection, that potentially covers the whole range of commitment and intensity, that contains characteristic norms and constraints, and that provides opportunities for recreation, personal growth and service to others.
>
> (Kaplan 1975: 26)

Confusion also arises over the indiscriminate use of the terms 'leisure' and 'recreation', which are closely related and often used interchangeably. The simplest distinction identifies leisure with time and recreation with activity. Recreation is activity voluntarily undertaken, primarily for pleasure and satisfaction, during leisure time, but it 'can also be seen as a social institution, socially organised for social purposes' (Cushman and Laidler 1990: 2). Whereas it is possible to conceive

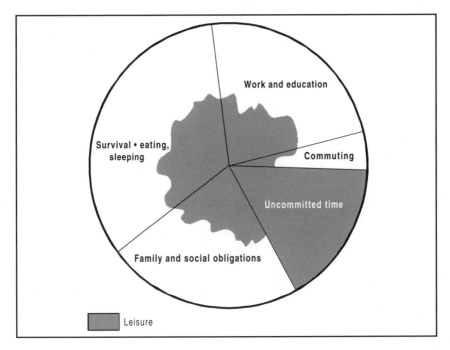

Figure 1.1 The diffusion of leisure time

Source: Department of Environment, Housing and Community Development (1977: 1)

some jobs as having a recreative element, the definition normally requires that no obligation, compulsion or economic incentive be attached to the activity. Recreation, therefore, contrasts with work, the mechanics of life and other activities to which people are normally highly committed. Certain activities are often thought of as inherently recreational. However, in a similar way to leisure, the distinguishing characteristic is not the activity or experience itself, but the attitude with which it is undertaken. To many professional golfers, for instance, perhaps golf is, or becomes, merely an occupation (though it could also be recreation); for the weekend golfer, presumably golf is looked upon as recreation and sport, even though at times it can involve much physical effort and frustration.

The concept of recreation, like that of leisure, is personal and subjective. Thus, value judgements as to the worth or 'moral soundness' of a particular activity often are inappropriate (Godbey 1981). Generally, recreation implies revitalisation of the individual, although purists would argue that *recreation* is, or should be, the culmination of recreational activity – 'the activity is the medium: it is not the message' (Gray and Pellegrino 1973: 6). If this argument were to be accepted, recreation could only be defined in terms of end-results, and potentially recreative activities which, for whatever reason, fail to 'revitalise' the participant, would be excluded. Rather than attempt to split ends from means, it would seem more useful to identify leisure as a process and recreation as a response. As Owen (1984: 157) puts it: 'Leisure has now come to be viewed as a process (Kaplan 1975) and recreation as an experience (Driver and Tocher 1974), which is goal oriented', with participation expected to yield satisfactions (London *et al.* 1977), and therefore physical and emotional rewards (also see Shivers 1967).

> In the specific case of outdoor recreation that element of reward may be stronger since participation will usually require the physical removal of the participant from the home or the workplace in order to engage in the activity in question. There is thus an additional cost in effort, time and/or money which must be part of the decision to participate.
>
> (Williams 1995: 6)

The term 'outdoor recreation' is more familiar in certain cultural contexts than others, but we do not agree with Mercer's (1994a: 4) view that, 'insisting on the distinction between "indoor" and "outdoor" recreation is as futile as emphasising the contrast between "urban" and "rural" leisure provision'. Regardless of indoor recreation developments, adaptations of such activities as cricket, soccer, tennis, athletics and rock-climbing, or of whether facilities have adjustable roofs, outdoor recreation is just what the category 'outdoor recreation' portrays – recreation that occurs outdoors in urban and rural environments. In this context, then, outdoor recreation raises significant resource management issues which indoor recreation activities do not.

Finally, discussion of tourism in the context of leisure and outdoor recreation is sensible. Tourism is one of the world's largest and fastest growing industries.

6

Much tourism is recreational, in that a good proportion of tourist activity takes place during leisure time, often outdoors, for the purpose of personal pleasure and satisfaction. Outdoor recreation overlaps with tourism in the distinctive characteristics and behaviour associated with each; tourism and outdoor recreation activity involve both travel and interaction with other people, and with the environment, in its widest meaning (see Chapters 5, 11 and 12). Some observers assign an emphasis on economic aspects and profit-making to tourism, while linking outdoor recreation primarily with noncommercial objectives (e.g. Gunn 1979). Unfortunately, others make a fundamental distinction between tourism and recreational travel (Britton 1979; Boniface and Cooper 1987). However, these distinctions create and foster an artificial gulf between tourism and outdoor recreation in applied and theoretical terms, leading to unnecessary obstacles to, among other things, understanding people's recreational motivations, choices, behaviour and experiences. In this book, tourism is considered within an essentially recreational framework, and 'may be thought of as the relationships and phenomena arising out of the journeys and temporary stays of people travelling primarily for leisure or recreational purposes' (Pearce 1987: 1).

Tourism also receives special consideration in this book (see Chapters 11 and 12). Attractions, facilities and services, developed for tourists in industrialised societies, are often utilised by local residents, and therefore will, in almost any case, impact upon local resident perceptions of, and attitudes to, a range of recreational facilities and services and, indeed, tourist activities. A more detailed discussion of tourism-related issues is provided in Chapters 11 and 12.

The significance of leisure and outdoor recreation

Three great waves have broken across the face of Britain since 1800. First, the sudden growth of dark industrial towns. Second, the thrusting movement along far flung railways. Third, the sprawl of car-based suburbs. Now we see, under the guise of a modest word, the surge of a fourth wave which could be more powerful than all the others. The modest word is leisure.

(Dower 1965)

Leisure

Leisure was once the privilege of the élite. Recently, it has been argued that leisure has become largely the prerogative of the masses. People's historical preoccupation with work as a means of livelihood appears to have been tempered by priorities geared, in part, towards the acquisition of more leisure. Developed countries are faced with the problems of adjusting to, and providing for, a society orientated perhaps as much towards leisure and recreation as it is towards work.

The dimensions of the leisure problem were discussed by Dower (1965), who described the leisure phenomenon as a 'fourth wave', comparable with the advent of industrialisation, the railway age, and urban sprawl.

7

The leisure phenomenon can be measured not only in terms of time availability but also in terms of activities engaged in, that is, how people spend their leisure time. It can also be measured in terms of consumer expenditure, that is, the extent to which people spend their money on leisure goods and services. The trends are unmistakable. Growth in participation in virtually all leisure activities since the Second World War has been dramatic... leisure has become a highly significant element of people's lives and of the economies of advanced industrial nations.

(Veal 1987: 2–3)

The growth of leisure, however, does not only bring benefits to individuals and society. The disadvantages of a leisured existence were foreseen (perhaps somewhat cynically) by George Bernard Shaw, who is reputed to have described a perpetual holiday as 'a good working definition of hell' (Gray and Pelegrino 1973: 3). This might well be applicable to unemployed people and retirees, to those living in remote rural areas, or to homeless people in inner city areas. Shaw's assertion reflects the apparent psychological inability of people to cope with the monotony and burden of a non-structured existence. Many people attempt to occupy time, which might have been utilised for leisure, with additional employment. This situation gives some substance to the notion that, for whatever reason, a life of leisure may not be the course of gratification it should be, or is not as accessible as many anticipated.

Work satisfaction does fulfil many human needs. However, for the majority, a reduction in work commitments must seem a highly desirable goal. At the same time, it is being realised that the fundamental consideration is not the overall amount of increased leisure gained, so much as the spatial and sectoral distributions of disposable time. Of practical importance in determining the recreational response, is whether this time is concentrated or dispersed. In Australia, for example, progressive reductions in contractual working hours have been introduced by way of a nineteen-day month, a nine-day fortnight or flexible work arrangements, in part, to improve recreational opportunities. Although the compression of leisure into standard packages probably suits the convenience of both employers and employees, concentrated periods of use place great pressures on the recreation environment, and, in particular, fragile areas (e.g. alpine and coastal areas). Therefore, it would seem desirable, in the interests of recreation resource management, to devise a system of more flexible work patterns incorporating extended, but staggered blocks of disposable time; a desirable goal to be sure, but complicated by social patterns and processes (religious beliefs; designated public holidays often incorporating long weekends; Christmas and New Year festivities; Easter; school vacations).

Despite growth in participation in outdoor recreation activities, there remain inequities. Many writers have noted that, notwithstanding individual differences, the extent and nature of leisure and recreational activities people engage in are related to their position in the socio-economic stratification of society and to the class structure (Jamrozik 1986: 189).

LEISURE AND RECREATION IN A BUSY WORLD

In Britain and in most West European countries, leisure participation generally, and sports participation particularly, is dominated by men, young people, white people, car owners and those in white-collar occupations. The participation rates of women, older people, ethnic minorities, and those in blue-collar occupations are generally lower.

(Glyptis 1993: 6)

Jamrozik (1986: 204) made similar, though more general comments:

The 'promise' of increased prosperity, and a life of leisure through technological innovation has not, so far, eventuated as expected. Material prosperity has become a reality but it is not shared, either within countries or across countries. On the contrary, inequalities in income and wealth are increasing and now extend to access to employment and consequently to consumption, including leisure consumption.

Not all sections of society in the developed or less developed worlds enjoy adequate access to leisure. A broad group on whom increased leisure appears to have had considerably less impact is the female component of the population. Women experience unequal access to, and participation in, leisure as an inevitable consequence of, among other things, 'sexist' policies in society.

The problem appears to relate, at least in part, to an unequal incidence of leisure time. While the recent picture is somewhat mixed, it has been a widely held view that men have more leisure time than women, especially women in the workforce (Cushman *et al.* 1996a). The burdens of domestic and child-care responsibilities fall inequitably on women, who make up an increasing percentage of the workforce in Western countries. These factors, together with economic and cultural constraints, might explain why women tend to be more active in home-based recreation activities than men. Even weekends 'become more a matter of overtime work for married women and more a matter of recreation for men' (Rapoport and Rapoport 1975: 13). Furthermore, men are more active in sports, while women tend to be more active in arts and cultural activities (Cushman *et al.* 1996a).

For the workforce in general, the same technological progress and social advances, which have permitted reductions in working hours, have imposed pressures on the way leisure is used, and on the extent and nature of recreational participation. LaPage (1970) suggested that non-work discretionary time, ostensibly available for recreation, is constantly eroded by the time necessarily spent in commuting to work and in travel to and from sites for social purposes. More than 30 years ago, Wilensky (1961: 136) deplored the fact that leisure was spent '... commuting and waiting – hanging on the phone, standing in line, cruising for parking space'; very little has changed. Urban sprawl and the concentration of much work in central business districts have, despite innovations and extensive public and private sector investment in transport technologies, resulted in lengthier commuting times for many workers in the western industrialised world. The development of communication technologies, such as the internet and the mobile

9

phone, means it is more difficult for people to remove themselves from the work place, which for many was once confined to an office and to particular hours. The picture has been further complicated by the effect of evolving social mores and changing lifestyles in urban and rural areas. Couples, where both partners are working, for instance, require more ancillary time in the home for necessary chores and maintenance, so that hours set free from work are taken up with domestic tasks.

Moreover, in a materialistic and sophisticated society, leisure without affluence often seems of little relevance. Preoccupation with material possessions can divert values away from the acceptance and simple enjoyment of leisure; thus the benefits of improved working conditions are often translated into money terms. Economic circumstances or personal inclination force a trade-off between more free time and increased disposable income, and can lead to the filling up of leisure hours with overtime or a second, or even third, job. This, in turn, curtails the opportunity for recreative use of leisure.

> One of the paradoxes of leisure is that while time and money are *comple-mentary* in the production of leisure activities, they are *competitive* in terms of the resources available to the individual. Some leisure time and some money to buy leisure goods and services are *both* needed before most leisure activities can be pursued.
>
> (Martin and Mason 1976: 62)

Bearing in mind the continuing emphasis on material possessions in modern industrialised nations, there is little reason to expect a reversal of this trend towards acquisition of leisure durables at the expense of leisure time. However, the pursuit of affluence is self-limiting to the extent that, ultimately, time is needed to make use of the possessions acquired, and that, beyond a certain level, marginal tax rates usually ensure that additional income becomes an 'inferior good' compared with disposable time.

Sociocultural factors, too, can have a bearing on the appreciation of leisure and its use for recreation. Contrasting attitudes and value systems mean that some individuals and societies continue to equate leisure with frivolity and wasted time. For others, a hedonistic orientation, which clearly ranks free time more highly than work, appears to welcome the emergence of a leisured society without any sense of guilt. Pearson (1977) relates this tendency to institutional arrangements biased towards greater amounts of disposable time, and to an environment with significant leisure potential. Certainly, the efforts of labour unions, and enlightened social reforms and legislators have contributed to a greater leisured existence for a wider cross-section of the population. Yet, even in Australia, there is evidence of the persistence of a puritan work ethic, especially on the part of those new settlers from a different cultural background, who have migrated to that country since World War II.

Such anti-leisure sentiment, displayed by apparently compulsive workers, is, of course, not restricted to Australia. Indeed, it could arise in any situation as a function of deficiencies in the leisure environment, rather than from a conviction of the necessity and desirability of work. These same deficiencies can inhibit recreative use of leisure time. Individuals slumped in front of the television set may be there, in part, because of their socioeconomic circumstances and inclination, and/or because of their lack of a more constructive outlet for leisure. The physical and mental demands of work in an automated society emphasise the importance of, and need for, challenging and satisfying leisure pursuits. In addition, the existence and apparent acceptance of a persistent core of permanently unemployed, for whom the provision of satisfying recreational outlets is an urgent task, must be acknowledged. Yet, the leisure environment very often cannot provide the opportunities needed for more positive use of disposable time, whether voluntarily acquired or enforced. Identification and remedy of such deficiencies are necessary for a fuller realisation of what leisure has to offer people and society generally.

The arrival of television coincided with a change to greater in-home leisure generally. Television is only one of several leisure goods that made the home a centre of leisure activities in the second half of the twentieth century. Hi-fi equipment and recorded music, videos, computers and computer games, the internet and other factors, singularly or in combination, have provided a variety of new products and activities in the home; products and activities which are being produced and distributed on a worldwide scale (Cushman *et al.* 1996a). Simultaneously, housing conditions (e.g. household design and technologies – insulation; air conditioning) have improved considerably so that there has been less incentive to get away from the house for entertainment and enjoyment.

Although the home has become much more important in leisure, other developments in technology (e.g. air transport; snow mobiles and off-road vehicles; trail bikes; walking boots) have either widened the scope of outdoor recreation activities or at least made it easier, more comfortable and speedier for people to venture further afield than ever before in the search for leisure experiences. There have also been important changes in the structure of national and regional economies. The leisure and tourism sectors have increased in significance as areas of personal expenditure and employment, and as aspects of public policy (e.g. see Carroll 1990; Henry 1993; Veal 1994). Clearly, the position of leisure in society is cemented in social, psychological, political and economic factors.

Outdoor recreation

The importance of outdoor recreation has been highlighted by Devlin (1992: 5, in Mercer 1994a: 4), who argues:

> People's recreational use of leisure time will almost inevitably at some stage include outdoor recreation. This is currently true for 90 per cent of those who live in Western countries, and, for many of these participants, it is a form, this form of recreation, which represents a very important part of their lives.

The 'leisure explosion' in the developed world has been paralleled by a striking upsurge in all levels of recreation activity. Institutional, technological and socioeconomic factors have been influential in this upsurge.

Much leisure activity, as noted above, is of course home-centred, perhaps also home-technology-centred (e.g. television, computers, videos), a feature which is being reinforced and cultivated by capitalist society. Nevertheless, participation in outdoor recreation in Australia, Canada, New Zealand, the United Kingdom (UK), the United States (US) and other industrialised nations has grown rapidly since World War II, and particularly since the 1960s, while participation in organised sports has seen some dramatic changes (see Cushman *et al.* 1996b).

Despite its unquestionable scale and significance in social and economic terms, sport remains a minority participatory activity – many more people actually watch sport than participate in it. The leisure activities with mass appeal are still those that are more informal, social and passive. Surveys for the Countryside Commission (1991 in Glyptis 1993) show that in 1990, 76 per cent of the population had visited the English countryside for purposes of recreation, generating over 1,600 million trips and 12,400 million pounds of expenditure. Countryside visiting attracts not only a large number of people, but also a high frequency of participation. In 1990, as much as 19 per cent of the UK population had visited the countryside within the past week, and nearly half had done so within the past month. At the other extreme, 2 per cent had never been to the countryside (Glyptis 1993: 5–6).

According to Veal (1994: 158–9), there will be increasing demands placed on the Australian outdoors (see Table 1.1). In examining demographic changes and their relationship to outdoor recreation participation, Veal argued that 'the facilities and activities which are more closely associated with older age groups (nature watching/sketching, walking for pleasure, golf and visiting national or state parks) show a higher than average growth rate'. As he pointed out, though, 'the projections do not take account of changes in incomes, occupation, education, mobility, tastes or the many other factors which could affect demand'. Moreover, 'In the early 1960s expenditure on recreation and entertainment accounted for around 2.5 per cent of total private consumer spending. This has subsequently doubled to 5 per cent' (Mercer 1994a: 2).

Common trends in modern Western societies can be noted. Land and water-based or related activities that have witnessed growth, include:

- golf;
- bicycle riding;
- walking/day hiking and backpacking;
- photography;
- nature study;
- horse back riding;
- orienteering, mountaineering, rock climbing and caving;
- off-road (four wheel) driving;
- rafting, wind surfing, water skiing, tubing and jet skiing;
- snow skiing/snow boarding, and cross-country skiing.

Table 1.1 Demand projections 1993–2001

	1993 estimate	2001 projection	Change	% Change
ACTIVITIES				
A. Picnic/barbecues	1,885,477	2,108,786	223,309	11.8
B. Drive for pleasure	2,607,860	2,928,039	320,179	12.3
C. Visit parks	1,418,591	1,590,336	171,746	12.1
D. Nature watch/draw	365,804	423,193	57,389	15.7
E. Walk for pleasure	3,672,938	4,159,726	486,788	13.3
F. Bushwalking/hiking	313,396	345,890	32,494	10.4
G. Fishing	464,799	513,159	48,361	10.4
H. Golf	568,541	648,145	79,604	14.0
I. Horse-riding	83,097	85,316	2,219	2.7
J. Shooting/hunting	42,740	45,956	3,216	7.5
K. Surfing/lifesaving	351,952	388,025	36,073	10.2
L. Swim/dive/water polo	2,138,946	2,348,234	209,288	9.8
M. Non-power water sport	182,152	193,647	11,495	6.3
N. Powered water sport	120,784	128,986	8,202	6.8
Total Activities	**14,217,078**	**15,907,440**	**1,690,362**	**11.9**
FACILITIES				
O. National/state parks	849,111	958,969	109,858	12.9
P. Park/playground	1,852,163	2,058,075	205,913	11.1
Q. Walking Trail	750,416	833,380	82,964	11.1
R. Beach	2,489,961	2,747,657	257,696	10.3
S. River/lake	872,230	974,559	102,329	11.7
T. Camping ground	263,048	278,640	15,593	5.9
Total Facilities	**7,076,927**	**7,851,279**	**774,352**	**10.9**

Source: Veal (1994: 157)

A number of interrelated events and social and political developments, arising from global, regional and local forces, have led to growth and increased diversity in outdoor recreation participation and tourist travel, and to the establishment of public and private (including voluntary) recreation organisations and programmes. The extent and nature of recreational participation and personal travel have been affected by many factors, including:

- population growth (including immigration in many countries/regions);
- changes in population characteristics – longer life spans and ageing populations;
- shorter working weeks. The regular working week has been reduced from an estimated seventy-hour, six-day week in the mid-nineteenth century, to around forty hours or less, spread over as little as four days, although overtime and second-jobs are common, and more households are dual-income;
- increased affluence and higher disposable incomes (although arguably becoming more concentrated in some countries), affected to some extent by growth in the number and proportion of dual-income households in several countries;

13

- increased holiday entitlements. The right to generous periods of paid annual leave has been established, with the addition, at least in countries such as Australia, of an additional holiday pay loading to enable workers to take better advantage of their vacations. Not only have work periods been reduced, but various peripheral activities such as travel time and lunch breaks may be incorporated into the paid working day, so that non-obligated time is increased;
- increased mobility (by way of the development and wider use of private motor vehicles, and the greater availability, speed and comfort of other forms of transportation, particularly long haul);
- urbanisation and suburbanisation;
- the influence of commercial interests (public relations and marketing) and technological developments in recreational equipment and infrastructure;
- the promotion of high-risk recreational activities;
- greater educational attainment;
- increasing attention to health and fitness programmes;
- growth in environmental and cultural awareness and interests;
- the age of retirement has receded to the point where 60 is the accepted norm and even earlier retirement is commonplace;
- a growing focus on human services and increased recognition of the needs of special groups and new roles for girls and women; and
- tourism development (Kraus 1984; Murdock, *et al.* 1991; Parker and Paddick 1990; Lynch and Veal 1996; Cushman *et al.* 1996b).

Participation in recreation activity is influenced by, among other things, socio-economic factors (see Chapter 3). Income and education, which are often reflected in occupation and correlate highly with car ownership, probably have the greatest impact on recreation. 'Men of substantial mental accomplishment have not usually lacked interesting ways of employing their time apart from toil. And it seems likely that they will be somewhat less susceptible to the management of demand' (Galbraith 1972: 359).

Demographic variables such as age, sex, family structure, immigration and concomitant cultural assimilation and diversity, are also important in explaining recreation patterns. Participation in recreational pursuits tends to decline progressively with age, although television watching, golf and bowls have higher participation rates among the older age groups than the young (see Cushman *et al.* 1996a). In short, the types of leisure pursuits and recreational activities undertaken, change throughout a person's life cycle (see Chapter 2). An important demographic aspect is the general ageing of Western societies, so that provision must be made for a less active, but growing segment of the population, with considerable leisure time.

Institutional, technological and socioeconomic forces operating at local to global levels, in combination and separately, have clearly influenced recreation patterns in the developed world. Growth in outdoor recreation and tourism, and the resulting escalating pressures on resources have necessitated both closer examination of planning and management of the recreational and tourist resource bases of countries and regions, and innovations in policy and planning approaches. Furthermore, recreation and tourism are becoming increasingly important elements in the relationship between the economic, environmental and social dimensions of countries, regions, cities and towns (e.g. see Mercer 1970; Cloke and Park 1985).

Nevertheless, much outdoor recreation research is generally disjointed (e.g. longitudinal studies are lacking), and is relatively scant in such countries as Australia and New Zealand, as compared to North America and the UK. Indeed, we know very little about the spatial and sectoral allocation and distribution of the benefits and costs of outdoor recreation. 'Research reported by Hendry (1993) in New Zealand and Hamilton-Smith (1990) in Australia, suggests that the most frequent users of local government recreation services also tend to be the most well-off in the community. Access and use by low income groups, ethnic minorities, Aborigines, the aged, persons with disabilities and women are more restricted' (McIntyre 1993: 33). For many in these categories, lack of status, money, mobility, ability and agility, access or awareness, can all inhibit the purposeful use of leisure and, therefore, knowledge of, access to, and participation in recreational activities. In short, the use of leisure, and the nature and extent of participation in outdoor recreational activities, vary spatially and temporally, and fluctuate, sometimes unpredictably, with changes in taste and fashion, and with other developments on the local, regional, national and global scenes. Clearly, an understanding of outdoor recreation patterns and processes requires an appreciation of such factors as:

- people's motivations, choices, participation and recreational satisfaction; and
- planning and policy-making.

In most circumstances, it might be assumed that the availability of more hours free from work would be regarded as a significant social advance. Yet, for large sections of society the acquisition of greater amounts of time for leisure, and therefore for recreation, is problematic, and is consequently emerging as a major social problem.

Leisure and recreational opportunity first became recognised as a cause for concern during the Great Depression of the 1930s. The concern continues, but has expanded. Conferences on the subjects of leisure and recreation have proliferated around the world. These gatherings have been organised and sponsored by a diverse set of organisations, ranging from academic bodies to professional administrators and marketing groups. Such meetings and conferences (e.g. world and national leisure and recreation congresses) are now commonplace, while associations facilitating research activity and dissemination (e.g. the World Leisure Research Association – WLRA; the Australian and New Zealand Association for Leisure Studies – ANZALS) have been established. Journals devoted to leisure, recreation and tourism issues have increased in number. Disciplines such as geography and sociology frequently include conference themes relating to leisure, recreation and/ or tourism. Courses in leisure, recreation and, in particular, tourism are widespread and expanding into the Asia-Pacific area and other regions. This book is, in part, an outcome of the need to synthesise an ever-increasing flow of ideas, approaches, conceptual insights and applied research concerning outdoor recreation management.

Approach and structure of the present book

For resource managers, in particular, the focus of interest on outdoor recreation is largely on active, informal types of recreation (i.e. those activities engaged in beyond the confines of a building, sporting arena or home). This is not meant to denigrate the use of free time for individual indoor pursuits such as reading and hobbies, or for formal, structured and institutionalised activities such as organised sports. It simply recognises that the really important resource issues arise with the allocation and use of extensive areas of land and water for outdoor recreation. This is where space consumption and spatial competition and conflict are most likely to occur; 'it is in this context that spatial organisation and spatial concerns become paramount' (Patmore 1983: 225). By considering outdoor recreation as a process in spatial organisation and interaction, the resource manager can focus on those aspects with spatial implications (e.g. imbalance or discordance between population-related demand and environmentally-related supply of recreation opportunities and facilities) (Wolfe 1964; Toyne 1974). Obviously, too, this is where the opportunities and the need for recreation resource management are greatest.

Outdoor recreation can be studied in many ways. It can encompass different disciplinary frameworks (e.g. economics, sociology, political economy, geography and law), and thus can incorporate a combination of theoretical and applied research approaches. This book does not claim to present any 'ideal' approach, but it does seek to fill a gap by bringing together many disparate and complementary ideas and studies concerning outdoor recreation.

Outdoor recreation is not the prerogative of all, as we may have been led to believe. People's accessibility to outdoor recreation opportunities is constrained by barriers linked with age, gender, class, income, race, a lack of facilities and opportunities (see Chapter 3), and inappropriate policy-making, planning and management. Outdoor recreation puts pressure on the physical environment, is an increasingly significant factor in the economic concerns of households, communities and regions, and is receiving higher priority in political arenas. Outdoor recreation presents great challenges to planners. In some respects, those challenges represent wicked tasks; tasks which have no definitive right or wrong answer, so that any planning, management or political response is open to challenge from various (sometimes unexpected) interests.

Several books and other sources of information on outdoor recreation and related resource/environmental management issues have been produced. Some of these are now somewhat dated in terms of their concepts, theories and case studies, and therefore in terms of their applied usefulness, while others have stood the test of time. These contributions are complemented by a growing number of leisure, recreation and tourism journals (e.g. *Annals of Tourism Research, Annals of Leisure Research, Australian Journal of Leisure Management, Journal of Leisure Research, Journal of Park and Recreation Administration, Journal of Tourism Studies, Journal of Travel Research, Leisure Sciences, Parks and Recreation*, and *Recreation Research Review*), conference and workshop proceedings, and the publications of such innovative government agencies as the USDA Forest Service.

This book has a wide catchment area, so to speak, and the authors were selective in their chapter foci and sources. Material for the book is drawn mainly from North America, Australia, Britain, Europe and Southeast Asia. Review questions and guides to further reading are included at the end of each chapter so that readers can explore issues in greater depth than discussed in the text. An extensive reference list is provided.

Guide to further reading

- *Definitions of leisure, recreation and tourism; the relationships between leisure and work, and between leisure, recreation and tourism*: Kaplan (1975); Patmore (1983); Murphy (1985); Fedler (1987); Mieckzowski (1990); Leiper (1995); Lynch and Veal (1996); Butler, *et al.* (1998).
- *Leisure and recreation participation trends and issues*: Jackson and Burton (1989); Murdock *et al.* (1991); Veal (1994); Cushman *et al.* (1996b). Students should consult the statistics compiled by national, state/provincial, regional or local governments, and other agencies.

Review questions

1 Compile, and then compare and contrast, a number of definitions of leisure, recreation and tourism. Can you identify (1) major flaws in such definitions, and (2) important temporal shifts in such definitions?
2 Discuss the significance of leisure and outdoor recreation in modern society.
3 To what extent should the study of outdoor recreation be viewed as not having defined disciplinary boundaries?
4 Why is it important to examine outdoor recreation in the context of resource management?

2

OUTDOOR RECREATION: MOTIVATION AND CHOICE

As with other aspects of human decision-making, explanation of leisure behaviour is complex. The unfettered personal connotations of leisure and the discretionary nature of recreation were noted in the opening chapter. An underlying dimension common to both leisure and recreation is discretion – freedom to choose and the exercise of choice. This discretionary element helps explain why observers find difficulty in explaining why people choose particular leisure settings and activities, and in accounting for recreation choice behaviour. It might be argued that the choice process is no more complex than that involved in, say, the selection of a new residence. After all, choice is subject to a range of influences and is not a completely random process. Nor is it unique to any individual. However, the unbounded nature of leisure and the subjective, even capricious characteristics of recreation decisions, make generalisation and prediction more challenging.

Motivation

Why do people choose to use their free time for recreation?
What motivates sky-divers or abseilers to take part in high-risk recreational activities?
Why do some city office workers devote much of their lunch breaks to intense physical pursuits?
How is it that certain individuals find great satisfaction in the isolation of wilderness recreation, while for others, leisure behaviour is associated with a stimulating social environment and the isolation of wilderness is abhorred?

The process by which a person is moved to engage in particular forms of behaviour has been the subject of speculation and research over a long period. Understanding leisure behaviour is no less complex. According to Iso-Ahola (1980), human actions are motivated by subjective, defined goals and rewards which can be either extrinsic or intrinsic. When an activity is engaged in to obtain a reward, it is said to be extrinsically motivated. When an activity is engaged in for its own sake, rather than as a means to an end, it is said to be intrinsically rewarding. Although the distinction is blurred and open to subjectivity, Iso-Ahola believes that leisure behaviour is chiefly motivated by intrinsic factors related to self-expression, competence and satisfaction, which, in turn, implies freedom of choice.

Recreation choice, however, should not be seen as unrestricted. Whereas individual motivations instil a propensity towards certain recreation activities, actual participation largely reflects the selection of the best alternative or compromise under the circumstances. Choice is bounded by any number of constraints, including physical capability, affordability, awareness, time restrictions and family obligations. The existence and intensity of these constraints vary among individuals and across socioeconomic, demographic and other groups.

A further source of complexity in the explanation of leisure behaviour arises from confusion over the nature of recreation demand and its relationship with recreation participation. In particular, there is an apparent inability to distinguish between the concept of demand in the broad, generic sense and its use to refer to existing levels of recreation activity. The latter, as indicated by numbers of participants or tourist visitation rates, is not a true measure of demand because it relates to observed or actual participation and behaviour, which is only a component of overall aggregate demand.

Recreation demand

The term 'recreation demand' is generally equated with an individual's preferences or desires, whether or not the individual has the economic and other resources necessary for their satisfaction (Driver and Brown 1978). Recreation demand, so defined, is at the preference–aspiration–desire level, before it is expressed in overt, observable behaviour or participation. In this sense, it is a propensity concept, reflecting potential or behavioural tendencies, and is detached from subsequent recreation activity. As one authority put it: 'Recreation demand is a conditional statement of the participation that would result ... under a specific set of conditions and assumptions about an individual... and the availability of recreation resources...' (US Bureau of Outdoor Recreation 1975: 10–22).

This broad notion of demand is supply-independent. It assumes no constraints on recreation opportunities or access to them. In these terms, recreation demand depends only on the specific characteristics of the population (e.g. age, income, family structure, occupation and psychological parameters), and not on the relative location of user groups, or the quality and capacity of facilities, or the ease of access.

However, actual consumption or participation in recreation activities is very much a function of the supply of those opportunities. Observed levels of leisure behaviour may conceal frustrated demand, which can only be satisfied by the creation of new recreation opportunities or by increasing the capacity of existing facilities (e.g. by management strategies encompassing land or water acquisition, hardening the landscape, and interpretation). If opportunities are less than ideal, people will actually participate less in recreation than their theoretical level of demand would indicate.

In the real world, recreation demand rarely equals participation. The difference between *aggregate demand* and *actual participation* (or 'expressed', 'effective', 'observed', 'revealed' demand) is referred to as *latent demand* or *latent participation* – the unsatisfied component of demand that would be converted to participation if conditions of supply of recreation opportunities were brought to ideal levels (Figure 2.1).

The confusion between recreation demand and participation, and the implications of such misinterpretation, have been noted by a number of authors (e.g. Knetsch 1972; 1974; Elson 1978; Lipscombe 1986). It is not enough simply to look at what people do and interpret this as reflecting what they want to do; it also reflects what they are able to do. Participation data are important, but they must be interpreted in terms of both supply and demand variables. Knetsch (1972) points out, for example, that if the participation rate in swimming in a given area is found to be very large, relative to that in some other area, it may be almost entirely due to greater availability of swimming opportunities. Adoption of attendance figures as a measure of demand confuses manifest behaviour with recreation propensities and preferences.

Nor is the problem merely one of semantics; the planning implications are clear for adjusting the supply of recreation opportunities and estimating the probable effect of alternative policies and programmes. It is important for planners to have answers to questions concerning for whom, how much, what type and where, in regard to the introduction of new recreation facilities. Equating demand with

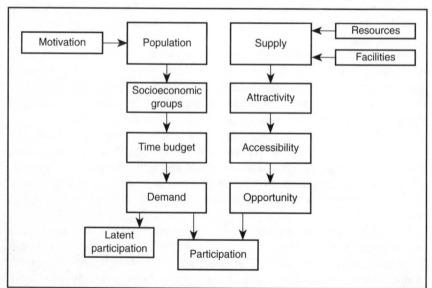

Figure 2.1 Recreation demand and participation

Source: Adapted from Kates, Peat, Marwick and Co. (1970: 1.1)

existing consumption or participation rates, can lead to the assumption that people will want only increasing quantities of what they now have, thereby '... perpetuating the kind of facilities already existing in the areas already best served and further impoverishing already disadvantaged groups' (Knetsch 1974: 20).

Another problem in relying upon past (observed) participation to guide future decisions is that observed activity patterns reveal little regarding satisfaction or the quality of the recreation experience. As Stankey (1977: 156) pointed out:

> ... when opportunities are available, particularly at little or no cost, they will be used. But use should not lead us to automatically assume people are satisfied with existing opportunities or that alternative opportunities might not have been even more sought after.

A deeper understanding of the true nature of recreation demand would throw light on the reasons for non-participation or under-participation in specific areas and activities, and reduce mis-allocation of resources (e.g. see Vining and Fishwick 1991). It should ensure also that any induced demand as a result of additional recreation investment is directed towards remedying these deficiencies. The supply of appropriate opportunities can release latent participation and translate it into effective demand, and can also be used to manipulate and redirect demand from one area or activity to another. Mercer (1980a) gives several examples of induced, substitute or diverted demand as the result of creating new resources and of improvements in access and technology. It should be noted, of course, that heightened levels of participation can just as readily be achieved by improvements in awareness, and by education, training and similar triggers (see below).

Awareness of the factors generating recreation demand and the relationships between its various components are important in recreation planning and resource management. That said, it is obvious that most attention in the social sciences has been devoted to recreation behaviour *per se* (i.e. to actual participation or effective demand). It is in the spatial and temporal expression of demand and the use made of specific sites and facilities where many resource problems exist. Whereas these patterns of use are derived in part from underlying preferences, they reflect also the availability, quality and effective location of recreation opportunities. Explanation of revealed recreation behaviour, therefore, must be sought in terms of the interaction between recreationists and the resource base, and in terms of the processes by which outdoor recreation sites are chosen. With respect to the latter,

> It is gradually becoming clear that human decision making cannot be understood by simply studying final decisions. The perceptual, emotional and cognitive processes which ultimately lead to the choice of a decision alternative must also be studied if we want to gain an adequate understanding of human decision making.
>
> (Svenson 1979, in Vining and Fishwick 1991: 114)

Recreation participation

A simplified representation of the factors which influence the decision to participate in recreation is set out in Figure 2.2. Once again, a broad distinction can be made between the potential demand or propensity for recreation and the supply of opportunities to realise these preferences or desires. The variables can be grouped into the demographic, socioeconomic and situational characteristics which generate a propensity to recreate, and those external factors which facilitate or constrain the decision and the choice of activity and site.

Demographic characteristics

The size, distribution and structure of the population are of crucial significance in explaining recreation patterns. Age, sex, marital status and family composition or diversity, have all been recognised as affecting recreation preference.

At the aggregate level, important demographic considerations are the overall size, structure and distribution of the population. Although population growth rates in Western countries remain low, significant shifts of population are taking place internally. One of the most widely publicised of these has been the migration from the Frost Belt to the Sun Belt States of North America. Whereas part of the attraction of the Sun Belt can be found in the outdoor recreation opportunities available, rapid, unplanned growth in these areas threatens the very qualities newcomers seek.

At the disaggregated, individual or family level, a good deal has been written on the effects of age and the progression of life from one phase to another through

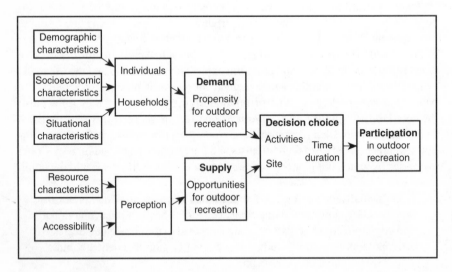

Figure 2.2 The decision process in outdoor recreation

Source: Pigram (1983)

what is known as the life-cycle. It has been suggested that although sharp lines of division cannot be drawn, certain preoccupations and interests predominate at specific stages in the life-cycle. With regard to recreation, not only are preferences influenced by age, but also by an individual's physical, mental and social ability to participate.

It is clear that the recreational importance of each phase is closely related to the family framework and to other 'life event' phases in an individual's 'life career' (Mercer 1981a). Apart from the family setting, these include the broader cultural background, government policies and the mass media. Mercer emphasises that the average life span subsumes and obscures major traumas such as illness, divorce, bankruptcy and the so-called 'mid-life crisis'. Moreover, during any life episode, recreation opportunities may be constrained by relative poverty, immobility and lack of time.

The implications of the family life-cycle approach are that recreation requirements can be expected to vary from individual to individual and between different people at different stages of the cycle, with important consequences for the planning and management of recreation space and resources. What is perhaps more important for current policy considerations, is that significant demographic changes are taking place within the family life-cycle, and that these, in turn, will generate altered priorities in recreation policy, planning and development.

In several countries of the Western world, the most dramatic demographic changes are shifts in age structure, stemming from the post-war 'baby boom' and the subsequent 'baby bust'. As these ripples move into maturity and beyond, their influence is reflected in recreation patterns so that resource managers need to be alert if a rapid and appropriate response is to be made. The changing status (some would say 'demise') of the family in modern society is another factor affecting individual and community participation in recreation. The prevalence of working couples and the freeing of women from many pre-existing constraints are gradually blurring sex-related differences in recreation participation. Childless couples, unmarried couples living together and greater numbers of elderly people living alone or in public and private community housing, all contribute to the growing complexity of 'family' life to which recreation planning must adapt.

Projected aged dependency ratios in OECD countries (see Figure 2.3) demonstrate the ageing profile of populations in Western industrialised countries. Proportionally, a smaller and smaller workforce is going to be required to support a growing aged dependent component of the population. This ageing profile also underlines the need for greater provision of suitable recreation opportunities for older, active people. In Australia, for example, those aged 65 years and over are expected to make up over 20 per cent of the population by 2025, compared with approximately 12 per cent at present (Borowski 1990).

The emergence of a significant elderly and retired component in the population, for whom greater longevity, improved health care and better financial provision generate a new set of leisure opportunities and requirements, takes on greater significance because of the high concentrations of older people in particular areas.

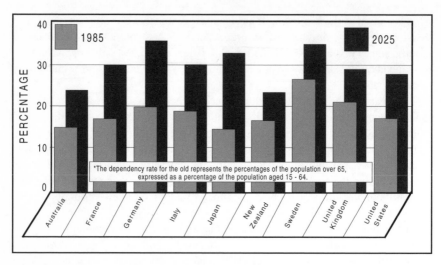

Figure 2.3 Projected aged dependency ratios in OECD countries

Source: Adapted from World Population Prospects, United Nations; National Australia Bank, in Freeman (1992: 10)

Mercer (1980b) identified several localities in Australia, in particular the Gold Coast of southern Queensland, as geriatric colonies, with above average numbers of retired people. However, retirement migration in Australia is perhaps not yet as pronounced in its regional effects as on the South Coast of England, popularly known as 'Costa Geriatrica', nor as in Florida, where the aged make up a significant percentage of the population. As Mercer (1980b) pointed out, such ageing of the population can occur very rapidly, and when accompanied by the departure of youth in search of employment or excitement, can give rise to a succession of strains and imbalances in the community.

Socioeconomic characteristics

Among the factors which influence the desires or inclinations of individuals for recreation, are social relationships and social structure, education, occupation and income. Recreation is a form of social interaction, and the way in which a society is organised affects recreation behaviour. For instance, interaction within and between families, peer groups and ethnic communities helps mould many facets of human behaviour, including goals and motivations for use of leisure.

Levels of education, too, whether considered in formal, structured terms or as incidental improvements in awareness and knowledge, must have a pronounced influence on actual recreation behaviour (also see Chapter 3). Indeed, the emphasis on advertising and marketing in the leisure industries reflects this relationship, while the efforts made by commercial enterprises to convince patrons of the quality of their attractions are themselves a form of education, and are also a facet of mass consumption and recreation. However, Mercer (1977) questions whether this correlation is causal when it comes to determining underlying propensities for recreation. The fact that the more highly educated person is likely to be more recreationally active may only reflect further correlation with a higher status occupation and reinforce income and class differences. As with so many of the factors impinging upon recreation demand, there is a degree of overlap, both with other influential factors and with the process of expression of demand through participation. Education contributes to knowledge, awareness and the development of attitudes and values, which, in turn, may generate aspirations and desires for recreation. At the same time, the acquisition of recreational skills through education can enhance opportunities for participation and for gaining satisfaction from recreation.

A similar problem occurs with income and occupation, each already highly correlated with the other. Undoubtedly, the amount of discretionary income available to an individual or family is a major factor affecting recreation participation, but does it help structure underlying recreation preferences? Do well-to-do people really prefer active outdoor recreation activities, or do their wealth and associated possessions merely open doors that are closed to the less affluent? Again, the former sharp distinction in attitudes to work and leisure between high and lower status occupations is becoming blurred. No longer can it be said with certainty that upper-class occupational groups show a preference for a more serious range of leisure pursuits, or view with disdain the thought of more mundane forms of recreation. Increased concern for conservation and environmental issues, especially among the 'baby boom' generation, has contributed to increased participation by a broad cross-section of the population in outdoor recreation, nature-based activities and use of national parks (Lacey 1996).

Situational characteristics

The third group of factors which impinge upon recreational choice is linked to some of those previously discussed and shows similar ambivalence. Under the category of situational or environmental factors could be placed:

- *Residence* – which incorporates such aspects as location, type, lot size and existence of a garden or pool, and which, to some extent, is a function of income and occupation. At a larger scale, the place of residence can influence recreation patterns. Obvious examples are coastal locations, winter sports areas, and large urban centres.
- *Time* – which also frequently reflects occupation, although this is changing with innovations in working conditions and the high incidence of unemployment. It is not merely the amount of time which is important, but its incidence in terms of usable 'blocks' at convenient periods (e.g. weekends). In general, self-employed persons have greater control over their time budgets and are, or should be, in a position to allocate more time to leisure. This, in turn, has the potential to widen the dimensions of recreation participation.

- *Mobility* – which, for most people, freely translates to car ownership or access to a motor vehicle. If a vehicle is not available, a person's recreation action space is obviously limited in terms of choice of site, journey, timing and duration of trip. Presumably, also, possession of a car generates a desire, or at least permits a propensity, for forms of recreation which otherwise could not be considered.

External factors

As noted above, some of the variables which are considered important in determining an underlying proclivity for recreation, can also be influential in the actual decision to participate. Several of the socioeconomic and situational factors, for example, appear to operate at various stages of the decision-making process. Furthermore, the role of resource-related characteristics is indicated in Figure 2.2. These characteristics have direct relevance to choice of recreation site, activities and travel, and are concerned with the opportunity to recreate (i.e. to activate latent participation).

Recreational opportunity depends upon the inter-related features of availability and accessibility of recreation resources or sites. The nature of recreation resources and their availability in functional terms, depends upon such things as quality, degree of development, carrying capacity, ownership, distribution and access. These, in turn, reflect economic, behavioural and political factors, which help shape public and private decision making for recreation provision.

Accessibility to recreation opportunities is a key influence on participation, and its several facets are examined in ensuing chapters. Its importance as the final deciding factor in determining the 'what' and 'where' of recreation participation is stressed by Chubb and Chubb (1981: 153): 'If all other external and personal factors favour people taking part in an activity but problems with access to the necessary recreation resources make participation impossible, the favourable external and personal factors are of no consequence.' Accessibility also helps explain the contribution of the travel phases to the overall recreation experience.

Recreation travel behaviour

Almost by definition, outdoor recreation implies that space, distance, and therefore time, separate recreationists from the sites and activities to which they wish to relate. A process in spatial interaction is stimulated as efforts are made to reduce spatial imbalance in recreational opportunities. The ease or difficulty of movement and communications are basic to the explanation of spatial interaction. Mobility and information diffusion thus become key elements in the spatial relationship between recreationists at the origin (i.e. place of residence) and the destination (i.e. the recreation site).

The friction of distance is important in all forms of recreation travel. For most movements, a distance–decay effect can be recognised, so that the strength of interaction declines as distance increases. Put simply, this means that recreation sites at a greater distance, or for which the journey is perceived as involving more time, effort or cost, are typically patronised less (a distance–decay effect). However, the effect of the friction of distance varies spatially, and with modes of movement and types of recreation activity. It can also change dramatically over time and space, with innovations in communication and transportation, and with advertising and promotion.

For some forms of recreation travel, the distance–decay effect may be heightened, manifesting itself in inertia or the reluctance to move at all. Alternatively, the reaction to distance may be in marginal terms. In most cases, the effect of distance will be negative, in that, beyond some point, further travel becomes less desirable; each kilometre offers more resistance or impedance than the last. Conversely, the effect may be positive, where the friction of distance is reversed; for some people and some occasions (e.g. ocean cruises), travel becomes so stimulating as an integral part of the recreation experience, that the further the distance, the greater the desire to prolong it.

The effect of travel and its key role in the satisfaction gained from the total recreation experience are important influences on recreation behaviour. The 'journey to play' can make or break the outing, and it is often the individual's perception of what is involved in the travel phases which is the crucial factor in the decision to participate or stay at home.

Recreation travel, in common with all aspects of recreation, is discretionary in nature in that it lacks the orderliness and monotony of, for instance, the journey to work. Yet, certain regularities can be discerned in recreation movement patterns in response to time–distance, connection, and network bias.

Time–distance bias, where the intensity of movement is an inverse function of travel time and distance, reveals itself in the distance–decay effect referred to earlier. Distance is constrained (or 'biased') by the time available and the type of recreation envisaged. Distance is also the basis for determining the extent of urban recreation hinterlands. In terms of travel distance, it is possible to conceptualise recreation traffic movements by a series of concentric rings progressively distant from the city to distinguish between day-trips, weekend trips and vacations. There is clearly scope for overlap between zones and such an arrangement may represent an oversimplification in an era of more sophisticated and efficient transportation systems.

Connectivity, and conversely barriers to movement, is another important aspect of transferability affecting the means or ease of spatial interaction. The presence or absence of interaction and the intensity of recreation travel are related to the existence and capacity of connecting channels of traffic flow. Recreational trip-making will respond positively or negatively to alterations in connectivity between origin and destination. An additional traffic facility such as a motor bypass, bridge or tunnel, can transform locational relationships by providing new or improved connections between places. Removal of linkages (e.g. destruction of a bridge) or impairment of capacity will lead to drastic alteration in patterns of recreation movement, and the resulting redistribution of traffic pressure can generate severe adjustments in dependent services and enterprises. Any number of examples exist where new communities and recreation facilities have sprung up and established sites have gone into decline because of alterations to pre-existing routes and modes of movement. Closure of railway lines, relocation of river crossings, construction of highway-motorway bypasses, even the conversion of streets to one-way traffic, can all have dramatic effects on recreation travel behaviour.

Finally, part of the explanation for regularities in recreation movements can be found in the characteristics of existing communication networks. Recreational travel is more likely where networks relate to shared information channels, a common transport system or the same sociocultural, national, political or even religious grouping. The huge volume of tourist flows based on group tours is but one example of the influence of network bias in promoting recreation travel on a large scale. The network effect, too, can be heightened by constraints on expanding links within or between systems, such as occur with national boundaries or language barriers.

Despite the regularities noted above, the essentially discretionary nature of recreation movements and the element of unpredictability put some difficulties in the way of developing an efficient and economic system of management for the special characteristics of recreation travel. Particular problems are the incidence of peaking, variability in participation and the heavy reliance placed on the motor vehicle. Patterns of recreation movement display daily, periodic and seasonal peaks and troughs, associated with time of day, weekends, vacations and suitable weather, especially in the summer season in coastal locations, and in winter in alpine areas. Some of these peaks are cyclical, and to that extent predictable. However, the problem remains of providing a transport system which can cope with short periods of saturation set against longer periods of under-utilisation.

The situation is worsened by the pervasive reliance on the automobile as the primary means of recreation travel. The motor vehicle ranks with television as the most powerful influence, positively and negatively, on recreation participation. The reasons are not hard to find. Use of the car allows for the unstructured nature of recreation (and other) trips, and provides for flexibility in timing and duration of the outing, and choice of route and destination. The car is readily available and is a good means of access to most sites, without the necessity for change of travel mode. It combines the function of moving people, food and equipment with shelter, privacy, a degree of comfort, and a relatively inexpensive means of transport.

The expectation of car ownership and its dominant role in outdoor recreation affect more than travel behaviour. The motor car is a fundamental influence on recreation landscapes and on the type and location of recreation facilities. As is noted in Chapter 11, the car has significantly affected the morphology and function of tourist areas, and has given rise to a completely new series of leisure activities and support industries.

In considering this close attachment of the recreationist and the motor vehicle, it would be wrong to assume that car ownership or access is universal. There will always be a social need to provide for the non-motorist in the community, if recreation opportunities for the less mobile are not to be severely restricted.

Given that recreational trip-making is largely unstructured and discretionary in nature, it is noteworthy that efforts have been made to isolate common variables influencing decision-making, and to use these to explain and predict recreation behaviour and associated patterns of movement.

Studies of trip generation are numerous, using models incorporating a variety of predictive variables to attempt to answer questions concerning: why particular forms of outdoor recreation are selected by different individuals and groups; why certain sites are patronised and others neglected; the expected frequency and duration of recreational trips; and the degree of substitutability between recreation activities and recreation sites.

One of the most popular and frequently applied techniques is some version of the gravity model, which has been used with success in forecasting visitor flows to recreation sites. Essentially, gravity models are based on the premiss that some specific and measurable relationship exists between the number of visitors arriving at a given destination from specific origins or markets and a series of independent variables, in particular, population and travel distance.

If these variables can be quantified with reasonable accuracy, predictions can be made as to the likely attendance at selected recreation sites from designated points or areas of origin (e.g. visitation rates to parks from surrounding regions, counties, towns or cities). If the actual, measured levels of attendance match the expected, then the model can be used to predict visits to proposed new parks, to indicate the need for greater efforts in publicity and advertising, or to assess the impact of improved accessibility on the propensity to travel.

The technique can also be applied to delineate the range or impact zone from which a site could be expected to attract visitors. In theory, if this zone was merely a function of the friction of distance, it would consist of a series of concentric zones surrounding the site, with numbers of visitors progressively declining outwards from the centre. However, distortion of the size and shape of the area is to be expected because of the kind of factors noted above. Variations in demographic characteristics, in conditions of accessibility and in the orientation and impact of promotional advertising within the hinterland, as well as competition from peripheral attractions, all help to explain why actual patterns of patronage depart from the theoretical.

Models of outdoor recreation participation can be developed at various levels of sophistication and application (i.e. local to national levels), but all must involve compromise and rest on certain assumptions, because of the complex nature of recreation behaviour. Caution is necessary, then, in the use of models and in the application of the results. In such an unstructured field of choice-making as recreation, where decisions are often more intuitive than rational, and more impulsive than considered, norms are not appropriate.

A further cause for concern in modelling recreation behaviour is the assumption that relationships between the several sets of variables remain constant. Yet, lifestyles and social mores change progressively, as do economic and technological circumstances, so that prediction is very difficult and value-laden (e.g. see Lee-Gosselin and Pas 1997). The dynamic nature of many inputs into recreation decision-making can be a source of miscalculation in planning. New trends and fashions, changing values, charismatic leaders and different policies by governments or other institutions, can all act as 'triggers' to release latent participation and bring effective demand more into line with overall demand.

Finally, the underlying element of choice in recreation means that individual participants or particular recreation pursuits should not be studied in isolation. Rather, the entire spectrum of leisure activities must be examined as a series of substitutes and complements that are capable of providing a variety of satisfactions, and that act as potential trade-offs for one another (Phillips 1977).

Substitutability and interchangeability are responses to the relationships between the experiences and satisfactions sought in outdoor recreation, and the geographic, social, psychological, economic or physiological barriers which prevent those expectations and satisfactions from being fully realised. The effect of these barriers is to stimulate replication of satisfactions by resort to some other activity. In short, the concept of substitutability implies that recreation preferences and propensities are much more elastic and open to manipulation than is generally accepted, making the recreation choice process that much more complex.

Recreation choice behaviour

Predictions regarding recreation behaviour would have greater validity if more was known about attitudes, motivations and perceptions affecting recreation decision-making. This would help explain: (1) why certain activities and sites are favoured; (2) why some recreation businesses are failures, while others provide satisfaction and even draw excess patronage; and (3) how and why alternative recreation opportunities are ranked.

The recreation choice process is influenced by people's perceptions of what recreational opportunities are available. In every decision-making situation, individuals evaluate selected environmental attributes against some predetermined set of criteria in order to arrive at an overall utility or preference structure (see Aitken 1991). A predisposition or propensity (i.e. demand) for recreation is translated into actual participation through a choice mechanism, heavily dependent upon perception of the recreation opportunity and experience on offer. Perceptions are personal mental constructs, which are a function of the perceiver's past experiences, present values, motivations and needs.

Perception operates over several dimensions and various scales in recreation decision-making, and initial mental constructs may be confirmed or revised as a result of further spatial search and learning. Information levels, as well as the ability to use that information (which may be governed by such factors as personality characteristics and aversion to risk), also help structure evaluative beliefs and mental images concerning the nature and quality of anticipated recreation experiences.

Information sources, and the credibility of the information itself, are key issues in the choice of leisure settings. The validity of some spatial choice models has been questioned because of the assumption of perfect information and the assumed ability of consumers to evaluate completely all alternatives (Roehl 1987). In reality, individuals typically consider only a subset of available alternatives. For example, in any choice situation, an individual's decision will be influenced by his/her awareness set. Larger natural settings (e.g. national parks close to urban populations), with distinctive characteristics, are more likely to be known and considered by potential participants. In an urban context, Roehl demonstrates that smaller neighbourhood parks, with fewer facilities, and designed to serve lower-order needs rather than community- or higher-order needs, are less likely to be in a consumer's awareness set.

Desbarats (1983) notes how the supposedly objective spatial structure of opportunities is narrowed into an 'effective choice set' comprising those (recreation) opportunities that are known to the individual and actively considered. Effective choice sets may represent only a small fraction of objective choice sets, because of the direct and indirect effects of constraints on behaviour stemming from the sociophysical environment. In particular, contraction of the initial choice set may occur because of lack of information about existing options. 'The better the information, the greater the congruence between effective and objective choice sets' (Desbarats 1983: 351). Both the quality and timing of information are important factors in recreation decision-making. Inadequate information and misinformation act as constraints in the process of discriminating between alternatives (Krumpe 1988). The implications for management and policy are obvious and are discussed further below.

Information also helps structure images of the environment to which recreationists respond. However, the cognitive processes involved in image formation are complex (Beaulieu and Schreyer 1985). The (objective) information flowing from an environment is filtered through the perceiver's set of preferences and values, and cultural interpretations of place meaning. The process is complicated by the personal nature of reactions to external stimuli and by the multifaceted characteristics of the environments being experienced.

Dissection is risky. In nature-oriented environments, in particular, it is difficult to reach consensus on what components – landform, water, vegetation, etc. – contribute most to the appeal of the landscape. In any case, these attributes must be mentally fused to complete the totality of the image, so that the whole is greater than the sum of the parts. 'A landscape is more than the enumeration of the things in the scene. A landscape also entails an organisation of these components. Both the contents and the organisational patterns play an important role in people's preferences for natural settings' (Kaplan and Kaplan 1989: 10).

Natural environments as recreation settings

The Kaplans applied this reasoning to what they see as human preference for natural environments. They believe that it is not only the dominance of nature in the scene which is appealing, but that it is also the spatial configuration of landscape elements which is important to people's reactions. Certain natural settings are favoured because of their openness, their very lack of structure and precise definition, their transparency, and the perceived opportunities to enter and move around. Wild environments, impenetrable forests and even built environments, on the other hand, may evoke less positive responses, along with feelings of insecurity.

Research by Driver *et al.* (1987) further attests to the importance of the natural setting in achieving the desired outcomes from leisure pursuits. In a wide-ranging study of wilderness users in Colorado, the most important 'experience preference domains' were linked to enjoyment of nature. Clearly, the natural environment plays a fundamental part in attaining the outcomes and satisfactions sought from participation in certain forms of recreation.

Given the widespread appeal which nature apparently holds for people, its importance in the experience of leisure should come as no surprise. Many of the benefits associated with natural settings are, or should be, fundamental to the realisation of leisure. The opportunity for self-expression and subjective freedom of choice, accepted by many observers as characteristic of leisure, appears to be sought more often in natural, than in created human-dominant landscapes. The intrinsic values derived from experiencing leisure are perceived as being more in keeping with the natural scene and with a minimum of social manipulation. In terms of 'effective functioning' (Kaplan and Kaplan 1989), the natural environment would seem to offer greater scope for personal satisfaction, through integration of mind and body in the leisure activity itself.

Undoubtedly, perceived environmental attributes are a powerful influence on recreation behaviour. This is borne out by Schreyer *et al.* (1985), who suggest that the most useful representation of the environment for the explanation of behavioural choice is at the macroscopic or holistic level, rather than at the attribute level. According to Schreyer *et al.* (1985: 16), 'People do not search for specific elements of the environment as much as they search for settings which will allow them to behave in the ways they desire ... which will allow for the attainment of the desired cognitive state'. They go on to stress the importance of the social milieu and the social definition of the physical environment to the totality of the setting in which recreation takes place.

By definition, outdoor recreation is resource-related, and increasing attention is being given to the setting in which action takes place, as a prime force in the satisfaction gained from the ensuing recreation experience. The assumption that recreation experiences are closely related to recreation settings is central to the concept of the recreation opportunity spectrum.

Recreation opportunity spectrum

In many ways, the recreation opportunity spectrum (ROS) is an application of behaviour setting analysis from environmental psychology (Barker 1968; Ittelson *et al.* 1976; Levy 1977). This approach suggests that all human behaviour should be interpreted with reference to the environment or behaviour setting in which it occurs. It is further suggested that, given knowledge of the behaviour setting for a specific recreation experience, such as a park visit, it should be possible to identify the human values and expectations associated with that experience. Examination of the human and non-human features of the behaviour setting should then indicate those contributing to or detracting from satisfaction.

As with all human behaviour, response to external stimuli is not always simple or direct. Environmental psychologists see people not as passive products of their environment, but as goal-directed individuals acting upon that environment and being influenced by it (Ittelson *et al.* 1976). All leisure environments affect recreation behaviour in some way; it is the dynamic interaction between the environment and users which is crucial to the outcome.

Within this conceptual approach, a recreation opportunity allows the individual to participate in a preferred activity, in a preferred setting to realise a desired experience (Driver and Brown 1978). The focus is on the setting in which recreation occurs. The ROS describes the range of recreational experiences which could be demanded by a potential user clientele if a full array of recreation opportunity settings was available through time. Clark and Stankey (1979: 1) define a recreation opportunity setting as:

> ... the combination of physical, biological, social and managerial conditions that give value to a place (for recreation purposes). Thus, a recreation opportunity setting includes those qualities provided by nature (vegetation, landscape, topography, scenery), qualities associated with recreational use (levels and types of use) and conditions provided by management (roads, developments, regulations). By combining variations of these qualities and conditions, management can provide a variety of opportunities for recreationists.

The basic premiss underlying the concept of the ROS is that a range of such settings is required to provide for the many tastes and preferences that motivate people to participate in outdoor recreation. Quality recreation experiences can be best assured by providing a diverse set of recreation opportunities. Failure to provide diversity and flexibility ignores considerations of equity and social welfare, and invites charges of discrimination and elitism (Clark and Stankey 1979). A sufficiently broad ROS should be capable of handling disturbances in the recreation system. These might stem from such factors as social change (e.g. in demographic characteristics) or technological innovations (e.g. all-terrain recreation vehicles) (Stankey 1982).

The ROS offers a framework within which to examine the effect of manipulating environmental and situational attributes or factors to produce different recreation opportunity settings. Clark and Stankey (1979) suggest that the most important of these 'opportunity factors' are:

- access;
- non-recreational resource uses;
- on-site management;
- social interaction;
- acceptability of visitor impacts; and
- acceptable regimentation.

Some of these factors are discussed in greater detail in later chapters. In particular, it should be noted that the weight or importance given to each will vary with individual site and management circumstances.

The range of conditions to which an opportunity factor can be subjected, and the way each can be managed to achieve desired objectives are shown in Figure 2.4. By packaging a recreation opportunity setting in some combination of the six factors described, a variety of recreation opportunities or options can be generated, and the ROS materially enlarged. In their scenario, the authors present only four generic opportunity types, arrayed along a 'modern to primitive opportunity continuum'. However, within each, there is scope for many complex combinations, thus providing even more diversity.

The ROS also allows an examination of opportunity settings with respect to the capability of potential users to avail themselves of the opportunities presented. Limited resources and, perhaps, lack of awareness or imagination mean that, generally speaking, the established recreation system caters for the majority, on the premiss, apparently, that everyone is young, healthy, ambulant, educated, equal and possesses the means to participate. The reality, of course, is very different. Reference is made in Chapter 3 to constraints on recreation because of age, lack of income and other factors. Racial and ethnic origins can be a disadvantage, particularly in inner cities and suburbs, where these minorities are often concentrated. Likewise, the spectrum of recreation opportunities for people who are disabled is likely to require special attention if real choice is to be offered.

Despite its inherent appeal as a means of facilitating choice in outdoor recreation, specific applications of the ROS approach have attracted some criticism. For some managers of recreation sites, the concept has been treated as a 'blueprint', from which little deviation was possible or desirable. In other situations, there has been a reluctance to amend the range of opportunity settings from that initially created, so as to allow some flexibility, in keeping with the dynamic aspects of the recreation environment and the preferences of users. Indeed, there appears to be relatively little consultation with potential visitors to identify preferred recreation settings. The approach is predominantly 'top-down', reflecting what management feels will be satisfying for visitors, and conducive to managerial convenience. The emphasis, too, has been on manipulation of the biophysical elements of recreation settings, whereas opportunities for social interaction are at least equally important influences on satisfaction and quality recreation experiences (Heywood 1989). Some of these shortcomings are brought out in examples of the application of the recreation opportunity spectrum concept in management situations presented in Chapter 6.

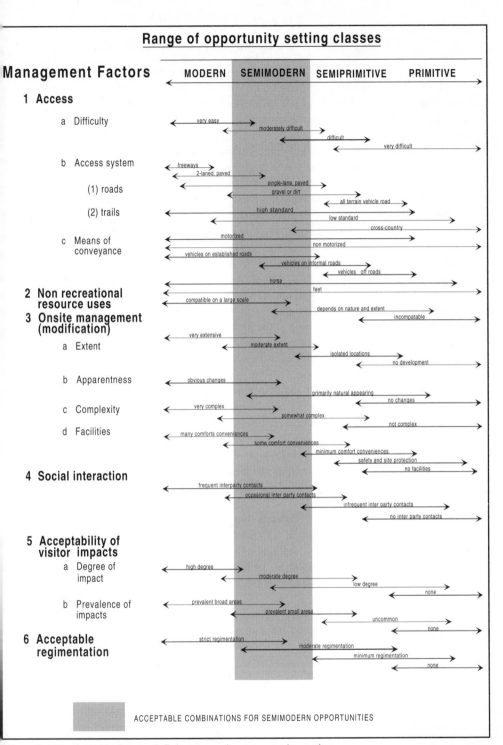

Figure 2.4 Management factors defining recreation opportunity settings

Source: Adapted from Clark and Stankey (1979: 15)

Incompatibility and conflict in outdoor recreation

Further complexity is added to the recreation choice process, when the issue of incompatibility is considered. Most often, this is seen as a problem between outdoor recreation and other forms of resource use. However, conflicts can arise just as readily between groups of recreationists, even when engaged in the same leisure pursuits.

The question of compatibility revolves around the degree to which two or more activities can co-exist in the use of a given recreational resource. Goodall and Whittow (1975) point out that the problem is linked with the resource requirements for particular recreational pursuits. Only where recreational activities have similar requirements is there a possibility of shared use of a site, or alternatively, of conflict. Noisy activities, such as those involving off-road recreation vehicles or power boats, conflict with fishing, bird-watching, use of wilderness and other activities requiring peaceful countryside locations. Nor is conflict necessarily confined in space or time; site disturbance can have a lasting effect and can spill over to adjacent areas. Goodall and Whittow stress that the incidence of incompatibility is, in part, a function of the activity, the manner in which it is practised and the characteristics of the site or the resource involved. Trails, rivers and other constricted linear resources are particularly sensitive to use incompatibility. On the other hand, timbered land may increase compatibility by reducing visual intrusion and noise penetration.

Conflict and compatibility involve a good deal more than simple one-to-one comparisons of selected recreational activities. According to Lindsay (1980), the conflict problem may be summarised as one of recreationists competing for the same physical, social and psychological space during the same time period. Thus, confrontation over use of recreation space should not be interpreted solely as inter-activity conflict. The complexities of human behaviour are such that conflict situations can develop between different types of recreationists engaged in the same activity.

Jacob and Schreyer (1980) believe that the key to conflict resolution lies in identifying the 'conflict potential' of recreation resource clientele, rather than in labelling certain activities as conflict-prone. It is not merely a question of skiers not getting along with snow-mobiles, or of 'motor versus muscle'. In this context, four causal factors are identified as conducive to conflict in outdoor recreation:

- *Activity Style* – Various personal meanings are assigned to an activity. For some, participation may be intense; the activity becomes the focus of life interest, with acquisition of status and achievement of a high quality recreation experience prominent goals of participants. As a participant becomes more specialised or 'involved', the potential for conflict increases with others not so committed or expert (McIntyre 1990).
- *Resource Specificity* – Some individuals attribute special values and importance to certain physical resources, and develop possessive, protective attitudes to favoured recreation sites – a common trait of skilled fishermen

36

and hunters. Tension can develop with lower status 'intruders', who do not share this appreciation and interpretation of site values, and who disrupt the exclusive, intimate relationship built up with a place. Once again, conflict has little to do with activities themselves, but can occur between divergent classes of resource users.

- *Mode of Experience* – The manner in which individuals approach a recreation experience can provide the ingredients for conflict. Jacob and Schreyer (1980) distinguish between 'focused' and 'unfocused' modes of experience. The latter is concerned with overall spatial relationships and environmental generalities, rather than specific entities within that environment. Thus, the 'focused' wilderness user, intent upon achieving an intimate relationship with specific aspects of the natural environment, has little in common with the 'unfocused', for whom merely being in the countryside is sufficient. The greater the gap between recreationists along this continuum, the greater the potential for conflict.

- *Tolerance for Lifestyle Diversity* – The suggestion is, that individuals deliberately choose recreation settings and associations which reflect their societal outlook and behaviour, and are unwilling to share resources with other lifestyle groups categorised as deviant, or merely different. Value-laden inferences are made about people indulging in alternative forms of recreation, stereotyped as 'less worthwhile'. Thus, the trail-bike, the power boat and the snow-mobile are seen as symbolic of a society that arrogantly exploits and consumes resources. Ethnic, racial and social class distinctions can also be the basis for lifestyle-based conflicts. Such people are often labelled 'out-of-hand' as 'inferior', so that even when pursuing the same activity and following the same rules, conflict still ensues, especially as the number and variety of people desiring access to recreation resources increase.

Jacob and Schreyer suggest that the degree to which these four factors are present, singly or in combination, represents the extent to which the potential for conflict exists. Conditions for conflict may just as readily occur in the mind and be part of the mental state and attitude of the participant, as in the nature of the recreational activities. The authors conclude with a warning for management:

> Unfortunately, the tendency to define conflict as confrontations between activities has left the sources of recreation conflicts unrecognised. In failing to recognise the basic causes of conflict, inappropriate resolution techniques and management strategies are likely to be adopted.
>
> (Jacob and Schreyer 1980: 378)

Summary

Many factors affect recreational motivation and choice, with much debate continuing about the forces affecting recreation decision-making at the individual, group and societal levels. This chapter explored the nature of recreation demand and participation, and the range of influences on recreation choice behaviour. The importance of accessibility and the travel phases in the overall recreation experience was stressed, and an overview presented of the recreation decision process. Reference was made to the factors affecting participation in recreation, in particular, the role of perception of recreation opportunity. The types of decision choices which confront individuals and groups, and how these affect people individually and collectively, are related to the concept of the recreation opportunity spectrum. Further complexity is added to the recreation choice process with consideration of compatibility and conflict between recreation activities and recreationists.

This review of the relationship between people and the leisure environment reveals some of the dynamics and complexities of the choice process in recreation behaviour. As stated at the outset, it is the unbounded, subjective nature of leisure and its expression in recreation activity which make explanation and prediction difficult. By definition, recreation is discretionary and any suggestion of obligation or compulsion must compromise the experience. Moreover, participants in recreation, as distinct from other forms of human behaviour, can exercise more control over decisions regarding what, where, and with whom, '… in the design of their desired products and thus the experiences they derive from participation' (Williams 1995: 32).

Finally, it is the interaction of such environmentally-related supply factors with demographic, socioeconomic and situational variables, or population-related demand factors, which generates opportunities to participate in recreation. However, recreation decisions depend not on actual objective opportunities, but on individual perceptions of those opportunities. These, in turn, depend greatly on formal and informal social and information networks, and on the personal characteristics and motivations of potential recreationists.

Guide to further reading

- *Recreation and travel motivation, choice and behaviour*: Mayo and Jervis (1981); Pearce (1982); Ibrahim (1991); Bammel and Bammel (1992); Garling and Golledge (1993); Ross (1994); Stopher and Lee-Gosselin (1997).
- *Tourist satisfaction*: Ryan (1995).

Review questions

1 Discuss the dimensions of recreation demand.
2 Critically review discussions of the relationship between recreation demand and recreation supply.
3 Present an overview of the factors affecting recreational motivation and choice. Generally speaking, is it enough to derive explanations of recreation motivation and behaviour from only one factor (univariate analysis), or should we consider the relationships between many factors (multivariate analysis)? Explain your answer with reference to appropriate studies.

4 What are the main demographic changes taking place within your country and local region? Identify recent recreation policy and planning responses to such changes at the national and/or regional level.

5 Identify a local recreation site where conflict between recreationists has arisen. Why does/did that conflict exist? Has the conflict been resolved? Why/why not?

3

OUTDOOR RECREATION: SPECIAL GROUPS AND SPECIAL NEEDS

Recreation need is characteristic of all human beings, but as Veal (1994: 189–90) so cogently stated:

> Every individual is unique and so could be said to have unique leisure requirements. In family settings and some organisational settings this uniqueness can be catered for, but human beings are social animals with interests, demands and needs in common... Classifying people into groups and considering their common characteristics and needs is not therefore to deny their individuality; in fact, it has been the failure of providers to consider the common needs of some groups which has, in the past, denied members of such groups their individuality. As a result of campaigns, regulations, research and the spread of ideas such as 'market segmentation' and 'niche marketing', some of these problems are now beginning to be overcome.

There are some for whom participation in, and the resultant satisfaction derived from, recreation requires that special services, programmes and/or facilities be provided to ameliorate or remove leisure constraints. These people are commonly regarded as having special needs.

Research on recreation non-participation and constraints to leisure is growing. Such research makes theoretical contributions to our understanding of leisure choice and behaviour, and makes practical contributions by providing information which will generate or affect service delivery by way of policy-making, planning, programming and marketing (Jackson 1990).

From this perspective, this chapter outlines the concepts of leisure constraints and recreation need, and considers factors which may act as barriers to recreational participation and satisfaction. It discusses possible approaches to the assessment of constraints and needs as a basis for future planning and programme development. As society becomes more complex and dynamic, as 'social services' of the state are privatised (moved to the private sector), and as much recreational need and tourist travel becomes more discerning, sophisticated and expensive, it becomes increasingly difficult for individuals, acting alone, to satisfy their recreational needs.

Leisure constraints, recreation need and human rights

'A constraint to leisure is defined as anything that inhibits people's ability to participate in leisure activities, to spend more time doing so, to take advantage of leisure services, or to achieve a desired level of satisfaction' (Jackson 1988, in Jackson and Henderson 1995: 32; also see Henderson *et al.* 1989: 17). The constraints associated with leisure and recreation participation have been studied by several authors in general terms, and in specific terms with reference to special groups, including people with disabilities, youth and adolescents, the elderly, and women with physical disabilities. In the main, two types of constraints have been identified – *intervening constraints* (those that come between a preference and participation and which thereby limit participation) and *antecedent constraints* (those that influence a person's decision and subsequently inhibit preferences).

Research on constraints to leisure behaviour and recreation participation is growing conceptually, theoretically and in practical application. Work on leisure constraints and barriers has a relatively lengthy history, with the Outdoor Recreation Resources Review Commission (ORRRC) studies in the 1960s (e.g. Ferris 1962; Mueller, Gurin and Wood 1962) receiving considerable prominence. Since the early 1980s, however, constraints research has proliferated, with perhaps the most notable early models put forward by Crawford and Godbey (1987). Their formulation of leisure constraints focused on intrapersonal, interpersonal and structural constraints (see Figure 3.1). In 1991, Crawford, Jackson and Godbey revisited that formulation. In its place, they proposed a hierarchical process model, in which the three types of constraints above (intrapersonal, interpersonal, structural) were integrated. They derived three propositions from that model:

- leisure participation is heavily dependent on a process of negotiating through an alignment of multiple factors, arranged sequentially;
- the sequential ordering of constraints represents a hierarchy of importance;
- social class may have a more powerful influence on leisure participation and nonparticipation than is currently accepted (i.e. the experience of constraints is related to a hierarchy of social privilege).

Crawford *et al.* believe that this more recent model may help to clarify some paradoxical findings that were not fully explained previously. In particular, they noted the more frequent reporting of structural constraints among people of higher socioeconomic status. They go on to point out, however, that this hypothesis is largely speculative, and that research should proceed in three main directions.

First, there is a need for intrapersonal and interpersonal-level data, with investigations encompassing the entire array of constraints – intrapersonal, interpersonal, and structural – simultaneously. This would permit testing of propositions that people negotiate through sequential levels of constraints, and that these levels represent a hierarchy of importance.

Second, the issue of social stratification needs to be examined from a dynamic perspective, particularly given the widespread and increasing disparity between

41

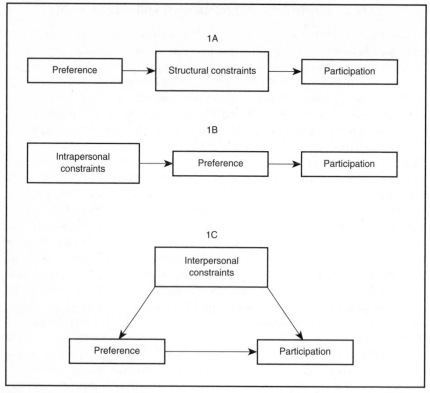

Figure 3.1 Crawford and Godbey's three types of leisure constraints

Source: Crawford *et al.* (1991)

the affluent and the poor. Longitudinal studies would reveal associated changes in recreation patterns and processes.

Third, we should move beyond examining constraints which result in non-participation, to investigate constraints which affect levels of participation. Clearly, then:

> Although for practical reasons it is often useful to conduct research at a high level of detail on separate parts of a system (in this case the system of leisure behaviour), there is a need to integrate leisure constraints research within the mainstream of leisure studies. Leisure researchers cannot afford to investigate the phenomena in which they are interested in isolation from other factors that influence leisure choices.
>
> (Crawford *et al.* 1991: 318)

Jackson, Crawford and Godbey (1993) went even further and suggested that leisure constraints negotiation (i.e. how a person decides to experience an activity despite the constraints encountered) is the key to understanding constraints. Whatever the case, constraints to leisure (whether they lead to nonparticipation or to less than optimal participation from the participant's perspective) stem from many factors, including biological, psychological, sociological, political and economic sources. For people to find leisure experiences or to establish desired levels of participation, planning and management to ameliorate or remove constraints are required. On the time dimension, free time has come to mean little if 'free time' is simply regarded as time not spent at work, or meeting other basic necessities (also see Chapter 1). Such is perhaps the case for the unemployed, the retired, those in public or private institutions (e.g. hospitals or prisons), or in circumstances where free time is a burden. Leisure activities for these people are not always, perhaps rarely, freely chosen or necessarily enjoyable, and may even involve physical and/or psychological stress. Under these circumstances, individuals may require additional services or information, for instance, by way of education in developing leisure decision-making and participation skills, as well as in identifying opportunities to seek satisfying leisure experiences.

The concern for special needs groups arises from increased societal awareness of and concern for, a more egalitarian society, based on human rights, social equality and accessibility to resources. Put simply:

> ... every person has the innate right to pursue his dreams and must be given the opportunity to fulfil his needs (within societal approval) as he has the capacity to achieve without artificial hindrance or restriction. The only limitations upon individual achievement should be biological potential and social acceptability.
>
> Shivers (1967: 131)

With respect to recreation, concerns for egalitarian recreational opportunity were given international prominence when the World Leisure and Recreation Association promulgated the Charter of Leisure (revised in 1981), which contains seven articles (see Table 3.1). These articles present an overriding ideal of equality of recreational access, extolling the virtues of leisure, and exhorting governments to make provision for leisure as a social service. Unfortunately, however, they do not declare access to leisure facilities and services as a human right (Veal 1994: 9), and this problem is manifested in recent planning and policy. For instance, in the United Kingdom, until 1995, when the Government's Disability Discrimination Bill was introduced in July that year and eventually became the 1995 *Disabled Rights Act*, there was no legal framework to protect people with disabilities from discrimination in seeking access to museums and country heritage sites. It was not until 1990 in the United States, that an *American Disabilities Act* was introduced, requiring all government, commercial and public premises to be readily accessible.

Humans are not created equally, nor do they share equality in life. Moreover, people, and public and private institutions have created (deliberately or otherwise) or contrived, artificial restrictions which may prevent individuals or groups participating in recreational activities which would otherwise be socially acceptable. Such restrictions are based on age, gender, race and ethnicity, religion, socioeconomic status, political affiliation, employment and location (e.g. remoteness).

43

Table 3.1 World Leisure and Recreation Association: Charter for Leisure

---

Article 1

Leisure is a basic human right. This implies the obligation of governments to recognise and protect this right and of citizens to respect the right of fellow citizens to leisure. This means that no one shall be deprived of this right for reasons of colour, creed, sex, religion, race, handicap or economic condition.

Article 2

Recreation is a social service of similar importance as Health and Education. Therefore, opportunities must be provided on a universal basis, reasonable access ensured, and appropriate variety and quality maintained.

Article 3

Ultimately, the individual person is his/her own best leisure and recreation resource; the primary role of governments, private agencies and groups are of a supporting nature, consisting of the provision of services where needed, with prime emphasis at the local level.

Article 4

Leisure and recreation opportunities should stress self-fulfilment, the development of interpersonal relationships, the fostering of family and social integration, international understanding and cooperation, and the strengthening of cultural identities. Special emphasis must be placed on maintaining the quality of the environment and on the influence of energy demands on future recreation resources.

Article 5

The development of recreation leaders, animators and/or counsellors must be undertaken wherever possible. The main tasks of these must include assisting people in discovering and developing their talents and helping them acquire desired personal skills for the purpose of broadening the range of recreation opportunities.

Article 6

The wide variety of leisure and recreation phenomena, including personal and collective experiences, must be subjected to systematic research and scholarly inquiry, with the results being disseminated as widely as possible to enhance the individual's knowledge of him/herself, to provide a stronger rationale for policy decisions, and to provide a more effective basis for program development and operation. All citizens must have access to all forms of information relative to the various aspects of leisure and recreation.

Article 7

Educational institutions at all levels must place special emphasis on the teaching of the importance of leisure and recreation, on helping students discover their leisure and recreation potential and on ways to integrate leisure and recreation into their lifestyles. These institutions should furthermore provide appropriate opportunities from which recreation leaders and educators can be developed.

Source: WRLA, in Veal (1994: 18)

In other words, it is society's definitions, perceptions and attitudes to such factors, and their relationship to recreational need which serve as one basis for inequality.

Recreation need is a multi-dimensional concept (Bradshaw 1972), and there exists a plurality of needs in any community (Hamilton-Smith 1975). Such needs are dynamic, individually and collectively. Taylor (1959: 107) outlined four uses of the term need:

- to indicate something needed to satisfy a rule or law;
- to indicate means to an end (either specified or implied);
- to describe motivations, conscious or unconscious, in the sense of wants, drives, desires, etc.;
- to make recommendations or normative evaluations. These are sometimes difficult to distinguish from the above three uses which are intended as purely descriptive statements.

There are several frameworks for assessing recreation needs. Mercer (1975) presented a typology of need comprising four categories based on Bradshaw's (1972) work:

- *felt need*: those needs which individuals have and which they want satisfied;
- *expressed need*: those needs which are expressed by people;
- *comparative need*: those needs identified on the basis of comparison of individuals or groups;
- *normative need*: those needs involving external assessments by experts, who identify a gap between what actually exists and what is desirable.

Each of the above dimensions lends itself to different methods of assessment (e.g. see Hamilton-Smith 1975), although Mercer (1975) views normative needs assessments with some suspicion. He argues that the 'experts' who make them are largely considered a 'small élite group in our society – the well-educated, well-to-do planners, politicians, engineers and academics'. This leads us to a critical point in identifying and assessing recreational need. The identification and assessment of recreation needs are value-laden activities, open to personal interpretation and subjective judgement, while any single measure of need will be inadequate, and a combination of approaches is needed. Values lie at the core of leisure and recreation public policy. As Simmons *et al.* (1974: 457) noted, 'it is value choice, implicit and explicit, which orders the priorities of government and determines the commitment of resources within the public jurisdiction'. These issues go to the heart of the structural constraints and problems identified by Crawford *et al.* (1991), and discussed earlier in this chapter.

As noted above, research concerning the recreational needs of special groups is expanding, in terms of both the types of groups studied and the depth of knowledge with respect to different groups (including variations within groups). Moreover, the normative aspects of recreational need, namely leisure and recreation policy-making and decision-making processes, which were largely ignored in the 1970s (e.g. see Mercer 1975), began to receive greater attention in the 1980s (e.g. see Henry 1993 and Veal 1994, for a detailed discussion), and deservedly so. Prescriptive–rationalist approaches to public policy, for instance, would see the decisions of government as being part of an inherently rational policy-making process, in which goals, values and objectives can be identified and ranked, after the collection and systematic evaluation of the necessary data (Wilson 1941).

However, this approach fails to recognise the inherently political nature of public policy, and the influences of values, power, institutional arrangements (including interest groups), and other factors (e.g. lack of monitoring and evaluation of policies and programmes) on the policy process. There are winners and losers with respect to any leisure and recreation policies and programmes. We need to know a lot more about who benefits and who loses out in terms of outdoor recreation.

People cannot always participate in the recreational activities of their choice. The satisfaction of recreational need requires individuals and groups to successfully overcome 'intervening variables' such as age, income and health status. The differential impacts of barriers to participation mean that some individuals and groups have more difficult barriers to overcome than other groups in society. For those individuals and groups unable to overcome the impediments associated with intervening variables, a case of special need may be identified. As a result, resources will need to be allocated to services, programmes and facilities, over and above those usually required. In this respect, governments necessarily play a crucial role. Recreation represents people's expression of the need to do things other than work, even though much recreation is institutionalised. That we are able to identify many people with special needs suggests that the institutional arrangements for recreational satisfaction are inadequate. The satisfaction of special groups' recreational needs thus requires institutional action. If this view is not accepted, then we run the risk of further disadvantaging these people.

The question of one's state of mind raises questions, too, about whether activities which may be seen by some as recreational, may be perceived very differently by special needs groups, who must be assisted to seek alternative opportunities during their leisure time. It is often the way in which a particular activity is perceived by an individual that will determine whether it is recreational or not (see Chapter 1).

Recreational choice and participation are affected by demographic, socio-economic and situational characteristics, external factors, and perceptions of recreational opportunity (see also Chapter 2). The availability of recreation resources in functional terms depends upon such things as quality, degree of development, environmental and social capacity, ownership, distribution and access. These, in turn, reflect economic, behavioural and political factors, which help shape public and private decision-making about recreation provision.

The special needs of some individuals and groups should be given due recognition in the context of the more usual recreation provisions of the community. In the past two decades, attention has been increasingly drawn to the problems facing various groups and individuals who might have special needs. We now turn our attention to some such special groups.

People with disabilities

Increasing public attention is being drawn to the problems facing people with disabilities or handicaps. Recent developments in legislation, policies and programmes in such areas as health, education, employment, facility design, and leisure and recreation, are evidence of changing attitudes and perceptions in society. The United Nations Year of the Disabled in 1981 was an important precursor to this situation, raising global awareness of people with disabilities.

A person with a handicap has been defined as one 'whose physical, mental and/or social well-being is temporarily or permanently impaired...' (Calder 1974: 7.3). It is perhaps proper to distinguish between functional disability as a result of primary impairment, and handicap which is determined by individual and societal reaction to limitations on social roles and relationships. Disability is a defined impairment, which becomes a handicap only when the disability prohibits activity in the pursuit of specific goals (see Dibb 1980).

In 1988, it was revealed that approximately 6.2 million adults (14 per cent of the population) in the United Kingdom had some kind of physical, sensory or intellectual disability. More recent estimates show that: nearly 1 million people are blind or partially sighted; 7.5 million are hearing impaired (of whom 2 million use a hearing aid and 55,000 use British Sign Language); 35 per cent of visually impaired people are also hearing impaired, while 14 per cent are mobility impaired (Blockley 1996). The last two statistics highlight a 'stigma on stigma' phenomenon (discussed below), which demonstrates (1) the diversity within singularly defined groups, and (2) the need for multivariate analysis in examining recreational access, motivations, choices and experiences.

The Australian Bureau of Statistics (ABS) (1993) provides an indication of the number of people with disabilities, as well as the nature of such disabilities, by way of a *Survey of Disability and Carers*. For the purposes of this survey, people were identified as having a disability if they possessed one or more specified limitations, restrictions or impairments, which had lasted, or were likely to last, for six months or more. A disability was defined as any restriction or lack of ability (resulting from impairment) to perform an activity in the manner, or within the range, considered normal for a human being (Australian Bureau of Statistics 1993: 6, in Lynch and Veal 1996). Disabilities encompassed such things as arthritis, ear disorders, mental disorders, respiratory, circulatory and nervous system diseases, disorders of the eye, head injuries, strokes, brain damage and other diseases.

The above survey distinguished between disability and handicap. The latter results from a disability linked to certain tasks associated with daily living, in relation to such activities as self-care, mobility, verbal communication, schooling or employment. These definitions of disability and handicap were based on the International Classification of Impairments, Disabilities and Handicaps, published by the World Health Organisation (WHO) (1980).

It has been estimated that in Australia in 1993, there were approximately 3 million persons (18 per cent of the total population) with a disability (Australian Bureau of Statistics 1993: 7, in Veal 1994), 250,000 of whom were classified as having some form of handicap. Handicaps were further identified as ranging from mild to profound, the latter necessitating help to perform one or more designated tasks. The prevalence of various disabilities, based on estimates of the ABS, are shown in Figure 3.2.

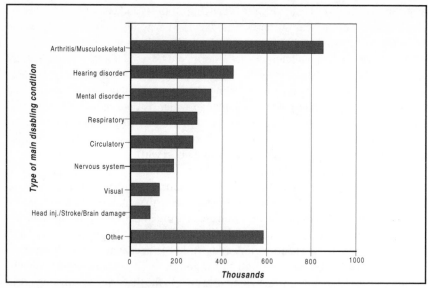

Figure 3.2 People with disabilities, Australia, 1993

Source: ABS 1993, Graph 2: 7, in Lynch and Veal (1996: 324)

The therapeutic value of leisure for people with disabilities

Leisure has therapeutic value and provides a means of integrating people with a disability into the wider community (e.g. see Shivers 1967; Patterson and Pegg 1995). Therapeutic recreation encompasses 'A process which utilises recreation services for the purposive intervention in some physical, emotional and/or social behaviour to bring about a desired change in that behaviour, and to promote the growth and development of the individual' (Gunn and Peterson 1978: 11). It promotes independent leisure for special groups through remedial, educational and recreational experiences that use various activity and facilitation techniques (Veal 1994). Nevertheless, integration is a complex concept and practice, which has been undertaken, in some instances, in haste, and in a piecemeal fashion with inadequate resources (see Patterson and Pegg 1995).

Recreational opportunity for people with disabilities

Developing a spectrum of recreational opportunities for people with disabilities should encompass three main principles: (1) strong leadership, which can overcome the stress and rigours of programme formulation, implementation, monitoring and evaluation; (2) appropriate assessment of the needs and skills of participants who are disabled (i.e. their physical, emotive, cognitive and social requirements); and (3) the means of integrating recreation programmes into the wider community.

Chubb and Chubb (1981) present a useful summary of the effects of disabilities on participation in recreation. The conditions and characteristics listed, range from left-handedness, allergies and aberrations of body size, through impaired manual dexterity and mental retardation, to physical disabilities, including sensory impairment. In the area of outdoor recreation, much emphasis has been given to this last category, especially to those affected by constraints on mobility and access, and by impaired sight and hearing. Recreation assumes great importance in the lives of such people, who often have a greater proportion of leisure time than most others. Yet, opportunities to participate, restricted in the first place by disability, are often worsened by building and design standards, and by regulations and requirements.

The dimensions of the recreation opportunity spectrum for people with handicaps are limited by 'environmental barriers', which are taken to include architectural barriers, transportation problems and societal attitudes (Calder 1974). Recreation participation and spectator opportunities for people who are handi-capped are seriously impaired by barriers of one kind or another, built into the design and construction of public and private buildings, national parks and playgrounds, and other recreation sites and facilities. Steps, gravel, escalators and narrow entrances, all effectively deny or restrict access for many classes of people with handicaps. Transportation, likewise, is often inaccessible to people with handicaps, because of unsuitable design, inadequate services or lack of appropriate facilities, especially space.

Technical approaches are only part of the solution. Attitudinal barriers within the community also have a marked influence on the ease with which people who are disabled can participate in recreational activity. Many individuals with disabilities are developing mature leisure attitudes and skills, and are no longer personally handicapped by their disabilities; they have developed adaptive skills that allow them to enjoy meaningful leisure experiences. Possibly the greatest handicaps they confront are the social barriers that prevent them from enjoying leisure and recreation activities. Such barriers include inaccessible facilities and services, the absence or lack of specialised policies, plans and programmes, and the attitudes of some sectors of the community, who discriminate against people who are handicapped as a minority group, and who, because of misinformation and misconceptions, stereotype people who are disabled as being incapable, unpro-ductive and in need of protection. The attitude of people who are disabled also has a bearing on their ability to make good use of opportunities. Problems of adaptation, education and retraining, especially where the onset of a person's handicap or disability is sudden (e.g. car accident or stroke), can reinforce the already difficult circumstances which tend to exclude these people from the normal leisure experiences enjoyed by the wider community.

Women

There is considerable debate and a growing field of research on gender and leisure (e.g. see Henderson 1994a; 1994b), and on the differences in leisure participation and constraints between men and women (e.g. see Harrington *et al.* 1992; Henderson 1991; Searle and Jackson 1985; Shaw *et al.* 1991; Jackson and Henderson 1995). More specifically, research has documented the similarities and differences in the leisure patterns and processes concerning men and women in the US (e.g. Blood and Wolfe 1960; Komarovsky 1967; Schneider and Smith 1973; Stafford 1980; Shaw 1985, 1992; Firestone and Shelton 1994; Hutchison 1994). In particular, the growing participation of women who voice their concerns, has added insight and significant depth to such research, which, in conjunction with feminist thinking, has developed foci concerning men/women comparisons and the barriers or constraints to women participating in recreational activities.

Women experience unequal access to and participation in leisure, as an inevitable consequence of societal attitudes, perceptions and public policies. Women's leisure is constrained by many factors, including:

* time limitations;
* lack of financial resources (e.g. a socioeconomic system that fails to reward women's labour equitably);
* increased participation in the workforce of industrialised countries by women, exacerbating the inroads into their discretionary time from domestic commitments and reaching the point where there may be little time left for personal pursuits;
* hegemonic constructions of heterosexual femininity – predominant inlfuences depicting women's heterosexual attractiveness as important;
* traditional family and societal arrangements that give men authority over women;
* a judicial system that trivialises male sexual violence against women;
* structural barriers and lack of broad acceptance of female participation in traditionally male activities (e.g see Henderson 1994b).

Women face considerable time and family constraints, and their discretionary time may be severely limited. They may, then, redefine leisure to signify a time when they can combine a leisure activity such as walking in the park or watching TV, with a family or domestic responsibility such as child-care or housework. These patterns are familiar in working-class contexts, where women's access to baby-sitters and household time-saving appliances is more limited, and in social-cultural milieus, where traditional views of a gendered division of labour and a 'woman's place' prevail.

Elderly

In several countries of the Western world, some of the most dramatic demographic changes affecting leisure and recreation demand, supply, and planning and programming, have been shifts in age structure. There is an increasingly significant proportion of elderly and retired people in the population, for whom greater longevity, improved health care and better financial provisions generate demands for new leisure opportunities and requirements. The ageing of populations of industrialised societies has caused government and non-government organisations to ensure that physical planning and service delivery of community recreation resources address the needs of the aged. Participation in outdoor recreation specifically, and recreation generally, often enhances the well-being, quality of life and physical and psychological health of the elderly, and can lead to reductions in social isolation and medical/drug dependence.

The extent and nature of participation in leisure and recreation change with a person's age (e.g. Singleton 1985; Hayslip and Panek 1989; Kelly 1990; MacPherson 1991). Generally speaking, participation in leisure activities declines with age, although there are variations according to one's 'income level, personality, interest, health condition, ability level, transportation, education level and a number of social characteristics' (Hayslip and Panek 1989: 425).

Much research has focused on the relationship between the leisure/recreation behaviour of the elderly and physical or psychological well-being (Iso-Ahola 1988; Coleman and Iso-Ahola 1993; Smale and Dupuis 1993), satisfaction (Kelly, Steinkamp and Kelly 1987; Losier, Bourque and Vallerand 1993; Delin and Patrickson 1994), constraints (Mannell and Zuzanek 1991), recognition and self-esteem (Tinsley, et al. 1987), increased coping skills (Coleman 1993), self-rated health (Delin and Patrickson 1994), and life satisfaction (Hayslip and Panek 1989; Kelly 1990; Hersch 1991; MacPherson 1991). The latter studies show that life satisfaction for older people who are not engaged in paid employment is very closely related to meaningful leisure and recreation participation.

Most forms of leisure, and indeed recreational participation, involve social interaction, which plays an important role in psychological well-being (Smale and Dupuis 1993), and offers many other benefits (see above). Interestingly, research has shown that older single adults, aged over 70, participate to a greater extent in organised social activities than do those who are of similar age and married (e.g. see Thompson 1992). More recent research in Australia indicated that:

> Overall, there appears to be considerable dependence upon relatives and friends as a source of social activities for those aged 60 and over. In addition, there is a trend among those who are single, and who may not have a range of family ties, to be reliant on a wider circle of social contacts.
>
> This is more pronounced for women. Of some question is the social support and network role of the organised groups in facilitating leisure of this age group. Only a small number of respondents (less than 3 per cent) indicated that they relied on organised groups for their participation in recreational activities.

(Simmons and Dempsey 1996: 41)

Leisure behaviour and recreation participation vary between the elderly and the rest of the population. Specific constraints such as lack of transport, poor health, insecurity (even fear), inhibit participation in community activities, so that home-based activities present a safer, more familiar and comfortable environ-ment. 'The elderly as a category are becoming younger, fitter and more affluent' (Veal 1994: 193), but disability is an important constraint for 45 per cent of the population aged 60–65, increasing by age to 83 per cent of those aged over 85.

Retirement, too, is becoming more common among women. Researchers, planners and policy-makers should be directing attention to gender differences in retirement and retirement recreational activities (Mobily and Bedford 1993), because 'it is abundantly clear that elderly women and men participate in different free-time activities' (e.g. Mobily *et al.* 1986).

Retirement impacts on a person's morale, and meaningful use of time in a person's later life becomes a significant adaptive task (Havighurst 1961: 310). Indeed:

> Staying alive requires effort on the part of the older individual to move beyond mere existence and in so doing he or she must be able to demon-strate a willingness to embrace risk, challenge and adventure. An important dimension of adventure is curiosity – the urge to know self as well as the mysteries of life that encompass our physical and social worlds. Samuel Johnson the author of the first dictionary is thought to have argued that curiosity is one of the most permanent and certain characteristics of a vigorous intellect. Youth do not have sole ownership on risk, curiosity, challenge and adventure. If they do, then it is because older people have relinquished these essential ingredients of a vital existence.
>
> (Seedsman 1995: 33)

So, there are marked variations among the elderly in terms of leisure and recreation. Despite these variations, useful generalisations have been made about the leisure needs of older people. These include the need to:

- render some social useful service;
- be considered a part of the community;
- occupy increased leisure time in satisfying ways;
- enjoy 'normal' companionships;
- be recognised as an individual;
- have regular opportunities for self-expression;
- attain a sense of achievement in leisure and other activities;
- access health protection and care;
- obtain suitable mental stimulation;
- acquire suitable living arrangements and family relationships; and
- achieve spiritual satisfaction (Hersch 1991).

Clearly, leisure is a realm of human activity for people of all ages. And perhaps 'the most important condition for good adjustment to the role transitions related to aging is the maintenance of meaningful activity' (Hayslip and Panek 1989; Parker 1979, in Williamson 1995: 63).

Youth

Leisure is a significant component (40–50 per cent) of the life of adolescents (Caldwell *et al.* 1992), and the central role of leisure activities is well-documented (see McMeeking and Purkayastha 1995: 360). According to Willits and Willits (1986: 190), leisure and recreational activities are 'not only ends, providing immediate gratification and enjoyment'. Rather, they are 'part of the learning process whereby the individual seeks to establish his/her personal identity... practices social and cooperative skills, achieves specific intellectual or physical attainments, and explores a variety of peer, family, and community roles'.

Leisure can provide an avenue for the expression and development of identity, autonomy, intimacy and personal growth. Leisure provides the opportunity for young people to hone or test skills and physical endurance, compete against others or better their own standards, and to broaden their general life experience (Iso-Ahola 1980). 'It is in this life phase that much searching is done as young people attempt to recast the identities which have been moulded for them by their parents, caregivers and other significant people and institutions in their lives' (Lynch and Veal 1996: 332). Participation or involvement in leisure activities in a person's adolescent years, in part, shapes the behaviour and attitudes that lead to more permanent patterns in later adolescence and later life (e.g. see Hultsman and Kaufman 1990), even to the extent that about 50 per cent of adults' ten most important recreation activities were begun in childhood (Kelly 1974).

Of course, some young people may choose not to participate in recreation activities. Nonetheless, access to, and participation in, leisure-based activities are influenced by many factors. Access to outdoor recreation activities varies among young people because:

- in urban areas, for instance, there is generally better access to art and cultural activities, sporting events, music (including discos and live bands) and eating venues (see Gordon and Caltabiano 1996: 37) than in rural (especially remote) areas;
- family, significant other adults and peers affect leisure choices and behaviour (e.g. see Snyder and Spreitzner 1973; Iso-Ahola 1980; Caldwell *et al.* 1992). For instance, parents are the major providers of advice and guidance during adolescence, to the extent that parental influence is important in early adolescents' decisions *not to join* an activity (Hultsman 1992; also see Youniss 1980). Parents may even be seen as a 'salient barrier to leisure' (Gordon and Caltabiano 1996: 37);
- access to public and private transport affects mobility in time and space (Hultsman 1993);
- lack of, or decline in, the number of volunteer leaders, has led to the collapse of some youth groups, yet such 'significant adults' strongly influence the recreational activities of adolescents (see Stephens 1983);
- employment, among other things, provides money and social contact, and a feeling of worth.

Constraints on leisure may lead to leisure boredom and, subsequently, deviant involvement (namely drug use and delinquency) (Iso-Ahola and Crowley 1991), and smoking and consumption of alcohol (Orcutt 1984). 'Because motivation is needed for active leisure participation, drug use might affect an adolescent's choices, when it comes to what kind of leisure activities he or she likes to do' (Gordon and Caltabiano 1996: 37). Frequent and/or prolonged drug use could cause physical debilitation, alienation from peers and family, or alter awareness and expectations of life events.

Unemployed youth is a problem which has manifested itself in many industrialised nations. Unemployment may be regarded as a manifestation of enforced free time, with leisure regarded as free time. The free time associated with unemployment is not the equivalent of leisure time. 'Unemployment imposes a number of burdens on individuals and people close to them, however, it also frees up large amounts of time which would otherwise be spent in the workplace or earning an income' (Lynch and Veal 1996: 340).

Unemployed people spend less time on outdoor activities and a great deal more time on home-based activity. If activities are expensive, they are largely curtailed. Activities may also be curtailed because of the social stigma of being unemployed, while diminished income appears to be an important factor in reduced participation in out-of-home entertainment, and in membership in clubs and associations (see Lobo 1995). Furthermore, 'Research on special schemes of public provision showed low participation levels. It is likely that the generally disappointing results of the schemes were due to the consequences of unemployment, namely, psychological, social and financial deprivation' (Lobo 1995: 26).

For McMeeking and Purkayastha (1995), an important consideration, and an issue warranting further research, is the extent to which leisure pursuits for adolescents are mediated by their experience of place. If we extend the accessibility of leisure and recreation opportunities to a person's opportunities to travel, then experiences of place may well be wider for those of higher socioeconomic status (individually, or through their family's wealth), those who are better educated, and those with greater social networks and access to marketing/travel information. As recent research has shown, 'the more leisure opportunities available to individuals, the more they want to participate' (Gordon and Caltabiano 1996: 41).

Stigma upon stigma

Just as the context of, and constraints to, leisure seem to differ somewhat between males and females, between people of different ages, and between people of different socioeconomic status, so differences occur within such groupings. Put simply, any understanding or explanation of leisure constraints must incorporate many diverse variables. What of single fathers? What of the growing number of men, who, either through economic circumstances or choice, decide to assume the role of primary care-giver to children, while the female partner, in a heterosexual relationship, pursues an income and career in the paid workforce? What of men who are labelled househusbands, and who soon find themselves occupying a status which has been the traditional preserve of women? How do these men manage to negotiate the values and practices of conventional masculinity (e.g. see Morrison 1994; Lynch and Veal 1996)? What of women with disabilities, where recent research (Henderson *et al.* 1993; 1995) indicates there is a magnification of leisure constraints for such women? Another study (Davidson 1996) demonstrated that women with young children do not have uniform holiday experiences or perceptions of those experiences.

Different characteristics may result in different leisure experiences among men and women: race, socioeconomic status, marital status, sexual orientation and physical ability (Henderson *et al.* 1995). The issue of sexual orientation also raises an important issue. While gay and lesbian studies appear to have gained increasing research legitimacy in some countries such as Australia and New Zealand, little attention has been afforded (1) the place of leisure in the lives of gay men and lesbians, or (2) the meanings attached to leisure by these groups (Markwell 1996: 42). According to Woodward (1993, in Markwell 1996: 43), 'sexual behaviour in general, and sexual pleasure in particular, has received insufficient attention in the leisure studies literature'.

Clearly, the opportunities for investigations concerning special needs groups are enormous. Specific data on manifestations of disability, gender, race and age (among other dimensions of special needs) are growing, but will never provide answers to all our questions in dynamic, modern, industrialised societies.

There is a clear need for continued questioning of the values that underpin recreational services and facilities, and, no doubt, recreation providers will perceive, and rightly so, many interests in any planning and development processes. If recreation is a fundamental human right, educators, planners and policy-makers must continue to probe the depths of accessibility in all its dimensions, and promote an egalitarian recreation ethic which fully accepts the recreational needs of people whatever their age, race, sex or sexual preference. However, this will only be possible if there is sufficient depth of understanding of constraints to leisure, accessibility to leisure opportunities, and the resources which the public sector and communities (e.g. associations and volunteers) are willing to provide.

Summary

The discussion of the recreation needs and opportunities of special groups, illustrates the broad potential for application of the recreation opportunity spectrum concept as a technique in recreation resource planning and management. However, it needs to be noted that interaction of people with resources is two-way. Understanding recreation behaviour and participation patterns, certainly calls for changes in personal and institutional dispositions involving attitudes and values, if we are to witness a more qualitative dimension to the human condition, and to the leisure part of human existence.

Guide to further reading

* For broad overviews and/or conceptual frameworks concerning *constraints to leisure*, see: Wade (1985); Crawford and Godbey (1987); Jackson, (1988; 1990; 1991; 1994); Jackson and Burton (1989); Crawford, Jackson and Godbey (1991); Jackson, Crawford and Godbey (1993); Veal (1994); Lynch and Veal (1996).
* For an excellent overview concerning *Women, Gender, and Leisure*, see *Journal of Leisure Research*, (1994, vol. 26, no. 1).
* The following suggestions for further reading serve to provide information on a wide range of conceptual and applied issues with respect to groups with special needs. The quite specific nature of many constraints- and special needs-related research means that the title of each work (see Bibliography) is often quite specific about the special group under investigation.

   Lopata (1972); Rapoport and Rapoport (1975); Deem (1982); Hendry (1983); Roberts (1983); Poole (1986); Ferrario (1988); Iso-Ahola and Weissinger (1990); Stokowski (1990); Atkinson (1991); Driver, Brown and Peterson (1991); Lockwood and Lockwood (1993); Sullivan (1993); Spinew, Tucker and Arnold (1996); Heit and Malpass (undated).

Review questions

1 Distinguish between the concepts of 'disability' and 'handicap'.
2 What is meaningful leisure? Does the concept of meaningful leisure take on different meanings for different groups, or is it a generic concept dictated by an individual's circumstances?
3 Apart from the special groups discussed in detail in this chapter, what other special groups can you identify? What makes those groups 'special'? What outdoor recreation planning and policy questions and issues do those groups raise?
4 Identify any policies which have been designed for a special group in your local area. Critically examine the extent to which the impacts and outcomes of those policies have met the needs of their intended audience.

4

OUTDOOR RECREATION RESOURCES

In a perfect world, demand for outdoor recreation activities would be matched by an ample supply of attractive and accessible recreation resources. Barriers to participation would be absent or negotiable, satisfactions sought would be realised, and quality recreation experiences would be the norm. A broad spectrum of recreation opportunities would be presented to potential participants, so that selection of desired opportunity settings was readily achievable, and real choice in the recreation experience was assured.

In reality, interaction between demand and supply factors is qualified by spatial, social/institutional/political, psychological, economic and personal impediments. These impediments prevent or inhibit satisfaction, and detract from the quality of the recreation experience. Thus, the *supply* of recreation resources in quantity and quality, and in space and time, is a fundamental element in creating and structuring fulfilling recreation opportunities. However, understanding of the factors which impinge on the adequacy of supply of recreation resources, calls, first, for consideration of some basic concepts underlying resource phenomena.

Resources – a functional concept

For many people, the concept of resources is commonly taken to refer only to tangible objects in nature which are of economic use (e.g. material substances, including mineral deposits, waterbodies, forests and agricultural soils). An alternative view is to see resources not so much as material substances, but as *functions* which such substances are capable of performing. In this sense, resource functions are created by human society through selection and manipulation of certain attributes of the environment. The physical existence of coal, iron ore or fertile soils does not constitute a resource; such elements *become* resources as a result of society's subjective evaluation of their potential to satisfy human wants relative to human capabilities.

This functional approach to resource phenomena was set out formally many years ago by Zimmerman (1951) and restated by O'Riordan (1971: 4), who defined a resource as:

An attribute of the environment appraised by man to be of value over time within constraints imposed by his social, political, economic and institutional framework.

In these terms, resource materials of themselves are inert, passive and permissive, rather than mandatory, prescriptive and deterministic. Creative use of resource potential requires the existence of a cultural and socioeconomic frame of reference, in which elements of the environment acquire a function as a means of production, or for the attainment of certain socially valued goals.

The existence of a body of water does not necessarily represent a resource in functionally useful terms. Indeed, in some circumstances water might be regarded as a hazard, and even dysfunctional to the utilisation of other more vital resource materials. Any number of attributes or constraints (e.g. size, depth, quality or accessibility) may inhibit the resource functions which water is capable of fulfilling. Creative use of resource potential requires that certain prerequisite conditions be met, among them:

- recognition of functional possibilities of resources;
- the will to exploit them; and
- technological know-how to put them to use.

The global environment offers many examples of materials with functional promise, but which must await the appropriate circumstances before being harnessed for human use. Consideration of resource phenomena in functional terms, also helps explain their changing roles and fluctuating values over time and space. To a marked degree, resource functions of environmental attributes are dynamic, reacting to changes in economic, social, political and technological conditions. Presently valued resources can lose their function as circumstances alter, and previously neglected resource potential may be put to use to meet new and complex demands. The salt resources of biblical times, and charcoal in the era of the industrial revolution, are just two substances which have only limited functional resource value in today's world. Exploitation of Australia's extensive deposits of uranium is subject to the political whim of successive governments and pressure from the 'green' lobby, so that just three mines are operating from a great number of potential resource deposits.

These dynamic characteristics of resource phenomena can be readily demonstrated by further reference to water resources, and to the range of functions identified with particular streams or waterbodies through time. For example, a river, perhaps initially valued merely as a convenient water supply, may subsequently acquire a function as a means of transport, a source of power or even as a waste disposal site. The emerging roles seen for water in outdoor recreation, and as a focus of environmental interest, are further evidence of the way in which changing perceptions of the resource are reflected in pressures to adjust its function.

Equally fascinating is the existence in space of contemporaneous, though contrasting, interpretations placed on an essentially homogeneous resource base. Again, water resources are a good illustration. The same physical attributes of a river valley, for example, can take on different dimensions in the minds of inhabitants aware and capable enough, and prepared to take advantage of, the opportunities offered. Different groups of people, occupying that same environment, may have literally different resources. For some, the valley and its waters represent, perhaps, a tranquil setting in which to carry on traditional farming pursuits; for irrigators and other primary producers (e.g. cotton farmers), the river is seen as providing the means of introducing intensive agriculture. Others may value the waterbody for active or passive recreation pursuits. Contrasting perceptions of what are taken to be appropriate resource functions help explain conflicts which arise over allocation and use. This theme will recur frequently in the issues considered in later chapters that deal with recreation resource use.

Recreation resources

As with the examples noted above, identification and valuation of elements of the environment as recreation resources will depend upon a number of factors (e.g. economics, social attitudes and perceptions, political perspectives and technology). Problems can arise in the identification process because, given the appropriate circumstances, most environments are, in some sense, recreational. Thus, resources for outdoor recreation can embrace a wide spectrum of areas and settings, ranging over:

- space itself (airspace, as well as subterranean and submarine space);
- topographical features, including tracts of land, waterbodies, vegetation and distinctive ecological, cultural or historical sites;
- the often neglected climatic characteristics of an area.

Hart (1966) used the term 'recreation resource base' to describe the total natural values of countryside or a particular landscape. He included in his definition such attractions as the view of a quiet agricultural scene, along with more tangible phenomena such as sites for picnicking, camping and boating. Recreation resources, then, embrace areas of land, bodies of water, forests, wetlands and other features of the natural or built environment in use for recreation. Current use identifies actual recreation resources, while the probability of use indicates potential recreation resources, rather than the characteristics of an area or site.

The process of creation, use and depletion of resources for outdoor recreation differs little from that in other areas of human activity, such as agriculture, forestry or mining. As Clawson and Knetsch (1966) put it:

There is nothing in the physical landscape or features of any particular piece of land or body of water that makes it a recreation resource; it is the combination of the natural qualities and the ability and desire of man to use them that makes a resource out of what might otherwise be a more or less meaningless combination of rocks, soil and trees.

(Clawson and Knetsch 1966: 7)

Recreation resources include natural attributes of the environment, as well as facilities and attractions such as sporting complexes and theme parks. This continuum is implicit in Kreutzwiser's (1989: 22) definition of a recreation resource '... as an element of the natural or man-modified environment which provides an opportunity to satisfy recreational wants'. As with resources in general, the supply of recreation resources depends, initially, on human recognition or perception of the environment as capable of satisfying those wants. However, society must also wish to use the environment for that purpose and have the ability, appropriate technology, organisation and administrative arrangements, to create an attractive, accessible and functional environmental setting for recreation.

Again, in common with the functional approach, recreation resources are not static or constant, but take on a dynamic character varying in time and space. Resources can become redundant. Changing economic, social and technological conditions can reveal new recreation potential in previously neglected areas. Natural resources are cultural appraisals, and what is recognised as a recreation resource by one group of people at one period of time may be of no conceivable use or value to them or others in different circumstances.

The renowned surfing beaches of the Australian coastline, for instance, have really only achieved prominence for outdoor recreation in the past half century, with the relaxation of attitudes to public bathing. To the Aboriginal inhabitants of the continent, they were an important source of food, whereas the early European colonists found the surf a formidable hazard in coping with the isolation of coastal settlements. Moreover, the gleaming sand itself, which to most Australians is an integral and attractive component of the recreation resource base, represents a very different kind of resource function for the rutile miner or the building contractor. 'The coast is not one resource, but many' (Patmore 1983: 209).

Contrasting perceptions of environment help explain conflicts concerning recreation resource utilisation. Forest and wild land recreation, for example, is largely a product of the conservation movement of the twentieth century, and claims on countryside and water resources can conflict with more traditional uses of rural land (see Chapter 8). Further attention will also be given, in later chapters, to the potential for conflict resulting from differing perceptions of the resource functions of water resources and the coastal zone. In the same way, scenic roads, walking tracks and trails of various kinds represent important resources for popular forms of outdoor recreation, but not all of these uses sit comfortably or compatibly with other demands made on such linear resources or adjacent areas. Even the extensive network of public footpaths, so much in demand by ramblers or hikers through the English countryside, can bring recreational users into conflict with neighbouring landholders.

It is important to recall from Chapter 2 that conflict in outdoor recreation need be neither resource-based nor activity-based. Conflict can just as readily arise from the attitudes and mindsets of recreationists, as from competing claims on a common resource base, or incompatible recreation activities. At the same time, many forms of outdoor recreation do not require exclusive use of land or water, but lend themselves to multiple use, in harmony with other resource functions.

Outdoor recreation and multiple use of resources

Outdoor recreation often imposes relatively non-aggressive and benign claims on the resource base, so that it is possible to envisage and actually plan for situations of multiple use. Forest lands and waterbodies are the most common examples of outdoor recreation existing as a compatible partner with the primary role for the resource. However, given the right circumstances, recreation activities can also coexist with agriculture and grazing land (Swinnerton 1982). Although more common with publicly owned resources, opportunities for multiple use can also be found in areas in private ownership.

From a social perspective, multiple use makes a lot of sense, especially where resources for outdoor recreation are limited, or where prevailing conditions limit their recreation resource potential. In economic terms, multiple use is justified if the combined benefits arising therefrom are greater than those from a single use, and are sufficient to cover any additional costs. This is generally accepted, in the sphere of forest management, to include outdoor recreation. In the US, for example, the *Multiple Use Sustained Yield Act of Congress*, provides for recreation as one of the main objectives of national forests. In Australia, managers of public forest lands have been slower to endorse recreation use alongside timber production. It is only in recent years that outright opposition to outdoor recreation in state forests has changed to guarded tolerance, and, now, to commitment to recreational use of forests and specific inclusion of recreation opportunities in management plans.

With water resources, the situation can be more complex. There are many different ways in which streams and waterbodies can function to satisfy recreation wants, and there are different forms (sometimes overlapping) of ownership and management of water resources. Expanding resource potential through multiple use is a challenge, given the diverse interests and requirements of recreational fishing, swimming, boating and passive shore-based recreation (see Table 4.1). Management approaches, based on multiple use of water resources, are explored further in later sections, along with the issue of operation of water storages to provide for outdoor recreation opportunities.

Recreation resources in the built environment

Up to this point, the emphasis has been on examining recreation resources as attributes of the natural environment, many of which are publicly owned and managed. However, a significant part of the recreation resource base comprises

Table 4.1 Examples of water-dependent and water-related recreation activities

Water-dependent activities

Aesthetic appreciation of water	Powerboat racing
Beachcombing	Rafting
Canoeing	Sailing
Crew racing	Shell collecting
Driftwood gathering	Shellfish gathering
Fishing	Small boat cruising
Houseboating	Snorkel or scuba diving
Ice fishing	Surfing
Ice hockey	Swimming
Ice skating	Voyages in cruise ships
Wading	Model boat sailing
Waterfowl hunting	Playing in water
Waterskiing	

Activities that are frequently water-related

Beach games	Pleasure driving
Birdwatching	Relaxing
Camping	Rock or fossil collecting
Hiking	Seasonal homes
Nature study	Sightseeing
Painting and sketching	Snowmobiling
Photography	Sunbathing
Picnicking	Walking

Source: Chubb and Chubb (1981: 314)

components of the built environment, which provide for incidental and perhaps opportunistic forms of outdoor recreation. Some of these come under what Ibrahim and Cordes (1993) call 'private recreation resources', which include private residences, second homes, clubs and organisations of various kinds, shopping centres, and industrial sites. In addition, plazas, malls, school grounds and parking lots, can all offer recreational opportunities in urban settings.

To these should be added purpose-built facilities and attractions which play an important role as recreation resources. Whereas a good proportion of these (e.g. urban parks, sporting facilities and community recreation centres) is the responsibility of government at various levels, many are commercial operations offering diverse attractions and services such as food-and-drink outlets, sports venues, accommodation and theme parks. As pressure on 'natural' recreation resources grows, these created or 'artificial' additions to the resource base will help take the pressure off the natural environment. The success of theme parks, for example, supports the notion that substitution of the distinctive (physical) attributes of a recreation setting might be possible without impairing satisfaction, so long as *functional* similarity is maintained (Peterson *et al.* 1985). Given that the desired attributes of nature can be identified and replicated, or simulated in a less pristine setting, pressure on authentic, nature-oriented environments may be relieved. Moreover, less demanding types of recreation might well make do with more tenuous links with nature. A bush barbecue, for example, does not necessarily have to be sited in a national park.

Ditwiler (1979) takes the consideration of substitutability further by questioning whether particular resources or environments are necessarily a prerequisite for the leisure experience desired. He argues that the experiences people seek from a natural setting, for example, could well be obtained from an artificial environment designed to include those characteristics of the natural environment required for the purpose. If Ditwiler is correct, and many supposed wilderness recreationists are more interested in diversion, excitement or challenge than in nature *per se*, it should be possible to substitute the utility inherent in specifically nature-oriented settings by creating artificial environments. Examples of such substitutions are already numerous, for example, in simulated settings such as Disney's Epcot Center, Florida. Despite scepticism and, perhaps, resistance from 'purists', there could be a useful role for technological ingenuity in helping to alleviate pressure on the natural resource base.

Space, location and accessibility

Identification, assessment, allocation and use of elements of the natural environment as recreation resources will depend upon a number of factors – technological, socioeconomic, political and perceptual. Physical characteristics are, of course, fundamental; water must exist for water-related tourism. However, such variables as space, location and accessibility have a direct bearing on functional effectiveness. Whereas some of these aspects may be offset or foreseen in the creation of 'artificial' inputs to the recreation resource base, they remain important in the ongoing function of all recreation resources to provide satisfying quality recreation experiences.

Outdoor recreation necessarily has its focus in space-consuming activities. Certainly, it is in the spatial distribution and frequent locational imbalance of leisure opportunities where much of the resource manager's interest and emphasis are focused. Space, then, is a critical resource for outdoor recreation, and certain kinds of activities require space with specific attributes, dimensions and qualities (Chapter 6).

As recreationists increase in number and mobility, there is greater pressure on recreational activity space, on service-space for ancillary facilities, and on access-space (e.g. parking areas, routeways). These latter considerations can have a marked bearing on the effectiveness of recreation resource space. Pressure on capacity, too, may stimulate multiple use of space, over time, for varied activities, day and evening, week and weekend and year-round rather than seasonal.

Conditions of location and access are basic to the definition of certain types of recreation space. Indeed these aspects were a major part of the rationale supporting the early classification of recreation areas as user-orientated or resource-based (Clawson and Knetsch 1966). For some outdoor recreation areas, isolation, to various degrees, is vital to maintain the individual experience sought. In this respect, an inherent characteristic of resource-based recreation areas is their locational immobility; they are site-specific to a particular environmental setting. Wilderness areas, for example, are largely delineated by their remoteness and difficulty of access, so that their primitive natural qualities will not be impaired by over-use.

On the other hand, valuation of the spatial element for user-orientated recreation resources depends very much on the location relative to population concentrations. In heavily populated areas, an adequate supply of readily available recreation space is especially valuable. Yet, urban centres often have to 'make do' with fragmented pockets of relict land and water, otherwise useless for the purpose of economic return. In many cases, these are legacies from past sub-divisions, where the statutory proportion of an estate set aside for open space has been selected on the basis of its inherent poor quality, either locationally or for residential purposes. The result is recreational space which is not appropriately sited for present needs, nor flexible or adaptable enough in quality to cater for the changing character of dynamic urban populations.

Awareness of deficiencies in current allocations of land for outdoor recreation has beneficial connotations in terms of planning for the more effective selection and location of new recreation areas in both intra- and extra-urban environments. Some constructive efforts in this regard have been made towards the provision of an improved environment for human recreation needs in new towns in Australia, Britain and the US. In such decentralised communities, the emphasis is on co-ordinated planning for the new city, together with its surrounding region. The maintenance of environmental values as a basis for recreational amenity, and provision of a wide range of accessible sites and settings, are seen as essential strategies.

At a finer scale, the relative disposition of spatial components into access and space and viewable space, may contribute to a fuller realisation of the recreation experience. The complementary nature of these fundamental spatial elements can be illustrated with reference to Banff, Alberta, Canada. The mountains and lakes (viewable space) in this vicinity rate low as occupiable space, but greatly enhance the appeal of the valley landscape, within which the resort and its access routeways are situated.

A similar distinction is made by Gunn (1988) in his study of the planning of tourist regions. Gunn suggests that a tripartite approach should be adopted in the design of tourist attractions. Stress is placed on the spatial relationship between the prime element, or *central attracting force* (e.g. a waterfall) and its essential setting, or what Gunn calls the inviolate belt. The function of this *setting*, or *entering space*, is to condition the visitor's anticipation in an appropriate fashion, so as to enhance the subsequent recreation experience. The third conceptual element is the *zone of closure*, or outer area of influence, containing service centres, circulation corridors and transport linkages. This functional component completes the tripartite concept and contributes to the wholeness of an attraction.

The example of Canada's Niagara Falls illustrates the importance of a complementary environmental setting in adding to, or detracting from, a prime tourist attraction. Niagara Falls is categorised as one of the 'wonders of the world', with the immediate vicinity being enhanced by an attractive reserve. However, only the most insensitive visitor could remain indifferent to the garish vulgarity of the outer approach to the Falls, marked as it is by inappropriate and poorly designed services and facilities. It is almost as if human beings had set out to mask the natural splendour of this magnificent feature with a veneer comprising the worst aspects of landscape design and commercial display.

A comprehensive approach to the planning and design of recreation space, therefore, should give attention to the regional background, the approach and means of access, and the immediate setting, all of which complement, or detract from, the satisfaction visitors and users derive.

Resource-based recreation areas are, by definition, located at the site of the prime element or attraction. Yet, even here, scope exists for manipulating the attributes of resource elements to enhance or inhibit their recreational function. In so doing, the 'location' of recreation resources and hence recreation opportunities is essentially being arranged to meet management objectives. A forest or a water-body, for example, may have several areas with recreational potential. However, it is the selective provision of access and facilities that will determine the location of those sites to function as recreation resources. In the same way, and on a larger scale, an agency may attempt to correct imbalances in the location and spatial distribution of visitors to national parks by strategically allocating ancillary facilities to selected sites (see Chapter 10).

Accessibility is a fundamental concomitant of locational aspects of recreation environments and of the functional concept of recreation resources. The presence or absence of roads, trails, parking space, boat ramps, airports or helicopter pads, can all impinge on the functional effectiveness of the recreation resource base. However, it is difficult to generalise when conditions of access for special groups, such as children, the elderly and people with disabilities, are considered. Moreover, it is not so much a question of physical access as of legal, institutional, and perhaps socioeconomic constraints on movement into and through recreation space. Such situations raise complex questions regarding public rights and access to common property resources; some of these are examined in Chapter 8.

The question of what constitutes a recreation resource and what factors add to, or detract from, the quality of the leisure environment can best be answered by a systematic assessment of resource potential. The task begins with identification and classification of elements of the recreation resource base.

Classification of recreation resources

An important initial stage in the resource creation process is an inventory and assessment of the quantity and quality of resource materials; those presently valued as resources, and those which may function as recreation resources in different socioeconomic and technological circumstances. Such stocktaking is necessary before the significance of stocks can be evaluated.

However, inventories themselves are of doubtful value; what is required is more than a simple listing of resource materials. The resource elements must be described and classified according to some recognised and agreed system in order to determine categories of resource deficiency and surplus as an input to recreation planning.

Classification of recreation resources can be approached from several angles. One of the earliest systems was devised by Clawson *et al.* (1960), who distinguished between recreation areas and opportunity on the basis of location and other characteristics, such as size, major use and degree of artificial development. Under this system, recreation areas were arranged on a continuum of recreational opportunities from user-orientated through intermediate to resource-based.

The Clawson system of classification has been widely applied, although the terminology can be confusing. All recreation areas must be user-orientated to some extent if they are to satisfy the functional concept of resources. Exclusion from the resource-based category of urban and near-urban recreation sites reflects a narrow interpretation of the term 'resources'. Obviously, there is scope for considerable overlap, and city parklands can be just as much 'resource-based' as remote wilderness. Moreover, large national parks, such as those which ring the city of Sydney, Australia (see Chapter 9), and which are close enough for casual day visits, actually qualify as user-orientated recreation areas. Obviously, too, there can be interchangeability over time, as resource-based areas, for example, come within the recreation opportunity spectrum of an increasingly mobile and affluent user population.

Various adaptations of the Clawson system of classification of recreation resources have been devised, using combinations of location, physical attributes, facilities, and type of recreational experience and use. An example presented in Table 4.2 encompasses both resource-based and built recreation facilities. Chubb and Chubb (1981) distinguish between the following classes of recreation resources:

- *Undeveloped Recreation Resources*, including land, water, vegetation and fauna;
- *Private Recreation Resources*, taking in 'personal resources' such as residences and second homes, private organisation resources (e.g. clubs), resources of quasi-public organisations (e.g. conservation groups), and farm and industrial resources;
- *Commercial Private Recreation Resources*, including shopping facilities, food-and-drink outlets and sports facilities, amusement parks, museums and gardens, tours, stadiums, camps, and resorts of various types;
- *Publicly Owned Recreation Resources*, covering local and regional parklands and sports facilities, state and national parks and forests, trails, tourist facilities and institutions;
- *Cultural Resources* in the public and private sector, including libraries and facilities for the Arts;

Table 4.2 A taxonomy of leisure facilities

Facilities not existing primarily for leisure	Resource-based facilities adapted for leisure	Built facilities adapted for leisure	Built facilities designed for passive leisure	Built facilities designed for active leisure
Agricultural land	Woodland parks	Historic houses	Museums	Marinas
Commercial woodland	Urban/rural parks	Ancient monuments	Galleries	Leisure/sports centres
Watercourses	Golf courses	Redundant churches	Libraries	Dance halls
Water masses	Beaches	Warehouses/industrial buildings	Arts/community centres	Squash/tennis centres
Private dwellings	Cruising waterways		Cinemas	Gymnasia
Workplaces	Public footpaths		Restaurants	Swimming pools
Streets	Canal towpaths		Hotels	Holiday camps
Moorland/mountains	Watercourses		Shopping malls	Snooker halls
Reservoirs	Water masses			Sports stadia
	Watercourses			Playgrounds
	Water masses			All weather sports pitches
	Zoos			Sports clubhouses
	Theme parks			Theme parks
	Open air museums			Shopping malls
	Holiday camps			

Source: Adapted from Ravenscroft (1992)

- *Professional Resources*, which can be divided into two broad areas: administration (the organisation of recreation systems, policy making and provision of financial support) and management (research and planning, design, construction and maintenance, resource protection and programming).

Clearly, such a listing of recreation 'resources' is all-embracing, consisting of a broader recognition of resource phenomena than those typically associated with outdoor recreation. Whereas all of the 'resources' included in the classification can undoubtedly function to satisfy recreation demands, the intention, here, is to focus on those elements of the natural and built environment (whether in the public or private sector), which are in use, or which have potential for use as *outdoor* recreation settings.

At a finer scale, classification systems have been developed for specific categories of recreation. Gold (1980) classified recreation space on the basis of location, function, capacity and service area, ranging from 'home-oriented space', through 'neighbourhood' and 'community space' to 'regional space'. Such classifications have much in common with Mitchell's (1969) hierarchy of urban recreation units, and have merit in exposing and correcting deficiencies in providing urban recreation opportunities.

Recreation resource capability and suitability

Increasingly, static classifications, specific to a particular time period, are being replaced by resource capability assessments of the potential of an area for a specified use. The assessment may be for one purpose or a combination of purposes. Perhaps the most ambitious and exhaustive scheme for classification of recreation potential has been carried out in Canada as part of the Canada Land Inventory (CLI), a comprehensive project to assess land capability for five major purposes – agriculture, forestry, ungulates, waterfowl and recreation. The inventory has been applied to settled parts of rural Canada (urbanised areas are excluded), and is designed for computerised data storage and retrieval as a basis for resource and land-use planning at local, provincial and national levels.

In marking the first 25 years of the Canada Land Inventory in 1988, Environment Canada highlighted the various features of this unique planning tool (see Table 4.3).

Despite its long life, the Canada Land Inventory continues to be relevant to a variety of land use and resource management problems across the country. Key questions in regard to outdoor recreation are: how much good quality land is available for this purpose? where is that land located in relation to potential users? The example 'New Parks with Water Access?' demonstrates how CLI information can assist in the planning of new recreation opportunities.

Again, the Canada Land Inventory can help planners choose among options where potential conflict exists. Targeting of areas of conflict can be indicated by a simple comparison, using overlays of single sector maps of the same areas. These

Table 4.3 Canada Land Inventory highlights: 1988

- One of the largest land inventories ever undertaken in the world, the CLI covers about 260 million ha of southern Canada for five resource sectors.

- Prime agricultural lands in British Columbia comprise only 5% of the Province's total area, and are under continuing pressure from competing uses. In the 1970s, the CLI was instrumental in helping this province and others move quickly to designate agricultural lands for protection.

- Acid rain is a critical threat to Canada's environment. Analysis using the CLI has demonstrated that more than 70% of Eastern Canada's prime resource lands receive acid rain in levels threatening sustainable productivity.

- Waterfowl are an important economic, recreational and ecological resource. In support of the North American Waterfowl Management Plan, the CLI has helped screen out areas where habitat maintenance might conflict with agricultural production.

- The CLI led to the development of the Canada Geographic Information System, a first-of-its-kind technical accomplishment in handling resource information.

- Not simply a valuable planning tool of the past, the CLI, established 25 years ago, can make a vital contribution to efforts to understand and resolve urgent environmental problems.

Source: Environment Canada (1988: 1)

Example: *New Parks With Water Access?*

Suppose that park planners wanted to determine the potential for establishing, within an hour-and-a-half drive of our major cities, more parkland on the shorelines of lakes, rivers and oceans. While the eventual decisions will be based on such factors as land ownership, price, accessibility and current uses, the first step, obviously, is to find out 'how much' there is and 'where it is'.

The CLI outdoor recreation inventory provides exactly this information. Table 4.4 summarises data on capability Classes 1–3 inventoried shoreline within a 121 km (75 mile) radius of selected cities. CLI maps will show the location of these sites, as well. The figures indicate that the potential for new shoreline parks varies greatly from city to city. Sudbury and Ottawa top the lists, while Regina and Calgary have limited potential.

The CLI has not identified the precise location of new park sites. But it has enabled planners to focus their search very quickly, allowing them to target detailed planning on the most promising shoreline sites.

Source: Environment Canada (1978)

techniques were the forerunner of the much more complex computer analyses in use today. The example 'Ducks and Wheat?' highlights sites of potential conflict between competing resource functions for the same area of land either as prime agricultural land or a prime habitat for waterfowl and the basis of extensive and valuable recreation activity.

In this way, the CLI system permits inter- and intra-sectoral comparisons to delineate suitable locations for recreational development, and to define priority areas between competing uses, as well as opportunities for compatible multiple resource use. However, the methodology does have some deficiencies. Although the classification is designed to accommodate a wide range of (then) popular outdoor recreation activities, inevitably it excludes certain pursuits (e.g. hang gliding), which have emerged since, as a result of technological advances and increasing levels of specialisation in recreation activities. Nor does the scheme have much to say about the *quality* of a recreation experience. Recreation resource capability is equated with the *quantity* (or level of use), which does not always coincide with satisfaction.

In Australia, classification schemes focusing on outdoor recreation capability have been applied to Crown (public) lands. The classification can apply to a range of land categories, from 'remote natural' to 'urban', reflecting the area's physical, social and management characteristics relative to its capability to provide opportunities for land-based or water-based recreation activities (Table 4.5).

Resource capability is a measure of the feasibility of allowing a range of specified resource uses on an area of land, reflecting both the likely productivity and resilience of the site. Whereas resource *capability* is based on mainly natural physical attributes, resource *suitability* is a socioeconomic and political evaluation of the acceptability or desirability of a particular resource use.

In terms of the suitability of a resource for recreation use, relevant issues include community demands and expectations, government policy, and conflicting needs of different user groups. The concept of resource suitability has four basic components:

Table 4.4: CLI recreational capability classes 1–3: inventoried shoreline within 121 km of the centre of selected metropolitan areas

Metropolitan Area	Class 1 (km)	Class 2 (km)	Class 3 (km)	Total (km)
Halifax	—	29	433	462
Montreal	12	253	1,785	2,050
Ottawa	80	471	3,076	3,627
Sudbury	24	221	4,181	4,426
Regina	5	19	285	309
Calgary	4	59	132	195
Vancouver	69	456	1,588	2,083

Source: Environment Canada (1988: 4)

Example: *Ducks and Wheat?*

Suppose wildlife managers want to protect prime waterfowl habitat in a particular rural municipality. They know that such habitat is frequently located throughout good farmland. Their goal is to locate those prime waterfowl areas that do not conflict with agricultural production.

As a first step, they obtain the CLI agricultural capability and waterfowl capability maps covering their rural municipality. Then they identify all the Classes 1–3 agricultural land and all the Classes 1–3 waterfowl area. The comparison identifies the general location and extent of areas where there are overlaps, and areas where there are no conflicts with agriculture. This latter group will be candidates for protection (Figure 4.1).

Classes 1 - 3 Agriculture

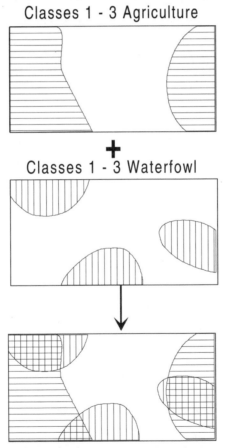

Classes 1 - 3 Waterfowl

Figure 4.1 Overlay of agriculture and waterfowl capability maps

Source: Environment Canada (1988: 4)

Table 4.5 Outdoor recreation capability classes

Classification activities		Physical setting	
Remote Natural	**Land based** Viewing scenery Bushwalking and hiking Camping Nature study, interpretive services Abseiling and mountain climbing	Size	1000 ha. or more (may be smaller if contiguous with natural areas; namely National Parks and State Forests etc).
		Access	Area at least 1km from all roads and tracks, which are usually open to motorised use.
	Water based Canoeing – rowing Swimming Fishing Surfing (body)	Man-made Modification	Essentially an unmodified natural environment which may contain a limited number of tracks. Buildings are very rare and isolated.
Natural	**Land based** Viewing scenery Picknicking Bushwalking, hiking, horseriding Camping	Size	No set criteria (generally over 20 ha) but may be isolated pockets in urban settings. Coast lands variable but contiguous with natural lands.
	Nature study, interpretive services Off-road vehicles Fossicking	Access	Area normally within 1km from primitive roads and tracks, which are usually open to motorised use.
	Water based Canoeing – rowing Sailing Swimming Diving Fishing Surfing (all)	Man-made Modification	A natural area with subtle-to-dominant modification. Serviced with primitive to sealed roads. Buildings are rare and generally isolated.
Rural	**Land based** Viewing scenery, arts and crafts Individual and team sports Picknicking	Size	No set criteria (generally over 5 ha) Coast land variable but contiguous with rural lands.
	Walking, bicycling, horseriding Sightseeing drives	Access	Area serviced by a formed road.
	Club and kiosk services Showgrounds and cemeteries Motorised transport (automobiles, 4-wheel drives, motorcycles) Camping and caravans; cabins Racing; games and playgrounds; fossicking **Water based** Canoeing – rowing; sailing; power boating; water skiing; fishing; diving; swimming; surfing (all); marinas; water sport.	Man-made Modification	A culturally modified environment reflecting past or current practices in agriculture, clear fell forestry, extractive industries, utility corridors etc. Area is serviced by formed roads to highways. Buildings are common and may range from scattered to small clusters, e.g. power lines, towers, resorts, marinas, pit heads and farm buildings.
Urban	**Land based** Viewing scenery, arts and crafts, bands Individual and team sports; picknicking Walking, bicycling and horseriding	Size	No set criteria (generally ¼– 20 ha) Coast land variable but contiguous with urban lands.
	Sightseeing drives; club and kiosk service Motorised transport (automobiles, 4-wheel drives, trains, buses, motorcycles) Camping and caravans; games and playgrounds	Access	Area normally accessed by foot, bicycle or, if necessary, vehicle.
	Water based Canoeing – rowing; sailing; power boating; water-skiing; fishing; diving; swimming; surfing (all) Wharfs, jetties; marinas; water sports	Man-made Modification	A structure dominated environment serviced with sealed roads to highways. Natural or 'natural appearing' elements may be present but buildings and building complexes are dominant, namely resorts, towns, industrial sites, residential areas.

Source: New South Wales Department of Lands (1986: 36)

Management setting (controls may be physical, e.g. barriers, or regulatory, e.g. laws)	Social setting (applies to a typical day's recreation. Peak days may exceed these limits)
Little or no on-site management control.	Frequency of contact with other users low on tracks and very low at campsites.
Management control is subtle-to-noticeable but in harmony with the natural environment.	Frequency of contact low-to-moderate on roads, tracks and developed sites. Low elsewhere.
Management controls obvious and numerous. Largely in harmony with the man-made environments.	Frequency of contact is moderate-to-high on roads, tracks, and in developed sites. Moderate away from developed sites.
Numerous and obvious on-site management controls.	Large number of users on-site and in nearby areas.

- *Economic Efficiency* – the allocation of resources to the use which yields the greatest financial return (e.g. if assessment shows that a parcel of land possesses a high capability for cultivation, preference should be given to cultivation, not an alternative use such as grazing);
- *Social Equity* – ensuring the equitable distribution of social benefits and social costs by adopting a particular form of land use. For example, one issue at the centre of past rainforest logging disputes in New South Wales, Australia, was the conflict between the tangible timber production and job security associated with the logging industry, and the more intangible loss to the National Estate of the rainforest resource itself;
- *Community Acceptability* – including such aspects as:
    - changing social attitudes (e.g. concerning environmentally incompatible land uses and external effects, such as adverse noise and visual impacts)
    - regional and community needs for land resources;
    - public participation and political influence;
- *Administrative Practicability* – the adequacy of existing infrastructure and services (e.g. accessibility to roads, education facilities, community services, power supply and sewerage) and the economic efficiency of providing services to undeveloped sites (New South Wales Department of Lands 1986: 50–1).

Recognition and classification of resource phenomena for outdoor recreation use call, first, for identification of the capability of the resource base to provide for a range of recreation experiences. However, this is only an indication of what recreation activities the area may support. From these possibilities, it is essential that the most desirable option/s or preferred resource uses be selected. Resource suitability takes account of physical capability, but focuses on choosing the recreational use which best satisfies community demand and government priorities.

Evaluation of recreation environments

Measurement of the capability and suitability of the resource base to support various forms of outdoor recreation is a more difficult undertaking than if the task is confined to classification or assessment for a single purpose. Yet, an area seldom provides for only one kind of recreation, and it is more realistic to consider several activities, with due regard for the complex relationships between outdoor recreation and other resource uses. The complexity of the task is typified by an early investigation undertaken in central Scotland to identify and evaluate recreation environments at a regional scale, on the basis of functional connections between different recreational activities, resources and users (Duffield and Owen 1970; Coppock, Duffield and Sewell 1974).

The approach adopted was to make four separate, independent assessments of the components of resource capability for outdoor recreation, and then to combine these into one single assessment. The components used were: suitability for land-

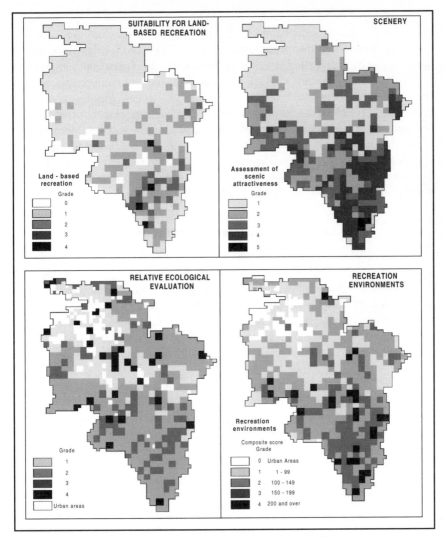

Figure 4.2 An assessment of the recreation resources of Lanarkshire

Source: Duffield and Owen (1970: Appendix 1)

based recreation; suitability for water-based recreation; scenic quality; ecological significance. The basic spatial unit was a 2 kilometre (approx. 1.25 miles) square grid overlay, covering the study area. Data on the four components were obtained from existing maps, aerial photographs and other published sources, and each grid square was evaluated separately for each category. The combined grid scores were then graded into six classes and mapped to indicate composite recreation environments (Figure 4.2).

Experimental techniques, such as used in the Scotland example, can always be subject to scrutiny, and adaptations to suit specific areas under study are advisable. The authors conceded that the assessment might be altered if other features and forms of outdoor recreation which attract visitors to the Lanarkshire region had been considered. The somewhat arbitrary choice of the four components, and the decision to give them equal weight, can also be challenged. Moreover, negative factors such as lack of access or the attitudes of management authorities, need to be examined at a subsequent stage of resource assessment and development.

Implicit in the discussion of recreation resource classification and evaluation has been the existence and even acceptance of, a strong element of subjectivity in the assessment process. One of the most difficult areas to contend with from the point of view of subjectivity is that of landscape evaluation.

Landscape as a recreational resource

Until relatively recently, landscape has been largely ignored as part of the recreation environment. However, growing concern for environmental quality has led to recognition of the scenic quality of landscape as a major recreational resource in its own right, rather than merely as the visual backdrop for other recreation pursuits. This, in turn, generated interest in systematic attempts to evaluate scenic beauty, and to examine the features of landscapes which contribute to their attractiveness and to their resource value in outdoor recreation (Robinson *et al.* 1976).

That said, difficulties remain with assessment procedures because of the intangible and multi-faceted nature of landscape, which does not permit precise measurement. The resource function can take on several dimensions, depending upon which senses are being satisfied and the characteristics of the population involved. These difficulties are compounded by the assessment of recreational values. Whereas most landscapes probably have some recreation potential, this fact is not easy to establish with any agreement, because of the personal nature of recreation and the subjective manner in which it is experienced. Generalisation and interpersonal comparisons are of doubtful validity, and the multiple characteristics of landscape make dissection and evaluation a risky undertaking.

Despite the essentially subjective nature of the variables involved, efforts are being made towards identification and measurement of scenic landscape values in response to competing resource users. In the ensuing discussion it is useful to distinguish between landscape *character* and landscape *quality*.

Analysis of landscape character is essentially descriptive and concerned with the attributes or components of landscape which constitute it as a visual entity – landform, water, vegetation, buildings and the like.

In contrast to landscape character, landscape quality is essentially a comparative, evaluative concept, subsequent to determination of landscape characteristics. Assessment of landscape quality is a three-phase process (Unwin 1975):

- *Landscape Description* – relatively objective inventory of landscape elements or characteristics (above), and classification of landscape types, without any scoring, ranking, or reference to quality;
- *Landscape Preference* – establishment of visual preference ratings or indices for landscape characteristics or types, based on personal value judgements or the opinions of panels of experts or representative populations;
- *Landscape Evaluation* – assessment of the quality of the particular landscapes under study in terms of the values or preferences expressed.

A variety of methods and procedures has been suggested to manipulate and rank landscape attributes in order to establish visual preferences. Criticism has been directed at particular approaches which claim to present an *objective* measurement of landscape quality. Such refinement is impossible with such an inherently subjective process, no matter what sophisticated analytical methods are used. Landscape evaluation can never be divorced entirely from subjective interpretations, and the best that can be achieved is some appropriate balance between operational utility and scientific elegance.

Apart from problems of subjectivity and the appropriate mix of landscape attributes, assessment procedures are unable to cope, as yet, with the internal visible arrangement and spatial composition of landscape (as opposed to its resource content). With all the difficulties and reservations regarding methodology, why bother evaluating landscape?

The answer can be found in the four broad objectives designated for landscape management by Penning-Rowsell (1975):

- *Landscape preservation* – identification of areas of landscape worthy of preservation and deserving of priority for conservation;
- *Landscape protection* – as a basis for development control decisions, to guide the direction of development, monitor environmental impact and provide for planned landscape change;
- *Recreation policy* – aimed at enjoyment of the landscape and realisation of its long-term potential for appropriate forms of outdoor recreation;
- *Landscape improvement* – identification of visual features which detract from landscape quality, so that such 'eyesores' can be removed or modified.

To these, could be added the requirement to satisfy a growing body of environmental law in many countries. In the US, for example, all government agencies must ensure that 'environmental amenities' are given appropriate consideration in planning decisions. In Australia, too, there is a growing commitment to landscape conservation, in both the natural and built environment. Over the years, several urban renewal projects have been delayed or abandoned because of 'green bans' imposed by trade unions, or as a result of the actions of conservationists anxious to preserve parts of the national heritage and landscape. More generally, though, environmental planning legislation in the States and Territories contains guidelines concerning environmental impact assessment and public participation in urban and rural development. However, environmental legislation can be thwarted by governments and developers. 'Secret dealings between State government officials and private developers have always been a feature of coastal resort development and land speculation in Australia, though there have been marked variations between the different States at different times in terms of the extent of the practice' (Mercer 1995: 173). In brief, development fast-tracking, statutory amendments, special deals and political favours are not uncommon in Australian tourism and related recreational development.

The national government in New Zealand is a key resource manager through the Department of Conservation, while the operational role of local authorities has been redefined through the introduction of the 1991 *Resource Management Act* (RMA). This Act represents an important attempt at regulating for sustainable development, which encompassed a redefinition of the structure and roles of local government. The RMA replaced many pieces of legislation with a more comprehensive framework for the allocation and management of resources (NZTB 1994, in Kearsley 1997). It outlines the responsibilities of central and local government, and is implemented by way of a hierarchy of policies and plans. However, perhaps the most significant change in planning methods, is in the explicit departure from prescriptive criteria for resource allocation (or 'zoning' approach), to a system focusing on the effects of activities rather than their intrinsic nature. 'The suitability of a particular use of land is determined by what its environmental outcomes might be rather than by what it is' (Kearsley 1997: 57).

Evaluation of recreation sites

At a finer scale, attention should be directed towards evaluation of the potential of the resource base to support a specific recreation activity or experience at a specific site. Evaluation at this level calls for a different approach from that used in broad regional assessment, especially where questions of land tenure, access and management can often be disregarded.

Site evaluation assumes knowledge and understanding of the detailed resource requirements for each type of recreation involved. The following is a list of the kinds of questions which might need to be answered:

- What kind of topography is most suitable for bushwalking, horseriding or trail bike riding?
- What river conditions are ideal for white water canoeing, trout fishing or bathing?
- What types of vegetation are preferred for orienteering or children's adventure play?
- What snow conditions are best for snow climbing, cross-country skiing or snowshoe walking?

- What characteristics make a rockface good for mechanical climbing, or a pool suitable for fly-casting? (Hogg 1977: 102).

For certain activities, conditions are necessarily more specific and closely defined than others, which are more flexible. Competitive activities, for example, are generally more demanding than less formal recreational uses of countryside. Physical and natural circumstances will be most important for some forms of recreation, whereas for others, social factors may need to be taken into account, and created facilities and infrastructure may be mandatory for effective functioning of the recreation resource base.

Recognition of recreation site potential involves synthesis of an 'identikit' specification (incorporating all the relevant site factors), which conforms most closely to the ideal. However, in the evaluation process, it is important to distinguish between *minimum* and *optimum* site requirements. The former represent a threshold or entry zone concept, in that they describe the set of obligatory conditions within a narrow range of acceptability that is essential if the activity is to take place at all. Unless such bare minimum standards are satisfied, the type of recreation envisaged cannot be accommodated. Optimum requirements imply a preferred situation such as might be demanded or experienced at an Olympic site of national or international repute. Such standards are obviously much more demanding and precise, and are applicable only to exceptional situations.

Once the necessary minimum site conditions have been established, a method of ranking or rating is needed which reflects the relative importance of each requirement, and which indicates whether it is considered an asset or a constraint. In an Australian study, Hogg (1977) lists the natural and cultural factors he considers important for overnight bushwalking or hiking (Table 4.6). Unless *all* site (route) requirements are held to be *equally* fundamental, the points allotted and rating scales used, need to be adjusted to reflect their greater or lesser importance. Without some weighting of this kind, serious deficiencies in more critical factors can be largely overcome by high ratings for more trivial aspects. Moreover, a zero rating given to indicate *total* unsuitability on the basis of a single vital factor can be swamped in the additive process by high ratings for more mundane requirements.

Thus, in Hogg's example, presence of quality campsites is assigned the highest maximum positive points value (25), whereas refuge huts are apparently considered of passing importance to bushwalkers and are lowly rated (5). On the other hand, the presence of hazards is judged to be a most serious constraint (–50), and some factors such as restrictions on access, are considered so critical as to be allotted the most negative value of –00. In Hogg's evaluation system, a recreational unit in which any *single* factor is rated –00 will also have a *total* rating of –00, and, therefore, is judged as totally unsuitable on the basis of that one factor, no matter how favourable other factors may be.

Actual evaluation of a specific site or recreational unit involves scoring the resource endowment according to the degree to which it satisfies each of the user requirements identified, and how it matches up to the conditions stipulated. This

79

Table 4.6 Factors affecting suitability for overnight bushwalking

Factor	Maximum value	Minimum value
Natural factors		
1. Topography (steepness and variability of terrain, length of uphill climbs)	15	–00
2. Rockiness of terrain for walking	0	–5
3. Weather characteristics during walking season	15	–15
4. Ease of negotiation of vegetation (denseness of scrub, fallen timber, blackberries, nettles, etc.)	0	–30
5. Presence and quality of campsites (ground suitable for pitching tents, firewood and drinking water, general environment)	25	–00
6. Extent of area	15	–00
7. Proximity to users	15	–5
8. Scenic quality (general attractiveness, variety, special features)	10	–5
9. Availability of drinking water between campsites	5	–8
10. Miscellaneous attractions (wildlife, swimming holes, historical features, etc.)	10	0
11. Undesirable features (snakes, leeches, mosquitoes, bushflies, etc.)	0	–00
Cultural factors		
12. Access to suitable starting points	20	–10
13. Tracks suitable for walking (related to 5)	15	0
14. Unnecessary or undesirable tracks and roads	0	–00
15. Refuge huts	5	0
16. Unnecessary or undesirable buildings	0	–00
17. Adequate track marking and signposting (including snowpoles)	8	0
18. Escape routes for emergency use	8	0
19. Hazards (e.g. mineshafts)	0	–50
20. Presence of conflicting recreational activities	0	–00
21. Presence of other conflicting land uses (e.g. logging, mining, grazing)	0	–00
22. Restrictions on access or certain activities (e.g. camping, fires)	0	–00

Source: Hogg (1977: 106)

gives an indication of the potential of a site for a specified form of recreation in terms of the presence or absence, and quality, of certain features. The evaluation, therefore, provides a kind of inventory and appraisal of the site's latent potential to supply particular recreation resource functions, although this does not mean that development of this potential will necessarily occur.

Evaluation is not always a straightforward field-checking procedure, and suitability scores should not be accepted without qualification. Some conditions need to be sustained over time, and others may only be ephemeral or present intermittently (e.g. wave conditions for surf-board riding). It is important for the assessor to be able to recognise in a low-scoring site, latent potential which can be realised if certain shortcomings are remedied by provision of additional features and sound management. Conversely, such insight is just as vital in the detection of inherent disadvantages (e.g. ground cover with low tolerance to trampling, or an unreliable water supply), which may become obvious with use and create a problem for subsequent management.

Evaluation of streams and routeways

Linear recreation resources such as streams and scenic routeways frequently call for specialised application of evaluation techniques. These methods seek to identify and measure or rank those physical, cultural and aesthetic attributes of a river and its environment which are considered significant when assessing its recreational value. Typically, the schemes divide the river into manageable segments, for analysis from maps, air photographs and on-site inspection. An element of subjectivity, again, is inevitable in judgements concerning the features to be assessed, the recreation activities envisaged, and the scoring and weighting procedures adopted. Most of the methods focus on relatively remote river resources, although some attempts have been made to develop and apply criteria for evaluating urban settings for recreational use, close to the centre of the river recreation opportunity spectrum.

Efforts should also be made to incorporate the concept of carrying capacity and the limits of acceptable change into evaluation and classification schemes, *and* to provide for user perceptions and public participation in the process. It is clear that there is still some way to go before development of an effective technique that will allow for the dynamic nature of the river resource and its potential users, and that is capable of being replicated in many different river situations.

One of the first systematic attempts to identify and assess scenic routeways was made by Priddle (1975) in Southern Ontario. Priddle's approach was also to break each selected routeway into segments, based on intersections or major changes in landscape. Each segment was then traversed and evaluated in terms of the distance that could be seen, the alignment of the road, and the scenic features and variety present.

The method was refined and developed further by Prior and Clark (1984) in an Australian study in the Hunter Valley region. In this study, the authors identified the essential components of a scenic road system and their relative importance, through a survey of likely users. The responses were used to develop a weighting system to evaluate standard and scenic quality in the valley. The method represents a rapid quantitative assessment of the major components of a rural scenic/ recreational route network to accommodate pleasure driving. Whereas some of the parameters used might be clarified, and additional aspects considered, evaluation of linear recreation resources in this way can provide valuable input to decision-making.

Summary

Classification and evaluation of resource potential are critical elements in recreation planning and management, but make up only one phase in the formulation of a rational strategy for recreational development. Comparative evaluation of resources provides valuable input to the process of informed, effective choice and decision-making. However, strict adherence to evaluation procedures and an over-rigid application of the findings, may obscure opportunities for substitution between sites of recreation activities with more flexible user requirements, and, hence, higher spatial elasticity. Scope should always be allowed for interpretation and sound judgement by management in the incorporation of assessment data into recreation resource development programs.

Guide to further reading

- The study of outdoor recreation and resources clearly has many dimensions. Liddle's (1997) work provides perhaps the most comprehensive coverage of the *ecological impacts of outdoor recreation and tourism*, giving detailed insights into *resource appraisal, capability and evaluation*. Readers should also consult Wall and Wright (1977); Cloke and Park (1985); Hammitt and Cole (1987, 1991).
- The need for *public participation* in recreation and tourism planning, including the development of visitor management strategies, is well recognised, e.g. see Murphy (1985); Haywood (1989); Dredge and Moore (1992); Ryan and Montgomery (1994); Simmons (1994); Hall and McArthur (1996).
- Shackley (1998) examines visitor management issues at *World Heritage Sites*.
- Sources concerning *recreation resource assessment*, include: Leatherbury (1979); Kane (1981); Cloke and Park (1985); Mather (1986); Hammitt and Cole (1987, 1991); Countryside Commission (1988); Thomson, Lime, Gartner and Sames (1995); Countryside Commission (1995); Fraser and Spencer (1998).

Review questions

1 How might an outdoor recreation resource be defined? Discuss the importance of considering resources generally, and outdoor recreation resources specifically, in functional terms.

2 Select a local outdoor recreation site. List the major uses and users of that site. Attempt to identify the current and potential conflicts which relate to that site. To what extent do you think multiple use of the site has been achieved and has been successful?

3 With reference to case studies, discuss the concept of accessibility in terms of recreation resources.

4 Resource assessment and evaluation require considerable subjective judgement in their application. What are the implications for decision-making? How, if at all, can subjectivity be controlled?

5

OUTDOOR RECREATION AND THE ENVIRONMENT

Identification of recreation resource potential, and classification and evaluation of resources for outdoor recreation are necessary, but only initial steps in the process of creating recreation opportunities. The real challenges for environmental management arise from human use of the recreation resource base.

In the final analysis, concern is for the quality of the recreation experience and the degree to which that experience contributes to the physical and mental well-being of participants. The quality of the recreation experience is largely a function of the environment in which it takes place, but there is nothing deterministic or inevitable about the relationship.

The primary concern of recreation resource managers is undesirable change in environmental conditions (Hammitt and Cole 1991). However, the mere presence of human beings in a recreation setting need not be the trigger for degradation, any more than 'a bull in a china shop' necessarily spells catastrophe. If the bull was led, docile, well-trained, and did not bellow, get upset, or otherwise disgrace itself and if the shop fixtures were well-spaced and the china secure and not irreplaceable, then perhaps the occasion would be without incident.

So it is with outdoor recreation; the use-impact relationship is not straightforward. A certain amount of recreation use in a particular environmental setting will lead to a certain level of impact, depending on a combination of factors, including the weather, the resistance and resilience of soils and vegetation to trampling, soil drainage, the extent and nature of recreational use, and management strategies (e.g. Stankey *et al.* 1984; Hammitt and Cole 1991; Liddle 1997).

Recreation–environment relationships

It is unwise to rush to conclusions about the impact of outdoor recreation on the environment, or to accept, without qualification, predictions of undesirable or irreversible consequences of human use. As noted above, the outcome is a function of the attributes of the environment, the nature and extent of the recreation taking place, and resource management strategies. The ensuing discussion focuses on the relationship between outdoor recreation and the natural, biophysical environment. However, the social setting can also be important in its effect on the recreation experience, and, in turn, can be transformed by the recreation activity taking place. This relationship is explored further under social carrying capacity.

Environmental attributes

Attributes of the biophysical environment differ from place to place. Geological and edaphic conditions vary, as do terrain, hydrology, fauna and flora. The biophysical characteristics of the natural environment can also be materially altered by ephemeral or transitory aberrations in weather and seasonal conditions. The simplest illustration of this, is the difference in effect on a recreation setting of the same type and volume of recreational activity in summer and winter. The ability of a site to recover over time also varies with the season and the weather.

Environments, too, differ in their ability to withstand use and to recover after use. Hammitt and Cole (1987: 23) make the distinction between *resistance* and *resilience.*

Resistance is the ability to absorb use without being disturbed (impacted); resilience is the ability to return to an undisturbed state after being disturbed. Resistant sites may or may not be resilient and *vice versa.*

Environments that are sensitive to disturbance may quickly reflect the effects of human incursion, but just as quickly recover after use. A rock surface is highly resistant, but once scarred by graffiti or other undesirable forms of 'recreation', the damage may be permanent. Attributes of resistance and resilience can also vary according to seasonal and climatic conditions. The best sites to use for outdoor recreation are those that are *both* resistant and resilient in the long term.

When comparisons are made between sites, it becomes clear that some ecosystems are more tolerant of recreation activity than others. Some areas are virtually indestructible, while others are so fragile as to permit only minimal use. Goldsmith and Manton (1974) suggest that the ecosystems or habitats most vulnerable to recreation impact are:

- coastal systems, such as sand dunes and salt marshes characterised by instability;
- mountain habitats, where growth and self-recovery are inhibited by climate;
- ecosystems with shallow, wet or nutrient-deficient soils.

A comprehensive survey of the ecological impact of outdoor recreation was presented by Wall and Wright (1977), who itemised the consequence of recreational activities for specific attributes of the environment. Most emphasis was directed towards soil, vegetation, water, and wildlife. However, the authors were able to show that even geology can be affected by certain forms of recreation, and that complex inter-relationships exist between types of recreational impact.

Recreation characteristics

A second group of factors which has a bearing on the recreation–environment relationship is related to the nature of the recreation activity and the characteristics of users. A recreation setting may well be able to withstand use by any number of sedate picnic groups, but would quickly deteriorate if subjected to an informal rugby match. Not only is the latter activity inherently harder on the ground surface, its concentrated nature has greater potential for impact than the typically more passive and dispersed pattern of picnicking. Moreover, rugby and similar forms of outdoor recreation can take place under rain-affected conditions, for example, where a natural playing surface is more susceptible to damage.

Some recreation activities, too, rely on certain types of specialised equipment, which add greatly to their potential for environmental disturbance and can allow users to penetrate deep into sensitive areas, not otherwise accessible. Off-road transport such as all-terrain vehicles, snowmobiles, dune buggies, and the increasingly popular four-wheel drive vehicles, have become a significant feature of the recreation scene. With this trend has come greater potential for degradation of the recreational environment and conflict with other users and uses. Much of the problem rests with the use of these vehicles in sensitive environments such as coastal sand dunes, arid zones, steep slopes, alpine areas and wetlands.

Off-road recreation vehicles can also be responsible for the spread of noxious weeds and the invasion of despoiled areas by exotic vegetation, normally unable to compete with indigenous vegetation. This is a particular problem in parts of Australia, where seeds from weed-infested roadside reserves are easily spread by tyres and mud on vehicles. Damage has also been experienced at sites of archaeological and scientific significance in coastal areas of Australia, where Aboriginal relics and middens have been destroyed or disarranged.

Quite apart from the physical effects of off-road recreation vehicles, the most persistent criticism is the noise associated with their use. Trail bikes and power boats, in particular, can be heard over great distances. Snowmobiles are criticised for being excessively noisy, and can add to site disturbance and adversely affect the chances of restoration. Other ancillary impacts attributed to off-road vehicles; are the spread of litter and the risk of fire in otherwise inaccessible areas. Finally, there are considerable hazards and risks with use of these vehicles; deaths and injuries are not uncommon.

Characteristics of participants likewise influence the interaction between recreation and the environment. Moreover, the attitude and behaviour of visitors can be as important as the pressure of numbers. Some recreationists act responsibly and leave a site in the same condition in which they found it; others are not so conservation-conscious, and make unreasonable demands on the resource base.

The problem is heightened by non-uniform patterns of recreation use and the manner in which participants distribute themselves within a site. Visitor pressure tends to be concentrated in space and time (Glyptis 1981). Gittins (1973) documented the differential intensity of recreational activity in Snowdonia National Park in Wales, where patterns of use vary with the time of year, seasonal conditions and popularity of certain features of the park (Figure 5.1). Clearly, at the time of the survey, the park was by no means crowded in the overall sense, but the intensity of recreational use and the potential for environmental disturbance were concentrated in a series of linear routeways and nodal points.

A comparable study was carried out by Ovington *et al.* (1972), on the impact of tourism at Ayers Rock – Mt Olga (Uluru) National Park in central Australia. The study established that, although contact areas for tourists within the park are restricted, each shows evidence of environmental change in terms of topography, soil, drainage patterns, flora, fauna, odour, noise and waste material accumulation. Ecological impacts included soil compaction and erosion, and destruction of vegetation and wildlife habitat. Even the massive monolith of Ayers Rock itself did not escape environmental damage from climbers, including the well-intentioned, but intrusive, installation of chain-railings and lines painted on the rock surface to assist visitors to reach the summit.

On a smaller scale, the pattern and extent of wear-and-tear by recreationists on campgrounds, picnic sites and sand dune vegetation have been demonstrated (LaPage 1967; Boden 1977; Slatter 1978). Various methods were used to record changes in ground cover and species composition, and these were correlated with the level of visitor use. LaPage used sequential photographs on a systematic grid system to reveal a progressive reduction in vegetative cover and number of species, closely associated with concentrations of use around fixed site facilities such as picnic tables and barbecues. With continued use, LaPage found a gradual rearrangement of plant species composition, leading to a relatively recreation-tolerant soil cover.

The ecological environment responds in different ways to visitor pressure, and the possibility of such beneficial changes should not be dismissed. Some observers consider that soil compaction around the roots of trees, for example, has a useful effect in terms of forest viability. Low intensities of trampling can stimulate plant growth, and the opening-up of forests with nature trails allows more light through the canopy and can contribute to an altered, but enhanced, recreation landscape. Types of impact and possible indicators for a park environment are set out in Table 5.1

Recreation impacts and assessment

Participation in outdoor recreation is increasing, for all the reasons noted in earlier chapters, and with it, the inevitability of environmental change and possible degradation. Such change is seldom sharp or catastrophic, but more usually, incremental and cumulative; the result of many individual actions. Impacts on the ecology of a recreation site often receive the most attention, and these, in turn, can detract from the quality of the recreation experience.

Comprehensive overviews and documentation of biophysical changes to the environment from recreation use are presented by Wall and Wright (1977), Hammitt and Cole (1987), Kruss *et al.* (1990) and Liddle (1997). Depending on the

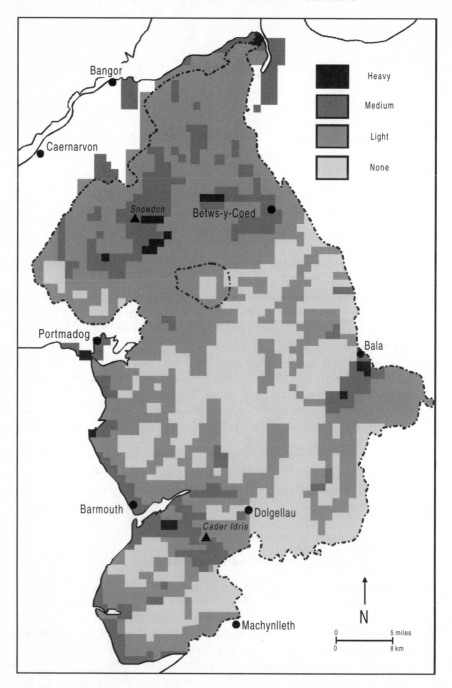

Figure 5.1 The intensity of recreational use in Snowdonia National Park

Source: Gittins (1973, in Patmore 1973: 238).

Table 5.1 Selected environmental impacts and potential indicators

Selected impacts	Potential indicators
Sewage discharge	Total phosphorus Faecal coliforms Streptococci
Solid waste disposal	BOD in leachate Air quality Wind blown litter
Accelerated erosion	Gullying Turbidity
Compaction	Bare soil Exposed tree roots
Vegetation disturbance	Area disturbed Special change
Wildlife disturbance	Habitat change Changes in animal sightings
Noise	Decibels
Traffic congestion	Delay times
Litter	Visual assessment
Introduced plants and animals	Species change Feral animal populations
Perceived crowding	Number of contacts Number of campsites Visitor satisfaction

Source: Adapted from Turner (1994: 135)

circumstances, outdoor recreation may affect the type and diversity of vegetation species, soil properties, wildlife populations, habitat, air and water quality, and even the geology of the recreation setting. Sensitive environments, such as parts of the coastal zone, are particularly prone to disturbance.

Despite the probability that recreation opportunities could be impaired by such changes to the ecology, visitors to a site appear to be more concerned about impacts that decrease its functionality or desirability (Hammitt and Cole 1991). Moreover, the *same* change can be seen as a problem or an advantage, depending upon the environment in question and its use for recreation. Hammitt and Cole (1991) offer the example of conversion of natural vegetation to introduced species of turf. In a pristine wilderness this change is considered undesirable; in an urban park the change to the playing surface may be beneficial.

The complexities of the relationship between recreation and the environment are matched by difficulties in detecting and identifying cause and effect. Not only does the effect on sites vary with the type and intensity of recreational use, visitors to a site vary in their reaction to change and to the presence of others. These differences are discussed further in the consideration of social carrying capacity.

The relationship between recreation and the environment can be direct or indirect; immediate or delayed. Although biophysical impacts of recreation may be easier to detect than disturbance to the social setting, precise measurement can be just as elusive. Wall and Wright (1977) point out that it is almost impossible to reconstruct the environment minus the effects induced by recreation, or to establish a base level against which to measure change – the environment is dynamic, with or without direct human intervention. The problem then arises of disentangling the role of recreation from the role of nature. Spatial and temporal discontinuities between cause and effect can further obscure the environmental impact of outdoor recreation. Erosion in one location may result in deposition elsewhere, and considerable time may elapse before the full implications are apparent.

Moreover, the recreation–environment relationship is reciprocal. Visitors have an effect on the environment which, in turn, affects users. For satisfaction to be maintained, environmental values must not be used up faster than they are produced. The capability of the resource base and the recreation setting to continue to provide for recreational use, raises the concept of carrying capacity.

Recreation carrying capacity

Like many concepts in outdoor recreation management, the term 'carrying capacity' is bedevilled by varying and sometimes conflicting interpretations. The concept of 'recreation carrying capacity' derives from the practice, in livestock and wildlife management, of referring to the estimated number of animals an area of rangeland or a given habitat can support. In its initial application in outdoor recreation, the concept was seen as a technique to limit use to the maximum number of visitors a recreation resource or site could tolerate, without damage to the biophysical or social conditions.

Most definitions of recreation carrying capacity attempt to combine this notion of protection of the resource base from overuse with, simultaneously, the assurance of enjoyment and satisfaction for participants. Thus, in broad terms, recreation carrying capacity involves both the biophysical attributes of the environment as well as the attitudes and behaviour of users. An early definition of recreation carrying capacity by the Countryside Commission (1970: 2) reflected this duality:

> The level of recreation use an area can sustain without an unacceptable degree of deterioration of the character and quality of the resource or of the recreation experience.

Leaving aside the vagueness of this definition, the Commission went on to identify four separate types of recreation carrying capacity – physical capacity; economic capacity; ecological capacity; social carrying capacity.

Physical carrying capacity is concerned with the maximum number of people or equipment (e.g. boats or cars), which can be accommodated or handled comfortably and safely by a site. In many ways, it is a design concept, as when referring to the capacity of a car park, a spectator stand or a restaurant. In other circumstances, it could relate to safety limits (e.g. for ski slopes or specific numbers for participation in sports). As will be seen later, restriction of the physical capacity of ancillary facilities can be a useful management tool for applying indirect control over visitor numbers. It is easier to limit boating activity on a lake, for instance, by deliberately reducing the physical capacity of on-shore facilities such as access points, boat ramps and trailer parks, than to regulate boats on the water surface.

Economic carrying capacity relates to situations of multiple use of resources, where outdoor recreation is combined with some other enterprise. Economic compatibility might be a better description, because the term is concerned with getting the right mix of resource uses, so that benefits and costs of recreation do not reach a point at which interference with non-recreational activity becomes economically unacceptable from the management viewpoint. This could happen, for example, at a domestic water supply reservoir, where recreation is permitted, but where the consequent costs of supervision, or of water treatment, cannot be justified. Similarly, with a farm or a forest, the demands and depredations of recreationists may push the costs of efficient production too high for economic management.

The final two components of carrying capacity – ecological and social – are of greatest relevance to outdoor recreation management and receive the most emphasis in the ensuing discussion.

Ecological carrying capacity

Ecological carrying capacity (sometimes confusingly referred to, also, as physical, biophysical or environmental capacity) is concerned with the maximum level of recreational use, in terms of numbers and activities, that can be accommodated by an area or an ecosystem before an unacceptable or irreversible decline in ecological values occurs. This concept has been the subject of controversy, especially in subjective judgements of what is 'unacceptable', or 'irreversible decline'. *Any* use of an ecosystem will result in some change, but over-restrictive management could negate the recreation resource function altogether.

It could be argued that an area's ecological capacity is reached when further recreational use will impact the site beyond its ability to restore itself by *natural* means. However, such a viewpoint ignores the essential flexibility of the carrying capacity concept and the scope for, even the presumption of, sound management practices to stretch carrying capacity beyond so-called natural limits (Godin and Leonard 1977). Technological and financial considerations are obviously also relevant to the question of irreversibility.

Any estimate of ecological carrying capacity must take account of:

- the nature of the plant and animal communities upon which the recreation activity impinges;
- the nature of the recreation activity and its distribution in space and time (Brotherton 1973: 6–7).

Several writers have warned against the misconception that capacity levels are somehow inherent or site-specific (Brotherton 1973; Ohmann 1974; Bury 1976; Manning *et al.* 1995). Bury is especially critical of the notion of a fixed, uniquely correct, recreation carrying capacity for a site, and suggests that the concept may be hypothetical in terms of managerial usefulness. He demonstrates the various components of biological and ecological carrying capacity, and the inter-relationships between them which inhibit generalisation. Bury gave the example of Big Bend National Park in southern Texas, which had about reached its *hydrologic* carrying capacity under 'existing standards of water use'. Lower standards, or elimination of certain forms of water use, could increase the hydrologic carrying capacity of the park. Similarly, capacity for sewage and waste disposal is, to some degree, a function of whatever mandatory regulation is adopted.

With all the contrasting physical characteristics possible within any particular site, development of strict measures of ecological carrying capacity, capable of general application, appears pointless and even counter-productive. As with the livestock grazing analogy, precise setting of the carrying capacity of a site at conservative levels, could be uneconomic and wasteful of legitimate recreational opportunities (Hammitt and Cole 1991). Over-restrictive limits could also reflect unrealistic management objectives in terms of maintaining the pristine integrity of a site. On the other hand, adopting levels that are too liberal for carrying capacity, may provide a short-term revenue windfall, but could lead to longer-term environmental degradation and ultimate closure of the facility.

In any case, the setting of carrying capacities is only one component of an overall recreation management programme, and must be accompanied by systematic monitoring of environmental conditions and the flexibility to respond quickly to indications of stress. Moreover, generalisation is not feasible. Each recreation site has a range of carrying capacities, depending upon the nature of the recreation activity, characteristics of participants, background environmental conditions and the management objectives adopted.

In particular, concern for ecological carrying capacity *alone* is inappropriate for outdoor recreation management. Recreation carrying capacity is evolving from its original emphasis on ecologically based use limits to an understanding of the complex relationships between environmental disturbance and participant satisfaction. The final test of whether a site measures up, rests with the minds of visitors, and their perception of, and reaction to, both the biophysical and social conditions of the recreation environment. Perception plays a key role in setting and managing the social carrying capacity of a recreation site.

Social carrying capacity

Outdoor recreation involves people, and the social environment in which recreation takes place has a good deal to do with the level of satisfaction experienced. Social carrying capacity (also referred to as perceptual, psychological or even behavioural capacity) relates primarily to visitors' perceptions of the presence (or absence) of others at the same time, and the effect of crowding (or in some cases, solitude) on their enjoyment and appreciation of the site. Social carrying capacity may be defined as the maximum level of recreational use, in terms of numbers and activities, above which there is a decline in the quality of the recreation experience, from the point of view of the recreation participant (Countryside Commission 1970).

The concept has much to do with tolerance levels and sensitivity to others, and, as such, is a personal, subjective notion linked to human psychological and behavioural characteristics. Put simply, social carrying capacity represents 'the number of people (a site) can absorb before the latest arrivals perceive the area to be "full" and seek satisfaction elsewhere' (Patmore 1973: 241). It is the least tangible aspect of recreation carrying capacity and the most difficult to measure. Not only does it vary between individuals, but also for the same person at different times and in different situations.

Bury (1976) suggests that visitor satisfaction is linked to the notion of 'territory' and 'living space', so that social carrying capacity is derived from the number and types of encounters with other humans in the recreation area. Manning *et al.* (1995) support the contention that trail and camp encounters, for example, are key variables in determining the quality of a wilderness experience. Bury (1976) makes the interesting point that it is not merely the *actual* number of times an individual meets other recreationists, but the *potential* number and type of such encounters which are important.

> ... recreation satisfactions may be impaired even before any encounters occur if the number and density of people *seem* higher than the visitor would prefer, or if the potential encounters seem likely to be more intense, or closer, than the visitor wishes.
>
> The condition may also be reversed – as when teenagers go to a beach to see, be seen, and interact with others. In this case, the desire of the visitor is for high densities of human use.
>
> (Bury 1976: 24)

The link between social carrying capacity and the type of recreational experience is illustrated graphically in Figure 5.2. The satisfaction derived from a wilderness experience is reduced, even at very low levels of use and social interaction – 'two's company and three's a crowd', indeed. The canoeists in Lucas's (1964) study of Boundary Waters Wilderness on the US–Canadian border, had no wish to see fellow humans. On the other hand, being the *sole* visitor to, say, Disneyland, would hardly be an enjoyable experience. In fact, the satisfaction gained from such essentially gregarious occasions increases with the level of use, at least until the point where crowding and congestion begin to irritate. It could well be the waiting and the queuing which then become exasperating, rather than the numbers of people in attendance.

Certainly, numbers of people alone do not cause visitor dissatisfaction. Reaction to crowding is variable and, to some extent, self-regulating. This makes any measurement of social carrying capacity just that much more difficult, because the non-gregarious individuals may be absent or may have redistributed themselves in space and time so as to avoid peaks in recreation use.

How a person reacts to the presence of others is influenced by underlying psychological factors such as personal values, goals, attitudes, expectations and motivations. The level of satisfaction is also affected by other events or conditions incidental to the recreation experience, for example, vehicle troubles or traffic problems on the trip, illness, or even the weather.

Social circumstances, too, help shape people's perception of a particular situation, and the way they receive and interpret information about a recreation environment. Human perception is, in part, a function of the psychological factors noted above, but is also a result of demographic characteristics and the socio-economic background of participants. Once again, it is not so much the size of the crowd, but similarities or contrasts in social status, behaviour or composition of the group which become a source of frustration and conflict (note the earlier discussion on incompatibility and conflict in Chapter 2).

Perception of the quality of a recreation experience also reflects the charac-teristics of the physical environment or situation in which the activity takes place. Site features such as location, size, configuration, terrain, vegetation, proximity to compatible activities and the type of support facilities, can all influence satis-

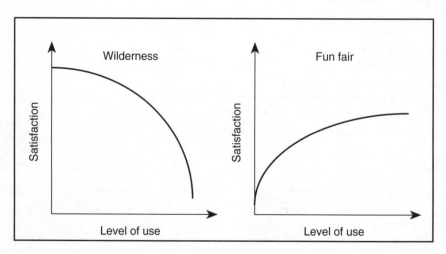

Figure 5.2 The effect of crowding on recreational satisfaction

Source: Adapted from Brotherton (1973: 9)

faction levels. In particular, they may affect the capacity of the landscape to 'absorb' users. It is the actual awareness of others which is crucial to social carrying capacity, so that any objective measure of the density of use may not be a true reflection of crowding. Out of sight *is* out of mind, and if others present are not visible because of certain site characteristics, social carrying capacity may be considerably enlarged. Bury (1976) points out that carrying capacity generally increases with increasing density of vegetative cover. Visitors cannot see or hear one another and so the area *seems* less crowded. Wilderness above the timberline, for example, has a smaller social carrying capacity than wilderness at lower altitudes, where participants are screened by both topography and vegetation. This associated notion of 'landscape absorption' has obvious implications for management of existing recreation sites and design of proposed sites and facilities.

In this discussion of recreation carrying capacity, most emphasis has been placed on ecological and social aspects. Consideration of these two components, separately, and sequentially, does not imply any order of importance, nor should it obscure the complex relationships between them. Both resources *and* people must be taken into account when considering carrying capacity. It is important for decision-makers to be aware of the dynamic, multidimensional nature of the capacity concept, in order to adopt a balanced approach to managerial responsibilities.

This comprehensive viewpoint was emphasised in a comment by Lindsay (1980: 216), who conceptualised outdoor recreation carrying capacity as:

> ... a function of quantity of the recreation resource, tolerance of the site to use, number of users, user type, design and management of the site and the attitude and behavior of the users and managers.

Thus, a competent recreation management programme would incorporate both environmental considerations *and* human needs and desires. It remains a matter of judgement as to when degradation of the resource base or deterioration in the quality of the recreation experience reach the point where action is called for. The difficult task of determining carrying capacities is matched by that of deciding when and where they are, or might be, exceeded, and, ultimately, the choice of remedial and pro-active management procedures.

Since the first rigorous application of carrying capacity to the management of parks and recreation areas in the 1960s, the concept has expanded from its original focus on resource impacts to include emphasis on the social setting and the quality of the recreation experience. More attention, too, is being given to the carrying capacity of the developed portions of parks, and to private lands, rather than exclusively to back country and wilderness settings (Manning *et al.* 1995). Despite its shortcomings, the concept of recreation carrying capacity is evolving as an important component of a more comprehensive and holistic approach to environmental management. When viewed in its proper perspective, recreation carrying capacity remains useful as 'an organizational framework for making rational judgments about appropriate conditions and public use of parks and recreation areas' (Manning *et al.* 1995: 337).

Recreation carrying capacity in review

Widespread application of the concept of recreation carrying capacity led to growing scrutiny of its effectiveness as a management technique. As noted above, the concept derived from use in the pastoral industries and wildlife management. Whereas the fixing of carrying capacities for recreation sites may appear intuitively simple, managing recreation use for people differs considerably from determining the forage requirements of cattle and wildlife. Even there, carrying capacities are not 'set in stone' but vary with climatic and vegetative conditions and other considerations. The same is true of recreation. The relationship between use and impact, typically, is not direct, and is affected by the type of recreation activity; its timing and distribution on the one hand, and the attributes of the environment where use occurs, on the other.

Although national park agencies worldwide continue to fix and observe carrying capacity levels for protected areas in their care, questions have been raised increasingly concerning the appropriateness of applying a supposedly simple model of range land management to the recreational needs and desires of people. Recent years have seen the adoption of a number of alternative approaches to monitor and manage the impacts of outdoor recreation on the environment.

As a result, the emphasis is changing from visitor levels as such, to the inevitability of change accompanying recreational use; from there to managing the amount and type of change resulting from human activity, on top of that occurring in nature. This approach reflects the view that variation in human behaviour is probably as influential as the actual number of visitors, in bringing about change to that environment. Moreover, it does not follow that change equates with degradation, or that impact is the same as damage. Some forms of impact can be tolerated by users and managers alike. Specific judgements are required on how much change or impact can be accepted before it becomes 'damage' and requires intervention by management (Turner 1987).

Limits of acceptable change

Out of this questioning evolved a more comprehensive and systematic framework for recreation decision-making, known as the 'Limits of Acceptable Change' (LAC). The planning framework based on LAC is essentially a reformulation of the recreation carrying capacity concept. The emphasis is on the ecological and social attributes sought in an area, rather than on how much use the area can tolerate. First tested in the Bob Marshall Wilderness Complex in Montana (Stankey *et al.* 1984), the system has received widespread endorsement as a rational planning approach to recreation and parks management.

Essentially, the Limits of Acceptable Change approach turns the recreation-environment relationship on its head, transferring the focus from the supposed cause (numbers of visitors) to the desired conditions – the biophysical state of the site and resource base, and the nature of the recreation experience. Moreover, change in nature is seen as the norm, and a certain level of natural variation in the environment is to be expected. It is when the rate of human-induced change accelerates, or the character of change becomes unacceptable, that managerial action may be called for (Figure 5.3).

The central question for recreation planners then becomes – how much, and what type of change can be tolerated? Whereas the response must necessarily be subjective, it needs also to be guided by reference to more than ecological criteria. Socioeconomic and political considerations can also be important elements of the consultative process in setting the Limits of Acceptable Change. A loose analogy can be drawn with the distinction often made between resource 'capability' and 'suitability'. Whereas a parcel of land may be judged *capable* of use, for example as a waste dump, from a biophysical standpoint, other factors, such as economic impacts and social pressures, may render it not *suitable* for such a purpose (also see Chapter 4).

It is important, therefore, that a systematic approach be adopted to establishing the Limits of Acceptable Change; one that reflects the natural conditions targeted, as well as economic, social and political realities. Establishing and implementing

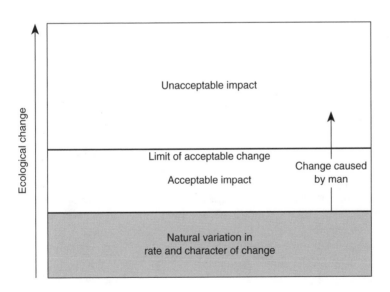

Figure 5.3 Model of acceptable ecological change in wildlands

Source: Hammitt (1990: 26)

the Limits of Acceptable Change management framework involves a multistage process. The following outline of each stage is adapted from Stankey *et al.* (1984):

Stage I

The first stage in the process is to undertake careful site analysis to establish base line data in terms of ecological, social and economic conditions and political circumstances. From this standpoint, area-specific issues and concerns can be identified and a better understanding gained of the recreation environment.

Stage II

For a range of proposed recreation opportunities, acceptable and achievable environmental and social conditions or thresholds are identified, and defined by a set of selected, measurable indicators.

Stage III

Existing resource and social conditions are inventoried as the basis for establishing the relationship between those conditions and what is judged acceptable for each class of recreation opportunity. Inventory data are also important for setting realistic and attainable standards or benchmarks for specifying acceptable conditions, or the limits beyond which change will become unacceptable.

Stage IV

Management actions which will achieve acceptable environmental and social conditions relative to the class of recreation opportunity are identified and evaluated (i.e. the measures necessary to transform the area from its existing condition to that desired). Where existing conditions are well within minimally acceptable standards, little action may be taken into consideration to select the preferred alternative.

Stage V

Finally, the management actions necessary for the chosen pattern(s) of recreation use are implemented and a monitoring programme established, based on the indicators selected in Stage II.

Monitoring is a particularly important part of the Limits of Acceptable Change process. It provides systematic feedback on the effectiveness of the management actions employed, alerting managers to the need to consider more intensive and rigorous efforts, or the use of other measures. It could also point to the need for revision of the standards and indicators specified. This could be the case especially where circumstances external to the recreation site have altered, e.g. changes in access or in contiguous land use.

By applying the Limits of Acceptable Change framework, it is technically possible to establish a rational basis for management intervention. However, Stankey *et al.* (1985) stress the wider context in which decisions have to be made. Recreation planning takes place in a political environment, in which different interests, views and values have to be accommodated. Management techniques based on the Limits of Acceptable Change approach are only part of the recreation planning process. Moreover, the subjectivity and judgement inherent in the identification of acceptable social and environmental conditions, in the setting of standards or thresholds, and in the choice of indicators, need to be balanced by opportunities for ongoing public participation at each stage in the process.

It is important to note that 'the LAC model has a close relation in the tourism planning field in a concept known as the Ultimate Environmental Threshold' (UET) (Mercer 1995: 171; for more detailed discussions of this concept, see Kozlowski *et al.* 1988). This concept refers to:

> The stress limit beyond which a given ecosystem becomes incapable of returning to its original condition and balance. Where these limits are exceeded as a result of the functioning or development of particular activities a chain reaction is generated leading towards irreversible environmental damage of the whole ecosystem or of its essential parts.
>
> (Kozlowski 1985: 148–9, in Mercer 1995: 171)

Areas where the UET approach has been adopted, include fragile mountain areas, and single islands and groups of islands (e.g. islands in the Capricornia section of the Great Barrier Reef) (Mercer 1995).

The limits of acceptable change approach: issues and concerns

As a recreation planning framework, the Limits of Acceptable Change approach was originally put forward as a means of rationalising recreation management in wilderness areas. Whereas it has had application outside this focus (see below), its wider potential is offset by 'lack of understanding of the capabilities of the LAC process… [and] poor or improper execution of the process' (McCool 1990a: 190).

As noted above, a strong element of subjectivity is present in the various stages of assessing acceptable levels of change from recreation use. Turner (1987) singles out the identification of indicators of acceptable conditions as a contentious issue, requiring professional judgement and experience, backed by community consultation. There is little agreement as to what constitutes useful generic indicators of recreation impact, so that it is necessary to derive site-specific indicators for particular environmental attributes at specific locations.

Turner (1987) sets out criteria for selecting environmental indicators (Table 5.2), and then applies these to a number of measures considered appropriate for the Australian Alps National Park. He notes, however, that few of the indicators adopted are entirely stable, even in the most undisturbed situation. The challenge for managers is to differentiate between recreation impacts and natural variations, and to identify base levels or reference points for particular indicators, outside of which environmental values provide an early warning of the need for intervention. This, of course, merely shifts the question for decision to the fixing of a reasonable, base level. It might be appropriate to adopt zero as the base level for soil compaction, for example as an indicator of soil condition. However, for other biophysical indicators such as contaminants in streams, 'normal' base readings could be above zero because of natural background levels.

Even then, a good deal of uncertainty prevails with respect to how an ecosystem might react to change in the longer term.

> The best way to handle such uncertainty is to plan on the basis that acceptable conditions will prevail for a certain proportion of the time. It is appropriate to say, for example, that phosphorous levels downstream of a ski village should be less than 40 mg/l for 95 percent of the time, rather than produce a blanket limit which will not be achievable... .
>
> (Turner 1987: 10)

The problem, noted earlier, of establishing generally accepted levels for indicators of the social impact of recreation, is equally contentious. Social impacts are obviously important in influencing the quality of the recreation experience, all the more so in remote areas of wilderness, where contact with other humans is usually not sought or welcome. Agreement on indicators such as the number of encounters might be possible, but specifying acceptable levels for such indicators is difficult when interpersonal attitudes and reactions are involved (Turner 1987).

Even the relatively straightforward requirement of monitoring the effectiveness of management actions with reference to the indicators specified raises a number of concerns. Given that the criteria for identifying valid indicators (Table 5.2) have been observed, there remains the question of sampling. Again, Turner (1987) notes as important considerations, the frequency of sampling, the spatial distribution of sampling sites and the need for replication in the interests of consistency. Systematic sampling is basic to monitoring procedures if the cumulative effects of recreation are not to go undetected.

Whereas the Limits of Acceptable Change approach is clearly not the panacea for confronting all recreation management challenges, it does offer the promise of 'more defensible decisions' (McCool 1990a: 191). In reviewing the potential strengths and weaknesses of the process, Knopf (1990) noted twenty possible strengths and only one perceived weakness. The shortcoming that Knopf identified was more concerned with the attitudes of those applying the approach, than with the process itself.

Knopf is concerned with what he sees as a certain kind of negative disposition implicit in the term, Limits of Acceptable Change:

> It seems that the LAC framework has the potential for feeding a certain kind of negative disposition that abounds in outdoor recreation management... that disposition has to do with an attitude that the primary

goal of resource management is to arrest the deterioration of environmental quality... people being construed as objects that impede quality
environmental management... that litter, form crowds, create noise...
trample vegetation... pests... messing things up.

(Knopf 1990: 207–8)

Some observers may agree with Knopf, and more than one parks manager has
been known to observe that parks management would be easy if it wasn't for the
people! Knopf's concern is that the Limits of Acceptable Change process has the
potential for encouraging the disposition that people are a problem, rather than an
opportunity, in recreation resource management. He contrasts two possible
statements introducing the process to make his point. The first is *negative*, stressing
the problem of resource degradation and the role of the step-by-step approach in
ameliorating the problems.

The second introductory statement emphasises the *positive* contribution of
outdoor recreation to human growth and development:

Increasing use of our outdoor recreation resource has presented even
greater opportunities for building peak experiences into people's lives.
This (LAC) plan summarizes a step-by-step approach taken to identify
new opportunities for serving our guests... and expanding our service
watershed considerably.

(Knopf 1990: 208)

Table 5.2 Criteria for selecting environmental indicators

Long-term significance	Indicator must detect changes that occur slowly but consistently, and must be able to detect trends over a 5-year period.
Short-term significance	Indicator must be able to detect changes in conditions which occur within any particular year.
Responsive	Indicator must detect changes early enough to enable a management response and must reflect changes that are subject to manipulation by management.
Detects amount of change	Indicator should be measurable and allow the amount of change to be assessed quantitatively.
Feasible	Indicator must be reliably measurable by field staff using simple techniques.
Economic	Indicator must produce meaningful information for managers at a minimum cost.

Source: Turner (1987: 20)

Perhaps Knopf is right in seeing the Limits of Acceptable Change approach as reinforcing a tendency towards a 'bunker mentality' on the part of some recreation managers. However, if applied sensitively and constructively, the process becomes a valuable recreation tool, offering a framework through which the nature of recreation management problems can be better understood and more effectively resolved (McCool 1990b).

Applications of the limits of acceptable change planning process

As noted earlier, the approach to recreation planning, based on the Limits of Acceptable Change, was first tested in developing a management framework for the Bob Marshall Wilderness Complex in Montana (Stankey *et al.* 1984). To demonstrate further how the system might be applied, a hypothetical case example is described by Stankey *et al.* (1985). The hypothetical area, Imagination Peaks Wilderness, is used to illustrate the flexibility of a management approach based on the Limits of Acceptable Change, rather than restricting and regulating visitors, except when and where necessary. Since this early work, the approach has attracted increasing attention as a decision-making framework for managers of wilderness and similar dispersed recreation settings.

A series of workshops organised by George Stankey and others in recent years in Australia, stimulated interest in wider applications of the Limits of Acceptable Change process in that country. Two of these applications will be described briefly, to illustrate the universality and versatility of the approach.

Wild and scenic rivers

In the State of New South Wales, the Department of Water Resources evaluated the applicability of the Limits of Acceptable Change planning system, as a conceptual context for integrated resources management on a catchment basis. In common with water resource managers everywhere, one of the principal challenges for the agency is how to resolve increasing demand for water between competing uses, including recreation. The Limits of Acceptable Change approach is seen as providing a rational, coordinating framework, within which to balance the needs of the aquatic environment with those of other resource users. In particular, the approach is considered relevant to the management of wild and scenic rivers, and wetlands. The following discussion of the Nymboida River draws, with permission, on reports of studies by Don Geering, Project Coordinator, former New South Wales Department of Water Resources, Sydney.

The Nymboida River typifies some of the issues that need to be faced when managing rivers in order to conserve essential environmental values, while allowing their potential for other uses to be realised. The river is part of the Clarence River system on the north coast of New South Wales, and its upper gorge is renowned for its wild and scenic characteristics, providing opportunities for rafting and white water canoeing. Concern is growing that the intensity of recreational use is compromising water quality along some reaches of the river; these being the source of urban water supplies, and habitat and breeding areas for threatened species of native fish. Riparian ownership and control are shared between various government agencies and private interests, and an area of wilderness overlaps part of the catchment.

The Limits of Acceptable Change approach was seen as a particular application of management by objectives in a complex resource environment, where a number of competing and conflicting interest groups have to be accommodated. The approach was combined with a procedure known as Adaptive Environmental Assessment and Management (Holling 1978; Walters 1986) to ensure that all environmental systems and all stakeholders, including recreational interests, were involved in an interactive manner in resource decisions. A modified adaptive procedure, known as Consultative Resource Planning, was developed to allow for even greater public involvement in specifying standards and measurable indicators of resource conditions, and in assessing the predicted impact, over time, of various management options, both positive and negative.

This approach enables both the agency and the catchment community to apply accepted criteria in order to determine when specified management objectives are being approached. Table 5.3 illustrates how the basic issue of water quality can be clarified, appropriate management action agreed upon, and responsibility allocated. The system links local knowledge and professional expertise; its flexible nature allows for fine tuning of levels of acceptable change, indicators and remedial measures as the approach is applied and evolves on site. Acceptable levels of disturbance for defined criteria within certain classes of river settings are set out in Table 5.4, with Class 1 being the least modified. Effective monitoring of these conditions means that the system should be self-correcting, responding to feedback on changes to criteria within acceptable limits for each setting.

Management of public lands

At the local and regional level, government agencies in Australia are frequently charged with preparing management plans for public lands and reserves so as to protect and restore the resource base, while assuring diverse and high-quality opportunities for outdoor recreation. Those objectives underpinned the planning approach for the conservation and development of Wallis Island Crown Reserve, part of a system of estuarine waterways on the central coast of New South Wales (NSW) (Gutteridge *et al.* 1988). The island covers an area of 880 hectares, some two-thirds of which is public land. The surrounding waterways and lake system are heavily used for water-based recreation and commercial fishing; the foreshores are a mix of residential and commercial development, and natural areas (Figure 5.4).

In drafting the management plan for the Island Reserve, the key concern of the resource management agency (NSW Department of Lands) was protecting important environmental features and processes, including wetlands and estuarine

Table 5.3 Nymboida River – water quality objectives and management options

Objective	That the water quality of the Nymboida study area remains of a standard that does not affect the recreational, scenic or urban water consumption potential of the resource.
Desired condition	That water quality be of a standard in its untreated form to conform with urban water supply criteria, during all flow conditions.
Management options	The Soil Conservation Service to investigate the soil erodability and sediment movement associated with agricultural practices within the Upper Nymboida catchment, and to recommend a strategy to overcome any identified problems.
	The Department of Water Resources to re-schedule monitoring at gauging stations in the Upper Nymboida catchment in order to investigate any water quality problems. The Forestry Commission to develop operational procedures for logging, adjacent to the Wild and Scenic River corridor, to ensure a minimum sediment load reaching the river.
	The State Pollution Control Commission to monitor water quality indices, particularly pesticide levels, within the study area.
	The Nymboida Shire to develop standards relating to waste disposal options within the Wild and Scenic River corridor, and advise landholders accordingly.
Performance review	The urban authority advises the management committee regularly of changes to water quality, using the monitoring information collected from the urban water supply system.

Source: Geering (1989: 4–5)

water quality. At the same time, it was recognised that increased demand for boating and outdoor recreation would place pressure on the natural environment, and that some low impact modification would be necessary for purposes of access and recreational use.

The planning approach taken was to define and allocate management classes (cf. recreation opportunity settings) for the Reserve, where environmental attributes were specified and certain types of recreation experiences provided for. Two management classes were designated – natural and semi-natural (see Figure 5.5), and within each, management actions were defined to maintain acceptable levels of development, relative to prevailing resource conditions and social considerations.

Table 5.4 Acceptable levels of disturbance for defined criteria within selected management classes of a river setting*

Criteria	Class 1	Class 2	Class 3
Land use	95% tree cover in sub-catchment; no settlement evidence or other built intrusions.	80% tree cover in sub-catchment; occasional grazing; clearing or logging; minor.	60% tree cover clearing for agriculture or forestry settlement, in a sub-dominant role.
Scenic value	Outstanding; typically includes deeply incised gorges and waterfalls.	High – including outcrops and rapids.	Moderate; no precipitous slopes, occasional rapids.
Access	No vehicular access within river corridor except for occasional four-wheel drive point sources.	Four-wheel drive access but limited primarily to point sources.	Occasional tracks or minor roads; possible lookouts on valley rim.
Hydrological modification	Essentially free-flowing with no modification.	Occasional minor regulation or water extraction.	Minor modifications, including regulation and extraction principally for agricultural uses.
Water	Unmodified; no increase in base load sediment or nutrients.	Partially modified; occasional minor deterioration (typically sediments 5 NIU for part of year and PO_4 and NO_3 levels around 1 and 6 mg/l respectively).	Modified; minor permanent deterioration (typically sediments 25 NIU for short periods and phosphate and nitrate levels generally exceed 1 mg/l and 6 mg/l respectively).

*Similar tables of criteria can be developed for alternative river and floodplain settings

Source: Geering (1989: 8)

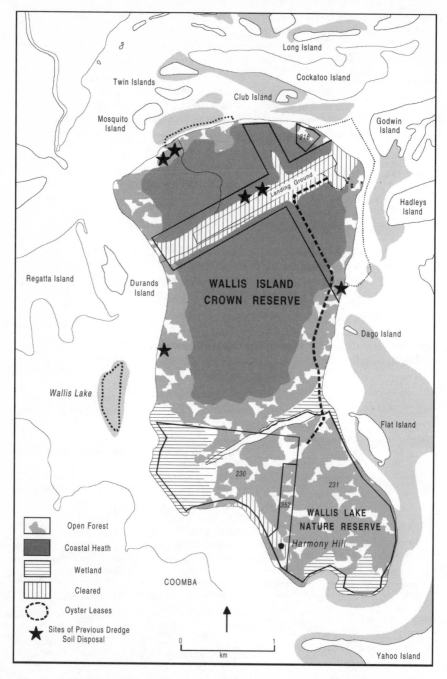

Figure 5.4 Features of Wallis Island

Source: Adapted from Gutteridge, Haskins and Davey (1988)

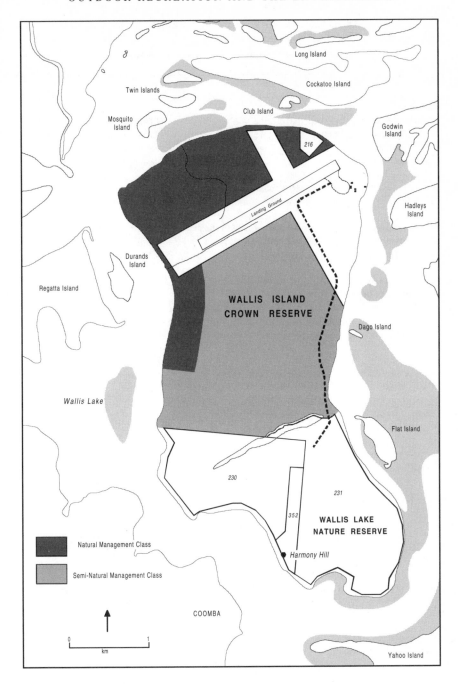

Figure 5.5 Management class allocation – Wallis Island

Source: Adapted from Gutteridge, Haskins and Davey (1988)

Table 5.5 Wallis Island Crown Reserve: environmental conditions for management classes – standards for each environment indicator

Indicator	Standard	
	Natural	*Semi-natural*
Resource		
Percentage of the blocks cleared of natural vegetation cover or not adequately regenerated	Nil cleared	Cleared area less than 5% of any 1 ha block except where necessary at any recreation development site
Presence of exotic species as dominant vegetation form	Minor component of herb layer only	Minor component of understorey only
Presence of seedlings or juvenile plants	All normal growth stages present – natural stand dynamics	Regeneration potential only Multiple age classes present
Length of shoreline affected by structures or other modification (permanent)	No shoreline affected	Less than 10% of shoreline modified
Number of clusters of buildings	No dwellings or structures	Dwelling structures 1 km apart (singly or in clusters)
Presence of services	Services not evident	Services not to be visually intrusive
Length of trafficable road/ presence of sealed roads	Walk trails or well-maintained fire trails only	Unsealed tracks/walk trails
Presence of mooring/jetty facilities	No shoreline facilities or disturbance	Limited facilities catering for lowest use only on a share basis. No permanent moorings
Social		
Number of encounters per day (peak period)	Less than 10 per day. Only few large groups per day	Less than 50 per day at defined use areas, 20 per day at other areas
Activities requiring mechanised access	No activities utilising motorised transport	No land-based activities. Only activity-powered watercraft
Management personnel presence per day	Occasional field patrols. Less than 1 per week	Daily patrols – resident manager
Visibility and enforcement of regulations	No outward or visible regulations	Visible and enforced regulations

Source: Gutteridge, Haskins and Davey (1988: 31)

Thus, the planning process for Wallis Island Crown Reserve represents a further application of the Limits of Acceptable Change approach. The process was driven by the concerns of the public agency and of the local community to develop appropriate classes of land management and use. The two classes were clearly defined by sets of indicators which relate to the resource and social conditions desired, and which specify acceptable limits to recreation use and development.

Selected indicators and standards for each are set out in Table 5.5 for both of the management classes. The plan of management also specified management actions required to achieve the designated standards.

Incorporation of the Limits of Acceptable Change into the planning strategy for Wallis Island Crown Reserve offers a framework for management of this public land, which will provide for a range of purposes appropriate to the island environment. Recreation developments are planned, including revenue-generating accommodation facilities within the semi-natural management class. Fire hazard control, foreshore protection and walking track maintenance, are other features of the management strategy to ensure that conditions for each of the management classes are maintained. Field checking and monitoring are also undertaken to assess whether standards and indicators are an accurate reflection of environmental conditions, and whether intrusive actions and associated change are approaching acceptable limits.

These examples from Australia of the application of the Limits of Acceptable Change process are a further indication of the relevance of this approach to outdoor recreation planning and the management of conflict with alternative resource uses.

Summary

The relationship between outdoor recreation and the environment in which it takes place, is complex. Clearly, the quality of the recreation experience will be affected by the recreation setting, whether natural or created, and the environment, in turn, will reflect the presence of recreationists and their activities. Whereas carrying capacity in its various forms remains an important aspect of managing the recreation–environment relationship, it is now realised that the concept must be applied with care. Generalisation is not possible, and the adoption of arbitrary limits to use ignores both the biophysical attributes of a site and contrasts in the nature and scale of recreation activity. A more positive approach concedes that some change is inevitable with recreation use; the challenge is to keep change within acceptable limits.

In addition, reference to carrying capacities is only one component of an overall recreation management programme. Systematic monitoring and feedback on environmental conditions, and a rapid and flexible response to indications of stress are essential elements in a more comprehensive and holistic approach to managing the reciprocal relationship between outdoor recreation and the environment.

Further reading

- *Recreation–environment relationships and impacts of outdoor recreation* receive thorough treatment in Wall and Wright (1977); Hammitt and Cole (1987; 1991); Kruss, Graefe and Vaske (1990); and Liddle (1997).
- *Recreation carrying capacity*, and recent qualifications and trends, are canvassed in Hammitt and Cole (1987; 1991); Kruss, Graefe and Vaske (1990); McCool (1990a); Manning, McCool and Graefe (1995); and Liddle (1997).
- The *Limits of Acceptable Change* concept was introduced by Stankey *et al.* (1984), and reviewed in McCool (1990b) and Knopf (1990).
- For discussion and applications of the *Ultimate Environmental Threshold* concept, see Kozlowski *et al.* (1988) and Mercer (1995). Where possible, readers should follow up applications of the above concepts by reference to case studies.

Review questions

1 How might seasonal conditions affect carrying capacities?
2 Explain the link between human 'action space' and social carrying capacity?
3 What basic variables influence ecological carrying capacity?
4 Why is social carrying capacity so difficult to measure?
5 What resource attributes have a bearing on recreational impact?
6 What is the significance of visitor distribution on the impact of outdoor recreation?
7 How can recreation management objectives affect the setting of carrying capacities?
8 What are some of the distinguishing features of the Limits of Acceptable Change (LAC) approach?
9 To what extent do you agree with Knopf's concerns about the LAC process?
10 What are some of the risks and advantages of bringing user preferences into the recreation management process?

6

RECREATION RESOURCE MANAGEMENT

Resource management has been defined by O'Riordan (1971: 19) as 'a process of decision-making whereby resources are allocated over space and time according to the needs, aspirations, and desires of man within the framework of his technological inventiveness, his political and social institutions, and his legal and administrative arrangements'. He argues that the emphasis in resource management should be 'upon flexibility and the minimisation of long-term environmental catastrophes, while maximising net social welfare over time', and that resource management is 'becoming increasingly concerned with the protection and enhancement of environmental quality and the establishment of new guidelines for the public use of such common property resources as air, water, and the landscape'. Thus, O'Riordan's definition and related perspective of resource management are closely aligned with the concept and application of carrying capacity, the Recreation Opportunity Spectrum and the Limits of Acceptable Change model.

Of all the resource management conflicts in the countryside, recreation offers perhaps the greatest opportunities for multi-functional resource use. Many recreational activities are compatible with other resource uses, while some are not and may have to be specially sited. Resource use conflicts are inevitable, but vary in their nature and extent. Innovative recreation planning approaches are called for.

The previous chapter examined the multi-faceted concept of carrying capacity, and the LAC model, in particular, giving some examples of their application and usefulness. This chapter examines a number of other approaches to managing resources for recreational use. It explicitly recognises that recreational resource management is a process which requires strategic planning for visitor management, generally, and for site selection, design, use, monitoring and evaluation, specifically.

The recreation resource management process

The primary aim of outdoor recreation management, presumably, is to bring together supply and demand to attempt to equate resource adequacy with human recreational needs and desires. In so doing, the manager must obviously have regard for the character and quality of the resource base, ensuring that capacity is not exceeded, and that environmental degradation is minimised. At the same time, the managerial role extends to visitor enjoyment and satisfaction. Action must be taken to reduce conflict and to maximise the quality of the recreational experience. These dual responsibilities hold, whether for the economic success of commercial enterprises, or for the protection of public investment in parks and recreation areas.

A first step in the management process is the establishment of management objectives. From these will flow the determination of carrying capacities or Limits of Acceptable Change, and the selection of specific management procedures. Modification of the system may well follow implementation of the management approaches decided upon. An important element in this phase is evaluation of the system, based on monitoring of its operation by managers and feedback from users.

An earlier model of the recreation management process is presented in Figure 6.1 and described in detail below. A set of objectives is delineated, first, with reference to the capabilities of the resource base. Information on resources should indicate which activities are physically possible, as well as some of the resource constraints on recreation opportunities.

Institutional, economic and other constraints, as well as personal circumstances, also have obvious implications for management, and set limits on the range of recreation opportunities possible. Legal restrictions and standards, administrative

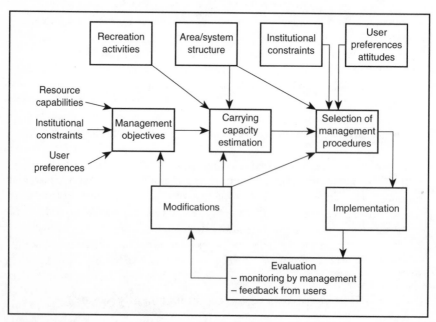

Figure 6.1 The recreation management process

Source: Adapted from Brown (1977: 194)

policies and guidelines, and budgetary and personnel considerations, can all influence the selection of realistic management objectives. In some countries and regions, problems for management stem from overlapping political and/or agency jurisdictions, each of which may have different interpretations of what is appropriate recreational use.

Ideally, management objectives should reflect user preferences if they are to receive support at the implementation stage. A good example is the 'battle' fought some years ago in Yosemite National Park between hundreds of young campers and a combined force of park rangers and state police. The dispute centred on what were seen as restrictive zoning regulations and excessive fees as demonstrators felt they were discriminated against by the park administration. A significant outcome of the confrontation was the admission that park planning authorities were out of touch with public opinion on many matters relating to national parks. As a result, provision was made for a much greater degree of public involvement in the park planning process for Yosemite and other sites (Mercer 1980a).

More and more people are demanding the right to participate in decision-making, and public involvement is increasingly seen as a necessary and desirable input to management (e.g. see Murphy 1985; Veal 1994; Pearce *et al.* 1997), although it is sometimes costly (due to time delays and consultation, and even court costs stemming from public challenges). This does not mean that the process must be totally democratic, in the sense that the user population 'calls the tune'. Identifying the population to be consulted is always a problem and, in any case, it would be foolish to disregard entirely the expertise of management in reaching a decision. As is often the case, compromise in the form of 'guided democracy' is probably the best approach.

User preferences regarding resource attributes, social characteristics of the recreation environment and preferred management approaches should certainly be canvassed, but interpreted in the light of managerial experience of what is desirable and possible. However, it is important that the process of public involvement be seen as more than just good public relations. Whereas user preferences are only one of many inputs to the formulation of management objectives, there should be clear evidence that they have been considered and integrated into the final decision. That said, it must be conceded that many factors inhibit and distort the clear expression and articulation of user preferences. Thus, a range of approaches may be needed to encourage participation, and combat apathy and indifference (Jubenville 1978).

Even when success has apparently been achieved in provoking a constructive response from communities affected by management proposals, there are risks involved in the process of public participation. Care must be taken to ensure that it is not only vocal pressure groups and politically active professional lobbyists who receive attention. The 'squeaky wheel syndrome' may not truly reflect majority preference. Conversely, the imposition of elitist managerial attitudes on users, could lead to management objectives which are unrealistic and unacceptable. A good measure of public participation is essential, but a balance should be struck between uninformed reactions, perhaps merely reflecting trends and fashions, and a more objective appraisal by supposedly detached management experts.

The second stage in the outdoor recreation management process (i.e. setting appropriate carrying capacities or limits of acceptable change, consistent with management objectives) should be related to the structure of the management area or system. As was noted in the previous chapter, limits on ecological and social carrying capacity are, in part, a function of the natural features of the site and the built facilities and amenities, and the recreation activities to be accommodated. It is also worth reiterating that carrying capacities, once set, are not inflexible but remain open to manipulation by management. Hence, the feedback loop in the chart from management procedures.

Several approaches to managing recreation sites are discussed below. In Brown's (1977) view of the management process, some of the factors which contribute to the formulation of management objectives also influence the choice of specific tactics or tools to achieve those objectives. It is not always a case of what managers may see as necessary, or desire, so much as what they are able to do.

Once again, institutional directives can act as constraints in the selection of management procedures. Obviously, also, the characteristics of the recreation area or site set limits on which combination of approaches is likely to be successful. Linear sites (e.g. walking tracks) call for a different approach from those with more regular dimensions, and remote forested areas with rugged terrain may not need such strict regulation as would more open, accessible sites with fewer natural deterrents on recreational use. Brown (1977) also stresses the value of knowing user preferences and attitudes with regard to the choice and effectiveness of specific management tools. Some people may respond kindly to appeal and inducements; others may react more favourably to direct regulation.

A desirable feature of effective management is flexibility, so that, if in the implementation of the management procedures, selected deficiencies are detected, the management process should allow for modifications. The need for adjustments may be discerned by management or become apparent in feedback from users. Subsequently, modifications can be made at various points in the system, and the management process becomes self-regulating.

Once objectives have been formulated and estimates made of carrying capacities, the primary task of outdoor recreation management emerges – that of selection, implementation and modification of on-site management procedures. However, effective management begins at an even earlier stage, with proper site selection, planning and design. If these preliminary considerations receive adequate attention, the most appropriate sites and the more resilient components of the environment (in terms of low vulnerability and high tolerance to visitor use) will already have been set aside for recreation, and developed so as to minimise management problems.

Recreation site selection

This is a most important step to which much of the ease or difficulty in subsequent operations can be attributed. Fundamental considerations are user access and the suitability of the resource base for the recreational activities envisaged. This should already have been established by application of the site evaluation techniques described in earlier chapters. However, the fact that a site apparently meets the basic suitability criteria laid down, may conceal shortcomings in specific resource attributes which will prove costly to management in later use. Where more than one site meets basic requirements, a detailed examination of site characteristics is needed to determine priorities for development. This examination should focus on the features of competing sites, which affect management's task of producing and sustaining worthwhile recreation values (McCosh 1973; Jubenville 1976).

Assuming that questions of location and convenience of access to potential users have been satisfied, many physical features of the site itself can impinge upon the quality of the recreational experience and, hence, the role of management. For instance, both the size of an area and its configuration are important. It is almost always helpful to have an area somewhat larger than required to allow rotation of use and provision of a buffer zone in order to segregate the site from adjoining developments. In most cases, too, a long narrow site is less efficient in terms of internal arrangement of attractions, facilities and services, than one of more regular configuration.

The nature of the terrain, the degree and direction of slopes, rock types and presence of rock outcrops, soil stability and compactibility, drainage and susceptibility to flooding, and availability of construction materials, can all have engineering implications for site development and maintenance. So, too, can the size, variety and density of the vegetation, and the extent and location of open space.

The importance to the recreation landscape of waterbodies of the right quantity, quality and dimensions has to be considered in site selection. Water is needed for drinking, sanitation and possibly irrigation, so that sources and suitability of water supply need to be determined, along with the costs of pumping, treatment, storage and disposal. Adequate estimates of water quality also require knowledge of groundwater, and of climate and weather patterns (e.g. precipitation, evaporation and snow cover) over an extended period. Other climatic factors which may have a bearing on decisions concerning a recreation site, include aspect, exposure to winds and seasonal conditions (e.g. length of shadows in winter and the incidence of high Spring pollen counts).

Finally, a site could have certain negative or undesirable features which could influence selection. For instance, Jubenville (1976) suggests that a hazard survey be carried out for each potential site, to identify possible hazardous conditions such as avalanches, falling trees, precipices, dangerous waters, poisonous plants and insects, and dangerous animals. Other annoyances, such as noise, dust, fumes, and aquatic weeds and algae can present problems for management; problems which, if foreseen, might be avoided, ameliorated or dealt with in a prepared and more systematic manner.

McCosh (1973) stresses the value of prior study and sound judgment in recreation site selection. Poorly chosen sites will become inefficient areas with problems that cannot easily be solved. Of course, some of the negative site characteristics noted may be offset by good planning and design. Use of design in this way as a compensatory device is fine, providing the cost is not excessive. However, it is preferable to implement design measures which are complementary to, and reinforce, the natural features of the site. The idea is:

> ... to utilize the features of the landscape to enhance recreational experience, minimize site maintenance and maintain natural aesthetics. Although fitting the development to the natural lay of the land may be more expensive and require more attention to detail, the resulting site should be more attractive, more able to handle large visitor-use loads and less expensive to maintain.
>
> (Jubenville 1976: 155)

Recreation site planning and design

Albert Rutledge (1971) put forward a set of design principles (or 'umbrella principles'), including:

- *design with purpose* – so that the appropriate relationships are established between the various parts of the recreation complex (i.e. natural elements, use areas, structures, people, animals and forces of nature);
- *design for people* – rather than to meet some rigid standards, or the impersonal demands of machines, equipment and administrative convenience. More attention to the 'why' of design would go a long way towards structuring outdoor areas to satisfy human behavioural needs;
- *design for both functions and aesthetics* – striking a balance of dollar values and human values with the achievement of efficiency, interwoven with the generation of a satisfying sensory experience.

Rutledge was writing of park design, but the principles and detailed procedures he describes have application in many other situations. For example, Lime (1974) has demonstrated the relevance of good location and design to the effective functioning of campgrounds. While it is probably true to say that aesthetics are often only considered after functional aspects have been satisfied, the two should go together in the design process, because attention to aesthetics can actually strengthen functional efficiency. In practice, functional elements of design tend to receive emphasis because of their more tangible nature – 'it works or it doesn't.' Aesthetics, on the other hand, are like beauty – very much in the eye of the beholder!

A further source of confusion can arise from overlap between the planning and design phases. In general terms, recreation site planning could be said to be concerned with the broad arrangement of site features, support facilities and circulation patterns necessary for the type of recreation envisaged. Design is related to micro-location and the moulding and fitting of the plans to specific topographic and landscape features of the site, while maintaining the desired positions of the facilities and circulation patterns (Jubenville 1976). For convenience, the criteria to be observed during both phases are considered together in the following discussion.

In the first place, planning and design of the recreation site should conform to known user preferences for given environmental conditions or situations (Christiansen 1977). Merely providing a picnic site is not enough. Service requirements, supporting facilities, equipment and site refinements should reflect the style and characteristics of participants. They should also be located to fit in with normal behaviour patterns, to minimise conflict and confusion, and to facilitate movement within the site.

A basic functional criterion of planning and design is that the recreation site and associated developments satisfy technical requirements (i.e. that they are useable in the sense of meeting standards of size, spacing and quantities). Operating needs and conditions are also important, and apart from meeting health and safety regulations, site developments should provide for the comfort and convenience of users. Rutledge (1971) illustrates the relevance of orientation to natural forces in the layout of recreation sites (e.g. the elevation and path of the sun's rays and the direction of prevailing winds), and stresses a common-sense approach to avoid unnecessary costs, and to provide for ease of supervision.

Recreational use of a site inevitably involves movement. The circulation system adopted can have a pronounced effect on efficiency of use, safety, satisfaction levels and supervision of visitor behaviour. Rutledge points out that the aim should be to get people where they want to go readily, and in doing so, not interfere with other activities. Therefore, the tasks are to anticipate flows, eliminate obstacles and confusion, and to provide unobstructed, well-defined, logical routes. Proper circulation planning and design can become an arm of recreation site management, not only in protecting the natural environment and visitors, but in promoting and facilitating desirable patterns of recreational use.

Sound site planning and design can minimise the task of supervision and the need for restrictive control measures over visitor behaviour. Public welfare should always be a concern, and if provision for visitor health and safety is built into a recreation site, many hazardous situations can be prevented. By definition, accidents are unplanned, but planning and design can go a long way towards eliminating the factors likely to generate emergencies. On-site control of vandalism and other forms of depreciative behaviour is also an important facet of visitor management.

Maintaining law and order at public recreation sites is also a serious problem for management. Depreciative behaviour can reduce or destroy the resource base and facilities, and interfere with the experience and satisfaction of other participants. Vandalism, acts of nuisance, violation of rules and crime, unfortunately, must all be anticipated. The monetary impact is staggering. In the US, the total yearly loss from vandalism alone was estimated 20 years ago at US $4 billion (Clark 1976).

The problem can at least be contained by prior attention to planning and design. Weinmayer (1973) believes that proper design can reduce vandalism by 90 per cent, and some observers suggest that much anti-social behaviour actually represents a protest against poor design and management of parks and other recreation sites (Gold 1974). So-called 'vandalism by design' is blamed for providing the opportunity for misuse by equipping recreation sites with objects, facilities and materials which invite disrespect and, ultimately, destruction. The inference is that opportunities for vandalism and other forms of undesirable behaviour can be removed or reduced at the planning and design stage. It is possible, of course, to attempt to devise structures which are vandalproof and virtually indestructible. It is preferable, and more positive, to provide sturdy, but attractive recreation environments which will be valued and protected by the users themselves. Site developments should be designed for easy maintenance and quick restoration if damaged. Rutledge (1971) suggests that thought be given to the clustering of potentially vandal-prone features, the opening-up of sites to external inspection, more adequate lighting, and the encouragement of higher levels of use, all as deterrents to anti-social acts.

Recreation sites which are properly selected and located, and which have had the benefit of thoughtful planning and design, should almost manage themselves; certainly, the task of management should be made much easier. Unfortunately, it is probably more often the case that managers inherit a poorly selected site, where little attention has been given to adequate development planning or design. Subsequent problems emerge, either because of overuse, deliberate misuse (above), or unintentional damage through ignorance and inappropriate use. Careful management of resources and visitors then becomes an ongoing concern.

Recreation resource management

Jubenville (1978) saw the managerial role in outdoor recreation as incorporating *resource management*, concerned with the reciprocal relationships between the recreation landscape and the visitor; *visitor management*, enhancing the social environment in order to maximise the recreation experience; and *service management*, involving the provision of necessary and desirable services so that the user can enjoy both the social and resource environment. Whereas each of these managerial roles is an important component of the overall recreation system, Jubenville considered visitor management to be fundamental, since it is the visitor who expresses demand for recreational experiences which require the other two elements. In the ensuing comments, provision of services will be regarded as a complementary, but ancillary, aspect of outdoor recreation management, and discussion will concentrate on resources and people. That said, it will soon become apparent that there is much scope for overlap between the two.

Recreation resource management implies close monitoring of the recreation site, to chart the rate, direction and character of change. It is vital that negative changes be detected early so that appropriate and positive management procedures be taken before site degradation proceeds to the point where the recreational environment becomes a source of dissatisfaction to visitors. Without a systematic monitoring and evaluation programme, management has no basis for comparison to determine change. Indeed, even before environmental deterioration or visitor dissatisfaction become evident, resource management procedures must be monitored and evaluated on a regular basis.

Resource management involves manipulation of elements of the resource base in order to maintain, enhance or even re-create satisfying opportunity settings for various recreational pursuits. In selecting the most appropriate course of action, the recreation site manager needs to balance concern for the resource base against other concerns, such as commercial considerations and the costs involved in loss of patronage. Leaving aside Jubenville's first suggestion – 'cut out and get out' – which is hardly a positive approach to management, other choices include:

- Site closure and rejuvenation through natural processes or cultural treatments. Site closure will certainly minimise recovery time and inconvenience, and may be justified for heavily deteriorated sites, especially where alternative opportunities are available.
- Rest and rotation of sites, or perhaps areas within a site, so that some recreation opportunities are always available.
- Leave open and culturally treat the site (i.e. keeping the site operating while implementing rehabilitation measures). If possible, this is clearly the ideal solution, but it can only succeed if treatment begins before site deterioration is well advanced.

Resource management procedures primarily involve technical and engineering-type actions, or landscaping techniques. Examples include various soil treatments and ground cover improvements such as irrigation; use of fertilisers; re-seeding; replacing or conversion to hardier and more resilient species; and judicious thinning of vegetation and removal of noxious species. These measures are aimed at increasing the durability of the biotic community, as well as inducing its recovery.

On-site patterns of recreational use can be influenced in various ways, including channelling the movements of visitors along selected paths (e.g. planting very dense and/or thorny bushes), or discouraging recreationists from entering a particular area (e.g. by fencing or the erection of some barrier designed as a 'people-sifter') (Seabrooke and Miles 1993). The effect may be discriminatory, but obstacles such as ditches and stiles, which prove a deterrent to some classes of visitor, are not insurmountable to all.

Vehicular traffic can be regulated according to mode and route, and many heavily-used sites no longer permit use of private vehicles; shuttle buses and other forms of communal transport are becoming more common in national parks (see Chapter 10). One-way traffic can be made mandatory, especially where parallel routeways exist, and separate trails can be designated for different classes of movement (e.g. skiers and snow mobilers).

Such action may be complemented by landscaping in order to enhance carrying capacities. This could involve the hardening or surfacing of intensively used areas such as viewing points; rotation of site furniture (barbecues or picnic tables) and movable facilities such as kiosks and shelters; rotation of entrances, trails and campsites; and provision of more effective waste disposal systems. As noted earlier, social carrying capacity can also be stretched. This can be achieved by many different management actions, including imaginative plantings to create more 'edges' or borders, or by breaking up the site with artificial mounds and buffers to boost the capacity of the landscape to 'absorb' visitors. By creating more levels or zones, a greater number of users can be accommodated on a beach. Lime and Stankey (1971) also indicate how recreational use can be redistributed, and carrying capacity increased, by improving access to previously under-used areas. Additional roads and trails, the installation of lighting, elevated pathways and bridges, and the elimination of hazards, are effective in redirecting visitor pressure.

With recreational waterbodies, capacity can be enhanced by providing more access points and ancillary facilities, and by manipulating the type and form of landscape features (e.g. addition of sandy beaches). Wildlife capacities, which indirectly impinge upon certain recreational pursuits, can also be built up by provision or improvement of habitats to encourage greater abundance and variety of animals, birds and fish. Wildlife populations will also respond positively to stable food and water supplies, control of diseases and pests (including predators such as feral cats), controlled use of biocides, minimisation of pollution, and reduction of fire and other hazards.

Recreation resource management is directed towards maintaining and enhancing the site as a viable setting for outdoor recreation. Ultimately, however, it is the reaction of the visitor to the site, which determines the success of the management programme. Ensuring a satisfying, high-quality recreation experience is the prime reason for developing an outdoor recreation management system. A specific procedure for visitor management which contributes to this aim, is the provision of information and interpretation facilties and services (see Chapter 10).

The rationale for the Recreation Opportunity Spectrum (ROS), carrying capacity, and Limits of Acceptable Change (LAC) were outlined in earlier chapters. The following discussion provides a brief overview of recent, innovative management approaches – the Visitor Impact Management (VIM) framework, and the Visitor Activity Management Process (VAMP). These approaches seek to address visitor management concerns, including those outlined above, while generally expanding on the principles underpinning carrying capacity, the ROS and LAC.

Visitor impact management framework

The development of VIM demonstrates the increasingly widespread view that recreational management requires scientific and judgmental consideration (e.g. see Hendee *et al.* 1978; Stankey *et al.* 1985; Shelby and Heberlein 1986; Vaske *et al.* 1995: 36), and that effective management of the recreation resource is much more than setting visitor use levels and specific carrying capacities (e.g. see Washburne 1982; Graefe *et al.* 1984; Vaske *et al.* 1995: 36) (also see Chapters 4 and 5).

The Visitor Impact Management framework resulted from a study by the US National Parks and Conservation Association (NPCA), which had two main objectives. The first objective was to review and synthesise the existing literature dealing with recreational carrying capacity and visitor impacts. The second objective was to apply the resulting understanding to the development of a methodology or framework for visitor impact management, that would be applicable across the variety of units within the US National Park System. A number of other goals underpinned the development of the VIM framework:

- to provide information and tools to assist planners and managers in controlling or reducing undesirable visitor behaviour;
- to suggest management approaches that build on scientific understanding of the nature and causes of visitor impacts;
- to consider impacts both to the natural environment and to the quality of recreation experiences, and to develop consistent processes for addressing such impacts.

(Graefe 1991: 74)

The review of the scientific literature relating to carrying capacity and visitor impacts identified five major considerations underpinning the nature of recreation impacts, which should all be incorporated into programmes for managing visitor impacts:

1 *Impact Relationships*: impact indicators are interrelated so that there is no single, predictable response of natural environments or individual behaviour to recreational use.
2 *Use–Impact Relationships*: use–impact relationships vary for different measures of visitor use, and are influenced by a variety of situational factors. The use–impact relationship is non-linear (i.e. it is not simple or uniform).
3 *Varying Tolerance to Impacts*: not all areas respond in the same way to encounters with visitors. There is inherent variation in tolerance among environments and user groups; for instance, different types of wildlife and user groups have different tolerance levels in their interactions with people.
4 *Activity-Specific Influences*: the extent and nature of impacts vary among, and even within, recreational activities.
5 *Site-Specific Influences*: seasonal and site-specific variables influence recreational impacts.

(Graefe 1990: 214; 1991: 74; Vaske *et al.* 1995: 35)

These five issues represent important considerations for the management of ecological, physical and social impacts (Graefe 1990).

In brief, the VIM framework is designed to deal with the basic issues inherent in impact management, namely: the identification of problem conditions (or unacceptable visitor impacts); the determination of potential causal factors affecting the occurrence and severity of the unacceptable impacts; the selection of potential management strategies for ameliorating the unacceptable impacts (Graefe 1990: 216). Given these basic issues, the VIM framework comprises eight steps (see Figure 6.2 and Table 6.1). Importantly, Graefe (1990; 1991) and Vaske *et al.* (1995) note that the task of managing visitor impacts is not over when management strategies are implemented, and that continuous monitoring and evaluation are necessary.

The Visitor Impact Management framework is based on high-level natural and social science research. As a planning framework it has the capability to deal with recreation impacts at a site level in a range of environments, and in conjunction with other planning frameworks within the management planning process. VIM has been applied in Australia (e.g. Jenolan Caves), Canada (e.g. Prince Edward Island), and in the US (e.g. Icewater Spring Shelter, Great Smoky Mountains National Parks; Logan Pass/Hidden Lake Trail, Glacier National Park; Florida Keys National Marine Sanctuary, Florida; Buck Island Reef National Monument, Virgin Islands; and the Youghiogheny River, Western Maryland) (e.g. see Graefe 1990).

VIM is a means of controlling or reducing the undesirable impacts of recreational use (Graefe 1991: 80). It has a sound scientific basis, and presents a systematic process for assessing visitor impacts by way of problem-solving. It is, in addition, a more detailed alternative to the concept of carrying capacity, and has potential for wider application in resource management (i.e. as part of an overall site or regional plan) (Graefe 1990; 1991), perhaps in conjunction with the LAC model.

The Visitor Activity Management Process (VAMP)

Tensions between resources and visitors led to the development of the Visitor Activity Management Process (VAMP) by the Canadian Parks Service (now Environment Canada). VAMP offers a fundamental change in orientation in parks management, from a product or supply basis to an outward-looking market-sensitive approach (Graham *et al.* 1988). Resource managers are thereby encouraged to be strategic in developing and marketing visitor experiences which will appeal to specific market segments.

> VAMP is a pro-active, flexible, conceptual framework that contributes to decision-building related to the planning, development and operation of park-related services and facilities. It includes an assessment of regional integration of a park or heritage site, systematic identification of visitors,

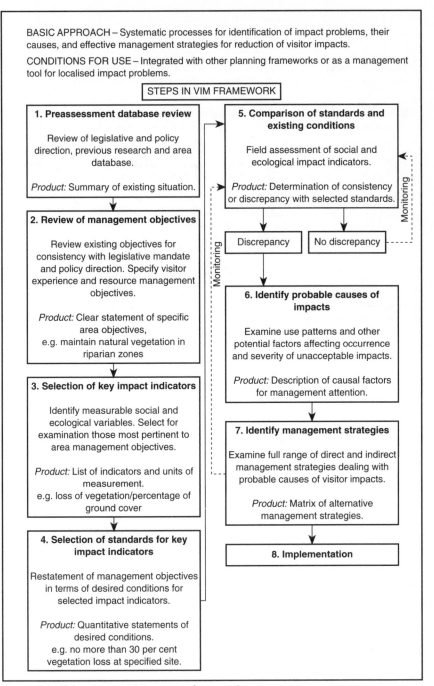

Figure 6.2 Visitor Impact Management framework

Source: Adapted from Graefe (1990: 218)

Table 6.1 Steps in the Visitor Impact Management framework

Step	Actions
1 The Preassessment Database Review	Compile and review relevant information that is available in order to gain initial perspectives of problems and issues.
2 Review of Management Objectives	Review current management objectives to ensure that they are stated clearly and specifically, and to define the type of experience to be provided in terms of appropriate ecological and social conditions. 'The objectives should be prioritized, since any single objective may lead to potentially conflicting goals' (Vaske *et al.* 1995: 38).
3 Selection of Key Indicators	Select measurable indicators that reflect the management objectives set out in Step 2: i.e. define what variables reflecting the planning objectives will be measured. Indicators will vary at different sites. 'Useful indicators include those that are directly observable, relatively easy to measure, related to the objectives for the area, sensitive to changing use conditions, and amenable to management' (Vaske *et al.* 1995: 39).
4 Selection of Standards for Key Impact Indicators	Decide how and when the Key Indicators identified in Step 3 will be measured, specifying appropriate levels or acceptable limits for the impact indicators identified in Step 3. 'The selected standards become the basis for evaluating the existing situation. This step serves the function of describing the type of experience to be provided in units of measurement compatible with available measures of the current situation' (Vaske *et al.* 1995: 39).
5 Comparison of Standards and Existing Conditions	Compare the existing situation with desired situations. The question to be resolved being: Is the area providing the types of recreational experiences identified in the management objectives, within appropriate maintenance of environmental conditions? Document problem situations.
6 Identification of Probable Causes and Impacts	Isolate potential factors (e.g. type of use, length of stay, group sizes, use timing and concentration, behaviour, and site characteristics) that may contribute to impact conditions. Identify the most significant causes of the problems identified in Step 5.
7 Identification of Management Strategies	Identify a range of alternative management strategies given some understanding of how the amount, type, and distribution of people using an area affect given impact indicators. The focus is on dealing with the causes of problems (see Table 6.3 in Vaske *et al.* 1995: 41).
8 Implementation	Implement the selected management strategies as soon as the necessary resources are available.

Sources: Graefe (1990; 1991); Vaske *et al.* (1995)

evaluation of visitor market potential, and identification of interpretive and educational opportunities for the public to understand, safely enjoy and appreciate heritage. The framework was developed to contribute to all five park management contexts: park establishment; new park management planning; established park planning and plan review; facility development and operation.

Graham (1990: 279)

In the same way as carrying capacity, ROS, LAC and VIM, the Visitor Activity Management Process uses information from both social and natural sciences to facilitate decision-making with respect to access to and use of protected areas (although it has the potential to be applied to a wider range of environments), and incorporates an evaluation requirement to measure effectiveness in outcomes and impacts (Graham 1990). It employs an overt marketing orientation to integrate visitor activity demands with resource opportunities, in order to produce specific recreation opportunities (Lipscombe 1993). A generic version of VAMP (see Figure 6.3) generally involves the following steps:

1 set visitor activity objectives;
2 set terms of reference;
3 identify visitor management issues;
4 analyse visitor management issues;
5 develop options for visitor activities and services;
6 provide recommendations and seek approval of activity/service/facility plan; and
7 implement recommended options (Graham *et al.* 1988).

Quite clearly, VAMP is 'issue-driven' (Hamilton-Smith and Mercer 1991: 58), and is flexible enough to incorporate process, planning and programme monitoring and evaluation. Since the late 1980s, VAMP has been applied to Canada's new park proposals and various park management plans (see Graham and Lawrence 1990). More specifically, VAMP has not been applied widely, save for a limited number of sites in Canada (e.g. Glacier National Park, British Columbia; Cross-country (Nordic) skiing, Ottawa; Mingan Archipelago; Point Pelee National Park; Kejimkujik National Park). VAMP is not a familiar planning approach in such countries as Australia (Lipscomb 1993), where long-run integration of visitor data is lacking, as most visitor management studies are carried out in isolation (Hamilton-Smith and Mercer 1991), and their findings rarely reported publicly. Nevertheless, according to Graham *et al.* (1988: 61):

VAMP continues to evolve and be modified as it comes into wider use, and as new concepts and techniques are developed... VAMP is a flexible management framework, undergoing transition and change. This frame-work must not become an end in itself, merely justifying design and

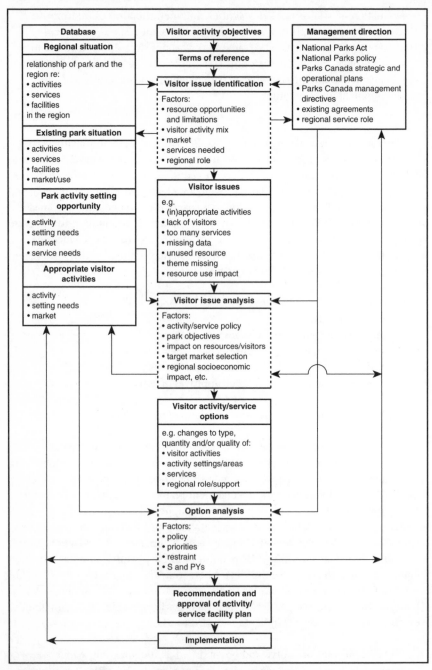

Figure 6.3 The Visitor Activity Management Process (VAMP) – A generic representation of VAMP

Source: Adapted from Parks Canada (1985, in Graham 1990: 278)

development of services, programmes and facilities. Case study research which highlights the effective [or otherwise] use of VAMP will be required... Further development of VAMP will require a supportive organizational environment, a condition not easily found in an era of constraint and pressure for more efficient management.

Over the past 25–30 years, a number of interdisciplinary or multidisciplinary planning and management processes and frameworks have been established, trialled, and variously described and evaluated, most on the precondition that they will help resolve, to a greater or lesser extent, visitor management problems. These advances in thinking, theory and application have gradually moved agencies towards more integrated approaches to environmental (including visitor) management. Perhaps not surprisingly, no single process or framework has received unanimous support among resource managers and researchers, as a means of solving the problems associated with visitor management. Put simply, a specific process or framework will be more effective depending on the scale at which it is applied (e.g. whether it is applied in a regional or single-site context), or on any number of other possible intervening variables.

Nevertheless, in addressing questions applicable to the four frameworks (LAC, ROS, VIM and VAMP) discussed in this and the previous chapter, a syndicate of conference representatives identified the following characteristics:

- LAC and VIM are mostly resource oriented and 'reactive' (i.e. they are frameworks which are applied after a resource input);
- ROS and VAMP are more oriented to visitor and use/activity, and include an interpretive activities component – a 'recipe' for potential interpretation;
- LAC and VIM applications can provide the opportunity for interpretive services and activities;
- ROS and VAMP applications result in the production of an 'interpretive prospectus-type' document. Including interpretation in the ROS and VAMP frameworks is automatic. LAC and VIM require a conscious managerial decision to include interpretation;
- none of the frameworks provides a superior time frame, and none appears to be more efficient than the other;
- improving the timeliness of the interpretive applications of any of the frameworks is primarily dependent on a quality data base, management priorities, and commitment to the interpretive element;
- all of the frameworks are information-gathering/decision-building processes rather than decision-making processes;
- inclusion of measurable standards and performance requirements would improve the use of the frameworks and their application to programmes such as interpretation (Pugh 1990: 354).

It would seem, then, that the application of recreation/visitor planning and management frameworks is largely a case of management skill in adapting and integrating components of one or more models or frameworks. For instance, the ROS was utilised effectively in an urban open-space study, despite its initial development and application in parks. As Hamilton-Smith and Mercer (1991: 57) noted, 'The possibility of using the ROS in any urban study [such as the Newcastle/ Lake Macquarie Open Space Study] should not be dismissed, but it is likely to require adaptation to the specific circumstances of the planning study concerned'. Perhaps, also, much hinges on the vagaries of politics, limited finances, and outdated organisational cultures and managerial frameworks, which can all thwart the best-laid intentions of innovative planners and managers.

Summary

The management of recreation resources is inherently difficult because of the extensive range of variables involved in the recreation–environment relationship. Furthermore, information concerning the impact of tourism on the environment is limited (see HaySmith and Hunt 1995). Despite these problems, several progressive technologies and planning approaches are being utilised in designing and managing recreation resources (particularly in protected and natural areas) across the globe. Geographic Information Systems (GIS) are being used, among other things, to identify ecologically sensitive areas and to plan tourist development. LAC is being utilised to monitor and set standards for acceptable levels of impact. VIM and VAMP call for monitoring and evaluation, the setting of clear and specific objectives, and the establishment of indicators and standards. These techniques, however, are not mutually exclusive. They can be integrated to identify, develop, implement, monitor and evaluate strategies for visitor management. Nor are these frameworks set in concrete. Political circumstances, resource availability and other circumstances will require organisations to adapt to their planning and management environments. This is best achieved by a strategic management approach, a subject discussed in Chapter 13.

Guide to further reading

- For detailed discussions on *VIM*, see Graefe (1990; 1991); Graham and Lawrence (1990); and Vaske *et al.* (1995); for *VAMP* see Graham *et al.* (1988); Ashley (1990); Reynolds (1990); Graham (1990); and Lipscombe (1993).
- Graham and Lawrence (1990), Hamilton-Smith and Mercer (1991) and Lipscombe (1993) contain discussions on the frameworks and applications of *carrying capacity, ROS, LAC, VAMP and VIM*, with Graham and Lawrence providing the most comprehensive coverage.

Review questions

1 Discuss the relationships between resource management, outdoor recreation and conservation.

2 What are the respective merits and limitations in applications of each of the following: the Limits of Acceptable Change; the Visitor Impact Management framework; the Visitor Activity Management Process; carrying capacity. Is any particular framework or approach better than the others?

3 Visit a nature-based outdoor recreation site (e.g. a national park). Conduct a resource inventory of that site. Make an assessment of the site's present condition. How might the site be improved? Determine what variables should be measured in order to monitor and evaluate recreational use of that site. If there is no current management strategy for that particular site, develop a management approach/framework to manage the site. If there is such a strategy, review that strategy and determine how it might be improved.

7

OUTDOOR RECREATION IN
URBAN AREAS

Most of the population of industrialised nations live in urban areas. Human beings are overwhelmingly social creatures and, as such, prefer to live together in communities created to serve individual and collective human needs. These communities can be of varying size and characteristics, but they have one feature in common – the potential to offer a wide range of functions to satisfy the needs of the population. One of these needs is provision for the recreational use of leisure.

Underpinning the attraction of urban places is the presumed availability of a diverse spectrum of recreation opportunities in a relatively limited and accessible spatial context. Part of the challenge of sustaining a livable urban environment, is to ensure the maintenance of a choice of quality leisure experiences through the existence of a spectrum of recreation opportunities, with the flexibility to adapt to the dynamics of a changing city landscape and evolving socioeconomic and political relationships. Indeed, 'the extent to which public outdoor recreation will flourish depends very much upon how planners and those in positions of power and influence face up to the challenge of urban restructuring – physical, economic and social – that is being posed by contemporary processes of urban change' (Williams 1995: 2).

The trend of urbanisation in Western industrial countries is well established. Some three-quarters of the population of the US and Britain live in urban areas and that figure is exceeded in parts of Western Europe. In Australia, also, another highly urbanised nation, life is centred on the cities, and most of the people are born, and spend their lives, within the confines of the built environment (Figure 7.1).

Despite this concentration of humanity, only in the past two decades has growing concern for quality-of-life issues, and in particular outdoor recreation, focused attention on the relative deprivation of city dwellers, and the need for more enlightened planning of the urban environment. This recent attention is not surprising. Many of the world's great cities are 'sick' – they are losing people and jobs; they have experienced or are experiencing declining fiscal solvency; they are less convenient, safe and attractive; and they are short on justice, tranquillity and general welfare. Increasingly, they are also seen as short on outdoor recreation opportunities.

Figure 7.1 Australia: Regions and population in 1992

THE OUTBACK

Population June 1992: 143 796
Growth 1991 to 1992: 265
Percent Australian Population: 0.8%
Net growth 1976 to 1992: 24 179 or 20.4%
Leading cities population in 1992:
*Kalgoorlie 27 182
*Alice Springs 24 500
*Mt.Isa 24 310
Comment: Virtually static in terms of population growth; growth is confined to service towns or areas of resource development.

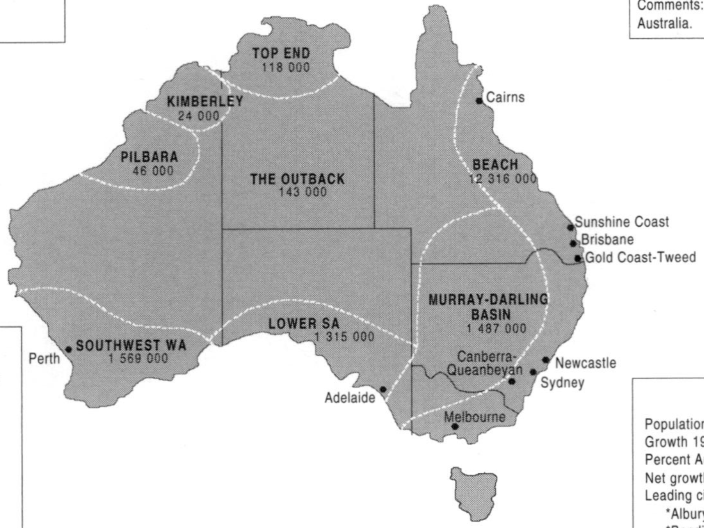

TOP TEN CITIES

Rank	City	Net growth 1991/92	
1	Brisbane	27 513	2.0%
2	Sydney	26 900	0.7%
3	Melbourne	21 200	0.7%
4	Perth	16 443	1.4%
5	Gold Coast-Tweed	10 375	3.4%
6	Adelaide	9 798	0.9%
7	Sunshine Coast	6 087	5.1%
8	Newcastle	5 800	1.3%
9	Canberra-Queanbeyan	5 558	1.8%
10	Cairns	2 298	3.2%
Total Top Ten		**132 472**	**1.1**

THE BEACH

Population June 1992: 12 315 773
Growth 1991 to 1992: 153 134
Percent Australian Population: 70%
Net growth 1976 to 1992: 2 512 463 or 25.6%
Leading cities population in 1992:
*Sydney 3 699 750
*Melbourne 3 177 900
*Brisbane 1 385 499
Comments: This is the new and fast growing urban heartland of Australia.

THE BASIN

Population June 1992: 1 485 537
Growth 1991 to 1992: 10 209
Percent Australian Population: 8%
Net growth 1976 to 1992: 162 126 or 12.3%
Leading cities population in 1992:
*Albury-Wodonga 90 578
*Bendigo 70 020
*Wagga Wagga 55 220
Comments: Comprises rich grazing and cropping land in inland Australia west of the Great Divide.

Map regions:
SOUTHWEST WA 1 568 000
PILBARA 46 000
KIMBERLEY 24 000
TOP END 118 000
THE OUTBACK 143 000
LOWER SA 1 315 000
MURRAY-DARLING BASIN 1 487 000
BEACH 12 316 000

Map cities: Perth, Adelaide, Melbourne, Canberra-Queanbeyan, Sydney, Newcastle, Brisbane, Sunshine Coast, Gold Coast-Tweed, Cairns

Parks and recreation are clearly not the highest priorities in urban development, but their roles in social welfare and urban renewal are becoming recognised. Enhancement of recreation opportunities in urban areas is now seen to contribute substantially to the quality of life of local residents, and to assist with the creation of a sustainable urban environment. For instance, increased attention is being given to the historical and cultural significance of park and recreation resources in the US (and elsewhere), with greater commitment to the restoration of historic buildings, facilities and designed landscapes. Julia Sniderman (in Dwyer and Stewart 1995: 607) of the Chicago Park District described her experience with park restoration in Chicago:

> The parks of Chicago are important cultural and historic resources. Documentation is the key to realizing the full historic value of urban parks. There is a vault under Soldier Field that was sealed off and forgotten for decades. When it was recently opened, whole archives of architectural plans for city parks were discovered. Documents like these are central to historic restoration work. Interpretation of restored parks is also important. Our recent restoration of Columbus Park is a good example. Seventh graders from the neighbourhood were trained as park docents to explain the key features of this Jens Jensen design to other kids and adults. One area in Columbus park was designed as an open-air theater; to highlight this feature, we presented a play there during opening ceremonies.

Urban restoration projects are occurring in cities around the globe, taking in such areas as remnant ecosystems (see Gobster 1994, in Dwyer and Stewart 1995) and waterfront development (e.g. see Law 1993; Craig-Smith and Fagence 1995; Williams 1995). Some areas, too, are utilising principles of ecosystems management, which emphasise relationships between physical, biological and social elements in the urban landscape (Dwyer and Stewart 1995), to underpin urban plans.

One of the first comprehensive attempts to document concern for urban recreation opportunities in the US, was the National Urban Recreation Study, undertaken in 1978 by the Heritage Conservation and Recreation Service. The primary objectives of the study were:

- to examine perceptions of needs and opportunities held by recreation users and administrators in urban areas across the country from the neighbourhood to the metropolitan level;
- to identify major problems of recreation and open-space providers in meeting needs;
- to explore possible solutions to problems with a wide variety of citizen and governmental interests;
- to identify a variety of open-space areas with potential for protection;

- to define a range of options for all levels of government, with emphasis on Federal alternatives which could assist or facilitate local, State and private efforts (US Department of the Interior 1978: 20).

The study concentrated on seventeen of the nation's largest cities, along with smaller towns and countries within their immediate vicinity. The sample field study cities were considered to reflect the dominant recreation issues and problems facing highly populated urban areas in the US.

The 1978 report established that no coherent national policy existed at that time for a balanced system of close-to-home recreation opportunities for all segments of the urban population. The study also found that recreational deprivation was not always a function of lack of facilities. In many cases, existing or potential recreation resources were not being fully utilised because of inappropriate locations or physical characteristics, deteriorating conditions, and poor quality management and programming.

Despite the broad spectrum of urban recreation issues addressed, the report was able to set common guidelines to indicate major directions for public action:

- conserve open space for its natural, cultural and recreational values;
- provide financial support for parks and recreation;
- provide close-to-home recreation opportunities;
- encourage joint use of existing physical resources;
- ensure that recreation facilities are well-managed and well-maintained, with quality recreation programmes available;
- reduce deterrents to full utilisation of existing urban recreation facilities and programmes;
- provide appropriate and responsive recreation services through sound planning;
- make environmental education and management an integral part of urban park and recreation policies and programmes;
- strengthen the role of the cultural arts in urban recreation.

Two decades later, and with allowances for scale and local circumstances, the findings of the 1978 report have relevance for other parts of the developed world. In particular, the report recognised the great disparity in the wealth of urban communities and the unevenness in resource endowment, both being factors which make a common strategy for addressing shortcomings difficult.

There are obvious differences in the physical and social geography of individual cities. Sydney, Australia, for example, is 'blessed' with a magnificent harbour and a wealth of accessible sandy surf beaches which provide unparalleled opportunities for water-related recreation. Many other coastal cities across the globe are likewise fortunate, whereas urban concentrations away from the coast typically present a different and more limited recreation environment. The presence of natural features and opportunities for contact with nature, within or close to the built environment, also enhance the potential for outdoor recreation. Sydney, again, is fortunate in being ringed with magnificent national parks only a short distance from the city's central business district (CBD) and periphery.

Climatic conditions also play an important part in the availability of a range of recreation opportunities in urban areas. The snowfields backing the city of Vancouver, Canada, go a long way towards compensating for restrictions on outdoor recreation in the city itself, because of otherwise pervasive rainy conditions. Cities in the tropics and subtropics can usually support more diverse forms of outdoor recreation than those where short summer seasons and severe weather can restrict activities.

A city with attractive natural features and an agreeable climate can take advantage of these for recreation; those not so fortunate may need to compensate by the creation of artificial environments. The provision of extensive facilities for indoor sports and other recreation activities in cities in the higher latitudes is, in part, a reaction to the severe winters of that part of the world.

Social differences between cities can also account for disparities in opportunities for recreation. Where cities are large, long-established and densely populated, diverse cultural features are more likely to exist, and these can be the basis for varied forms of recreation experiences, from participation in ethnic festivals and traditional celebrations, to the sampling of exotic foods and shopping for unusual products. On the other hand, a bland urban environment with an essentially monocultural population and a narrow social focus, can offer a strictly limited, and perhaps, predictable range of outlets for recreation.

The dynamics of urban recreation environments

> The urban domain is complex – towns and cities present an outwardly confusing mosaic of land uses into which recreational provision must fit and over which recreational activity must be superimposed. Patterns of opportunity are in most cases the product of lengthy periods of urban evolution in which physical growth, economic development and social change have combined to produce an environment that is dynamic, competitive and diverse.
>
> (Williams 1995: 14)

Urban areas are complex. They require planning approaches which are integrated (thus recreation and tourism are planned and developed in conjunction with other urban functions), flexible, and focused on 'the complementary function of the city and its region' (Jansen-Verbeke 1992: 33). Understanding the processes influencing urban development, and thus the nature and potential of outdoor recreation and tourism in urban areas, requires some understanding of global and local-scale developments, impacts and issues. Several important processes have shaped urban development, including:

- global economic restructuring;
- the physical expansion of built-up areas from compact, densely settled areas, with low overall populations, to modern post-industrial cities of highly populated but relatively low-density occupation, with residual zones of higher-density occupation in older, inner areas;
- increased social segregation as a result of mobility, preference and powers of different groups, and the ability of some to exert influence over emergent municipal authority;
- increased municipal regulation of development by way of attempts to control suburban sprawl, to plan new towns and to establish green belts;
- contemporary interests in greening the city and in encouraging environmental enhancement;
- urban redevelopment as a result of war damage, the establishment of urban development corporations, and the undertaking of high-profile projects (e.g. in dockland areas) to regenerate areas;
- urban redevelopment and restructuring, stemming from the need to address urban decay (Williams 1995: 15–6).

Given such processes and the problems that cities, generally, are encountering, it is not surprising that urban policies are now largely 'concerned with both winning economic growth and regenerating the core areas' (Law 1993: 23). The key elements in current urban policy comprise:

- emphasis on economic policies;
- emphasis on obtaining private investment;
- emphasis on property investment;
- public sector investment in infrastructure;
- public sector 'anchors', e.g. convention and entertainment centres, museums and art galleries;
- focus on the city centre;
- public-private partnerships;
- semi-autonomous agencies such as urban development corporations;
- flagship projects, e.g. Commonwealth and Olympic Games, Formula One Grand Prix, America's Cup;
- image and reimaging strategies (after Law 1993, in Hall, Jenkins and Kearsley 1997).

Tourism as an element of outdoor recreation has become an extremely influential element in urban planning in some areas. The potential of recreation and tourism as instruments in the policy of urban revitalisation is being increasingly recognised by local authorities and urban managers (Jansen-Verbeke 1992). Indeed, 'the growth of tourism as a form of economic development is having a major impact on the urban landscape of some cities, and reflects a changing attitude toward inner cities as well as a need to diversify repressed economies'

(Lew 1989: 15). Since the early 1980s, tourism has emerged as an important element in urban planning for such reasons as:

- economic globalisation and the consequent economic decline and restructuring of heavy industries and manufacturing in Western nations, led to a search for economic and employment alternatives by government at all levels, particularly in service industries;
- changes in transport technology have contributed to a decline of waterfront areas;
- tourism was seen as a way to rejuvenate and redevelop urban areas, often inner-city areas, which had experienced economic decline;
- in order to assist urban regeneration, governments have consciously sought to integrate tourism policy and development with cultural events and festivals, sports and leisure policies, and conservation of heritage, to help develop, market and promote urban regions and thus to attract the tourism and investment dollar (Hall *et al.* 1997: 199).

Of course, while it is possible to generalise to some extent about the patterns and processes of urban development and restructuring, there are some national and regional variations. Apart from the obvious physical, economic, social and political differences between cities which affect recreation potential, intra-urban contrasts develop over time in urban morphology, land-use patterns and socioeconomic characteristics. These can impinge on recreation needs and opportunities. Features such as the decline of the CBD; the establishment of satellite shopping complexes in the suburbs; the gentrification of inner city slums; the alteration of conditions of accessibility by the construction of new transport links; the emergence of ethnic enclaves within the urban system; the effect of political decisions on investment in recreation and sporting facilities; and ongoing changes in the economic, social and age structure of the population, can all have dramatic effects on opportunities for outdoor recreation.

The City of Sydney provides ready examples of the dynamics of urban recreation potential. The 'centre of gravity' of the city has now moved westward, with population growth, to the suburb of Parramatta, some 30 kilometres (approx. 18 miles) inland. Moving with it are the sporting and recreation facilities which have long dominated the eastern core of Sydney. These moves have been strengthened by the need to provide world-class facilities for the Olympic Games, scheduled for the year 2000. In turn, the availability of large areas of vacant land, once occupied by stockyards and industrial sites, gave further stimulus to the move westward and inland.

Any major city around the world can offer similar examples of the shifting and fluctuating nature of the spectrum of urban recreation opportunities, in response to changing physical, socioeconomic, environmental and political circumstances. Barrett and Hough (1989) point out that many municipalities in Ontario, Canada, have a legacy of parks and recreation space designed to meet the needs of earlier generations, yet, frequently they are inappropriate for the changing preferences, needs and lifestyles of the present population. They note that the social, economic and political context of urban communities continues to change at an increasing rate, but the planning response has not kept pace.

Pressure on the urban recreation landscape comes from a number of directions. Social concerns include the growing proportion of seniors and people with disabilities in the urban population – both groups with special recreation needs. Multicultural diversity in Canada, for example (as in Australia – see below), is placing a different set of demands on parks systems which were designed for a more homogeneous, predominantly white Anglo-Saxon culture. Urban recreation space also attracts the socially disadvantaged – the poor, the homeless, the transients and the unemployed. Apart from averting conflict with other users, the challenge is to develop programmes which encourage the disadvantaged to use recreation space more constructively. Again, the nature of recreation demand in cities is changing, with higher levels of environmental awareness, health and fitness programmes, and the emergence of a more varied array of leisure activities requiring specialised equipment and facilities.

At least some of the shortcomings in the urban recreation environment can be related to the above-mentioned dynamic elements in the character of towns and cities. An evolving pattern of urban growth and development can be recognised, marked typically by inner decay, suburban expansion or peripheral sprawl, and increasingly mobile and sophisticated groups of inhabitants. The implications are that any corrective measures proposed, must be adjusted for a particular geographical setting, area and population. Moreover, a distinction should, at least, be made between the inner city, the suburbs and the urban fringe.

The inner city

There is a good deal of evidence to suggest that the greatest deficiencies in regard to urban recreation space and facilities are to be found in the inner cores of large cities. Serious physical problems exist relating to the age, design and location of components of the recreation system. These are made worse when coupled with emerging recreation demands within rapidly changing urban precincts.

Population dispersion tends to take place from the centre, leaving behind both a diverse ethnic and cultural heritage, but typically, also, less affluent, elderly and otherwise disadvantaged groups. Any reverse movement of population is often representative of dissimilar and incompatible lifestyles, and merely adds a further dimension to the task of recreation provision. With gentrification of old inner neighbourhoods, basic deficiencies are aggravated by a new set of recreation demands from a diverse and rapidly changing clientele.

The inner city suburb of Marrickville, Sydney, for example, is experiencing a declining as well as a changing population. Recent census figures reveal that people born overseas made up 46 per cent of the population of the municipality. Of these, 87.5 per cent were from a non-English-speaking background. Major birthplace

Table 7.1 Marrickville municipality birthplaces

Country	Number	Percentage
Australia	44,063	54.0
Greece	6,033	16.1
Vietnam	4,386	11.7
Portugal	3,036	8.1
Lebanon	2,262	7.0
Yugoslavia	2,601	6.9
England	2,046	5.4
Italy	1,651	4.4
New Zealand	1,514	4.0
Philippines	846	2.3

Source: Australian Bureau of Statistics 1986

groups are shown in Table 7.1. In other inner suburbs of Australian cities, ethnic enclaves are becoming established (e.g. Vietnamese in Cabramatta, Sydney; Koreans in Campsie, Sydney; and Greeks in Coburg, Melbourne, supposedly the largest 'Greek city' outside Athens).

Some older core-city areas are fortunately able to retain a sound financial tax base with an established network of parks and open space. Other declining, fiscally-troubled core cities are forced to allocate a large share of their recreation budgets to operation and maintenance, at the expense of acquiring and developing new facilities and programmes. Those capital funds which are available for investment in recreation in older cities are typically spent on rehabilitation of ageing facilities.

In some of the world's larger cities, further difficulties are encountered in attempting to cater for the recreation needs, not only of residents, but also of commuters and visitors, all within the same inner-city environment. A good example is the City of Westminster, central London, where, apart from permanent residents who number about 240,000, some 500,000 workers commute daily, and where many millions of visitors are constantly present from all parts of Britain and the rest of the world. In this case, the Westminster City Council recognised that its particular responsibility was towards its resident population, especially in providing recreation opportunities close to home for this group.

London, in common with most of the world's great cities, developed without the benefit of a comprehensive recreation plan. By the time the need for planning was evident, many options were closed off by the massive social and dollar costs of acquiring recreation space. Skyrocketing land prices and finite funding sources placed any available recreation space beyond the reach of urban authorities. In downtown Atlanta, for example, the excessive valuation placed on a 1.7 acre (approx. 0.6 ha) site sought by the city, meant that its acquisition was only made possible by donation. The result is that traditional recreation activities requiring large expanses of land are now simply not possible in the densely populated neighbourhoods of most inner-city areas. Fortunately, though, for Londoners and visitors, Hyde Park, which was established and opened to the public in the 1630s, and Regent's Park, which was a central feature of John Nash's housing development in the early 1800s, remain as recreational assets. Clearly, historical developments are important, and so it is also worth noting that the number and accessibility of parks increased dramatically in the period between 1850 and 1880, when 111 urban parks were created in Britain, compared with 49 between 1820 and 1849 (Conway 1991). Interestingly, in that same period – and more specifically in 1879 – Australia's first, and the world's second national park, Sydney Royal (see Chapter 10), was established, mainly to serve recreational functions. It still serves as an important recreational asset within a dominant conservation management perspective.

In places where outdoor recreation opportunities are lacking, recreation needs can be met, in part, by providing indoor facilities or by innovative programmes to create additional urban recreation opportunities. Seattle, Washington, for example, transformed the air space over a ten-lane interstate highway into the 3.5 acre (approx. 1.4 ha) Central Freeway Park. Spanning the 'concrete canyon' on a bridge structure, the park offers an unusual retreat in downtown Seattle. Sydney, Australia, is another example of a large modern city where hectares of wasted space on the rooftops of city buildings have been transformed into sporting and recreational facilities for office workers and residents. New buildings are the prime target of this policy, with incentives for developers to incorporate use of rooftop space into their plans. In the US, Britain and Europe, gardens, swimming pools and recreation areas have been established on the rooftops of hotels, private homes and city apartment blocks. More specifically, more than 25 years ago, the US Bureau of Outdoor Recreation (1973) identified several successful space-conversion projects resulting in useful additions to the urban recreation resource base:

- in Baton Rouge, Louisiana, 35 acres (approx. 14 ha) of unproductive land beneath an elevated highway interchange was transformed into Interstate Park as a neighbourhood recreation area;
- in San Francisco, several park areas were developed on top of underground parking facilities (including Union Square), and in downtown Los Angeles two large corporations created a 2.5 acre (approx. 1 ha) rooftop park above a garage, as part of an urban renewal project;
- in Albuquerque and Honolulu, airport buffer lands were transformed into a community golf course;
- along the lower Rio Grande, the city of El Paso recognised the potential of the river's floodway, in developing a linear park incorporating recreation activities and facilities capable of withstanding periodic flooding;
- the surface of covered water storage facilities in Denver and San Francisco were developed for public tennis courts and sports facilities;
- in Washington DC, a sanitary landfill site was transformed into a useful and valuable recreation resource;

- in New York City, construction of a 30 acre (approx. 12 ha) park on the roof of a sewage treatment facility, provided a picnic area, baseball diamonds, tennis courts, trails, swimming pools and an ice rink.

Despite such initiatives, it is surprising that not all undeveloped areas of urban land and water are perceived as recreation resources or used as such. Many city authorities apparently lack the imagination and/or the means to capitalise on the potential of neglected areas such as floodplains, water supply reservoirs and catchments, waste treatment facilities, waterfronts, parking lots, service corridors and abandoned rights-of-way and railway lines. Especially valuable are strips of linear open space, where the edge effect promotes greater recreation use.

The inner city remains the focus of intense competition for space, for commercial and industrial premises, for transport and communication, and for high density/high rise residential purposes. It is important that provision for recreation space is not ignored in the redisposition of the land and water resources of the urban heartland.

The suburbs

If it is difficult to make useful comparisons of recreation provision between inner cities, it is impossible to generalise, regarding the recreation environment in the suburbs. Clearly, the dispersion of population from the core area referred to earlier, is stimulated by a perceived improvement in the quality of life, part of which is reflected in a better range and standard of recreation opportunities. However, the extent to which this is experienced depends upon the particular local mix of such factors as location, resource base, socioeconomic status, community spirit, and the affluence and initiative of the local government authority.

Suburbia has, or should have, one advantage – relatively newer facilities are likely to mean lower operating and maintenance costs, leaving more funds for investment in capital projects and acquisition of land. However, recently settled suburban communities with small, but rapidly growing populations and limited financial resources, are more often concerned with the availability of basic services than with the 'luxuries' of amenity provision.

Again, the very nature of the suburban environment, typified by a real sprawl of dispersed housing units and dormitory-style subdivisions, with heavy reliance on the motor car, makes it difficult and expensive to provide a full range of recreation facilities. Site design of many early subdivisions precluded the establishment of large open spaces for community recreation. Both private developers and public housing authorities appear to have given only minor consideration to this aspect, apart from labelling a mandatory minimum area as 'recreation reserve'. There seems little evidence of any comprehensive planning of recreation facilities as an integral part of emerging neighbourhood and community development patterns. Where tracts of recreation land are set aside, they are usually developed on an *ad hoc* basis, with scant regard for other than the immediate needs of the existing population.

Large modern cities typically spill over, unimpeded, into the surrounding countryside, in a process aptly termed 'metropolitan scatteration' (Wingo 1964). The rapidly diffusing residential frontier is allowed to outpace provision even for basic service needs, the urgency of which merely reinforces the traditional low priority given to recreation planning. Unimaginative land subdivision perpetuates the conventional grid street system, with associated large-scale alienation of potential recreation space. Little thought is given to the most appropriate size or form of the overall neighbourhood, or its relationship to the rest of the city.

Australian cities, in particular, are much less compact than their older European counterparts, with corresponding lower residential densities. Perhaps this has less to do with the relative scarcity of land than with Australians' preferences for maximising private space at the expense of public amenity.

The resulting featureless sprawl promotes a degree of introspection in urban Australia, or a tendency towards 'privatization', as Mercer (1980b) calls it. In the absence of local clubs or pubs, community meeting centres or sports complexes, cinemas, or even shopping centres in some cases, the inhabitants place greater emphasis on the home environment for their leisure pursuits. The sheer lack of public facilities forces households to maximise private space by way of compensation.

Moreover, the only practical means of transportation in a highly dispersed, rapidly expanding, low-density suburban area, is the private motor car. Those without a car are severely disadvantaged with respect to leisure options. The scale of metropolitan planning is geared to the car and not the human being. As the metropolis spreads, pressure to construct intra-metropolitan freeways to accommodate the car and overcome traffic congestion, also increases. Such freeways accelerate residential development towards the periphery. Ex-urban recreational opportunities are pushed further and further away from the centre of gravity of the population, to the detriment both of people living in the inner suburbs and of those without access to private transport. Thus, transport improvements, proposed as a solution to one urban problem, merely give rise to other problems, and recreation opportunities decline further.

It appears, then, that the modern city has let its inhabitants down as far as outlets for leisure in any communal sense are concerned. What seem to be lacking are the essential ingredients to create the 'village' atmosphere of earlier times – a setting which will generate a sense of togetherness, belonging and place. Features which once had an important recreational function as part of that setting have no place in present-day suburbs. The town square, the village green, the dance hall-cum-cinema, even the local 'pub' or bar in some cases, have given way to home-based recreation, centred on the television set, perhaps the backyard pool, and all manner of electronic gadgetry. The sterile facilities which often serve for community recreation purposes do nothing to offset urban alienation. It is difficult to identify with a slab of concrete or fibre-glass, and it is little wonder that the potential users attempt to humanise or deface these structures with graffiti. They see nothing wrong with vandalism of incongruous features to which they cannot relate and which apparently cannot satisfy their recreation needs.

Local authorities, which have the prime responsibility for recreation, face a deepening cost-revenue crisis, made worse by a general indifference on the part of the higher tiers of government to the problems of cities. This situation serves to underline the need for fresh initiatives in urban recreation planning. Part of this strategy should be a broader approach to the provision of leisure and outdoor recreation opportunities in the suburbs, with greater emphasis on self-help and community involvement. Out of necessity, planning bodies might come to realise that some of the deficiencies inherent in suburban life and living may be remedied by encouraging fuller utilisation and management of communal recreation resources.

The urban fringe

In 1981, the US Department of Agriculture estimated that some 3 million acres (approx. 1.2 million ha) were being converted each year to urban and built-up uses across North America. More recent observations give no basis for optimism that this trend is in decline. In these urbanising areas, local initiatives to direct development away from critical agricultural, environmental and recreational uses, are often weak or non-existent. The city periphery thus becomes the focus for some of the most urgent programmes for general living and open-space retention purposes.

One of the problems in discussing peri-urban recreation is to decide where suburbia ends and exurbia, or the urban fringe, begins. Yet, it is important to consider recreation opportunities in this transition zone, because mobile city populations readily incorporate nearby fringe areas into their effective recreation space. Despite the growing importance of home-based recreation noted earlier, the neighbouring countryside is increasingly perceived as an extension of life in the city. The tendency for people to seek natural settings to offset the pressures of an urban-industrial existence is well-documented, and is prompted, in part, by the urbanisation process itself. Janiskee (1976) explained the recreation appeal of extra-urban environments in the context of a push–pull model of motivation. Periodically, environmentally undernourished urbanites are 'pushed' from the city because of stresses imposed by their lifestyles. At the same time, they are 'pulled' into the more natural hinterland by the opportunity to experience compensatory alternative surroundings and activities. Apparently, urban dwellers, who have voted with their feet for city living, are not totally adapted to the urban environment. They have a physical and social need to seek novel, irregular and opposite situations, in exchange for routine, boredom and the familiar. This need is reinforced by growing awareness of what the surrounding countryside has to offer, together with enhanced means of making use of its recreation potential. Natural settings offer city dwellers the capacity for self-renewal in a different, specifically outdoor setting, inevitably leading them to the urban fringe and beyond for recreation. That said, the challenge remains of maintaining the essentially undeveloped character of the urban fringe so that its function as recreation space is unimpaired.

The task is given added urgency by the different perceptions held of the urban fringe. To land developers, it could be seen as a speculator's paradise; to urban planners it might represent a useful reserve of land for future urban expansion, or for the location of less compatible elements of urban infrastructure such as motorways, airports, waste disposal sites and noxious industries. Alongside these, is the potential to expand the city's spectrum of recreation opportunities by the creation of active and passive, large-scale and specialised recreation facilities.

Pullen (1977) argued strongly for the establishment of permanent areas of 'Greenspace' beyond the periphery of cities, as a means of guiding and containing urban development. Pullen saw Greenspace as a valuable resource for the provision of important social functions, including recreational activities. Despite the attraction of the concept, experience in major world cities suggests that the protection of a permanent zone of Greenspace is difficult in the face of compelling pressures to maximise economic use of valuable land.

Short of public acquisition and creation of formal park land, the zoning of land in the urban fringe, as reserve for recreation, should act as a further deterrent to development. This step could well be enhanced by the establishment of community forests on the outskirts of major cities in Britain (Countryside Commission 1989). The inspiration for the new community forests came from Bos Park near Amsterdam and the ancient stands of trees in Epping Forest, close to the heart of London. The North East of England now has two community forests – the Great North Community Forest, established in 1990 on 16,500 hectares (approx. 41,250 acres) of derelict urban land, and Cleveland Community Forest, established in 1991, covering 25,000 hectares (approx. 62,500 acres) around an estuary and on private agricultural land. Despite the desirability of these imaginative projects, in terms of adding to outdoor recreation opportunities in the urban fringe, their viability is uncertain given the long-term investment involved, along with financial constraints on new plantings (Wilson and Biberbach 1994). Moreover, recreational access is not always guaranteed by private land-owners reluctant to take on added management responsibilities.

This illustrates an important qualification regarding implementation of any plan to utilise the recreation potential of the urban fringe. The plan cannot succeed without the firm commitment of responsible public authorities, both financially and in terms of statutory powers over land use. Cooperation with controllers of private land and resources in the semi-urban countryside is also necessary if this resource is to fulfil its role as an integral part of the urban recreation environment.

Urban open space and recreation space

In the highly urbanised countries of the Western world, the city functions primarily as a place of residence and as a base for work commitments. The growing segment of life given over to leisure appears to find only restricted expression in urban environments. More and more people are looking beyond the city limits to find their 'activity space' for outdoor recreation in rural areas. However, for many urban residents, this alternative is not accessible for such reasons as lack of transport, time or money. These people must turn towards open space within the city for relief from perceived deficiencies in the urban environment.

Much of the dissatisfaction with urban living, and many of the concomitant social problems, can be traced to the apparent inability of the modern city to meet the basic needs of its inhabitants. One of the objectives of urban environmental and recreation planning is to produce a more satisfying array of amenity stimuli and responses. The range and intensity of amenity responses are, in turn, a function of the nature, characteristics and location of what may be called amenity precipitants. In an urban situation, a fundamental component of the amenity response system is again the availability of open space for recreation.

According to Gold (1988), an effective recreation experience in cities calls for opportunities to experience freedom, diversity, self-expression, challenge and enrichment. Servicing such opportunities provides much of the justification for providing open space within cities. In this context, open space is basic to the structure and function of the built environment in meeting human needs. Yet, various factors can affect its role as part of the urban outdoor recreation resource base. In the first place, it is too simplistic to equate open space with recreation space, since not all urban open space is equipped to function as recreation space. By way of example, modern, planned national capitals such as Brasilia are blessed with vast areas of open space, geometrically arranged, trimmed and manicured, yet devoid of any feature which would encourage, facilitate or even permit leisure activities. In many cases, any recreation function, apart from perhaps passive viewing, is specifically excluded by physical barriers, equally forbidding signage or other effective means of discouraging participation. 'Open space it may be; recreation space it is not' (Pigram 1983: 109).

This is not to deny that urban open space *per se* has value, apart from a potential recreation role. Demands for lower residential densities in affluent areas, and for extensive landscaped sites for public buildings and industrial estates, demonstrate a growing social awareness of space as a community asset. Added to this is the acknowledgment of what economists term the 'existence value' of open space and green areas within cities. Nearby residents can develop strong attachments for, even rather ordinary, local parks, which they may rarely use for recreational purposes.

However, satisfaction of the leisure needs of urban dwellers requires more than the existence of open space. In the provision of recreation space in cities, it is not a matter of how much, but how good that space is. In part, this will reflect the characteristics of urban open space in terms of size, range of facilities and accessibility. The importance of the natural setting in contrast to the surrounding built-up environment would also seem to be paramount. With reference to Sydney, Australia, McLoughlin (1997) argues persuasively for the retention of bushland within the urbanised area. Earlier in this chapter, the close proximity of a number of national parks to the City of Sydney was noted. However, despite this, McLoughlin identifies a range of complementary values for bushland, in or near urban areas:

- natural and cultural heritage values;
- habitat for resident and migratory species;
- aesthetic landscape values separating parts of the city, and as a screen for unpleasant urban structures;
- environmental protection values;
- recreational values for a variety of activities;
- scientific and educational values (McLoughlin 1997: 166).

McLoughlin also identifies threats to remnant areas of bushland, and the measures which need to be taken to minimise the impacts of city growth and development on this valuable element of urban open space.

In a study of urban parks in Melbourne, Australia, the attractiveness and variety of the vegetation, and the presence of waterbodies, were found to be important factors in accounting for variations in recreational use (Boyle 1983). At some parks, a strong preference was expressed for peace and quiet in relatively natural areas with few facilities. A significant number of respondents at two native eucalypt parks, for example, where minimal equipment has been installed, insisted that more facilities were *not* needed.

A similar preference for nature-dominant environments was revealed in a major study of inner-city parks in the City of Brisbane, Australia (McIntyre *et al.* 1991). Results of the study suggest that the natural setting of inner-city parks and green areas provides a venue for rest, recreation and release from tension for urban residents, as well as an opportunity to appreciate nature. The preference revealed for natural settings 'emphasises the need for the preservation of these "islands of naturalness" within the cityscape' (McIntyre *et al.* 1991: 16).

In the US, corridors of protected open space, known as 'Greenways', are managed for conservation and recreation purposes under a programme established by The Conservation Fund. Greenways often follow natural land or water features, linking nature reserves, parks and cultural and historic resources with each other, and with populated areas. Some are publicly owned, some are privately owned, and some are the result of public/private partnerships. Some are open to visitors, others are not. Some appeal to people, and some attract wildlife. Greenways, linking large natural areas, have also been developed and promoted in rural areas. According to The Conservation Fund, Greenways protect environmentally important lands and native plants and animals, simultaneously linking people with the natural world and outdoor recreational opportunities. Greenways can also preserve biological diversity by maintaining connections between natural communities; soften urban and suburban landscapes; protect the quantity and quality of water; direct development and growth away from important natural resource areas; provide alternative transport routes; and act as outdoor classrooms (for further details see the World Wide Web page at: http://www.conservationfund.org/conservation,greenway/htm).

The above discussion adds emphasis to the importance of matching park and protected area settings to the preferences of users. It also raises questions of multiple use, and of non-use or underuse of urban parks.

Multiple use of urban recreation resources

Implicit in several studies of urban open space is the waste involved in setting aside resources for some exclusive use. Public institutions, in or near urban communities, frequently provide opportunities for innovative recreation programmes. Despite additional surveillance costs and possible problems with anti-social behaviour, establishments such as schools, hospitals, child-care centres, health clinics, religious and cultural facilities, fair grounds, sporting arenas and even military bases, can all have significant potential in multiple use of cost-effective communal recreation space.

School properties, in particular, represent a sizeable part of readily accessible publicly-owned resources. They are usually well distributed within cities, and occupy strategic locations in residential neighbourhoods. Most have playgrounds or playing fields attached, and many have indoor gymnasiums and pools. Yet, aside from their primary role, they are often one of the least utilised public facilities, remaining empty when recreation pressures are greatest – after working hours, at weekends and during vacations. In many areas, opportunities also exist for reclamation or conversion of abandoned public buildings to provide indoor recreation centres. Key elements are diversity and flexibility: the opportunity for a range of recreation opportunities likely to attract a broad cross-section of the community, yet amenable to a change of function and orientation.

Children and play-space

> How do we create an environment that meets a child's need and urge to explore, test and experiment?

Play-space for children is a particularly sensitive issue in urban environments. Play which involves interaction with nature and natural processes is considered important for childhood development (Cunningham and Jones 1987). Whereas natural environments have been shown to have innate appeal for pre-adolescents, the absence of suitable, accessible and safe sites precludes this experience for many children (Cunningham and Jones 1994). In many cases, it is not the decision of the child which dictates the play location so much as the perception of the parent or guardian regarding what constitutes a suitable child-friendly environment.

According to Raymond Unwin (in Williams 1995: 18), the amenities of life have been neglected in that 'we have forgotten that endless rows of brick boxes, looking out upon dreary streets and squalid backyards, are not really homes for people'.

> The types of city Unwin and his followers wished to plan afforded not just those opportunities for leisure that were already associated with parks and gardens, but extended significantly the scope of the home and the suburban streets to support informal recreation and children's play. In time, the utility of the street as a recreational environment would decline in the face of the environmental onslaught of increased road traffic but in, say, the years between 1918 and 1939, the streets of leafy suburbia became an almost unnoticed, but significant recreation resource.
>
> (Williams 1995: 18)

More recently, playing on or near city streets, for example, has been generally frowned upon. Yet, with children, the streetscape tends to be popular for recreation purposes, despite obvious hazards. In fact, it appears that the busier the street may be, the more appealing it is. The unstructured nature of city streets and footpaths, with its clutter and ever-present element of danger, apparently offers an exciting and challenging contrast to conventional playgrounds. Rather than attempting to counteract this appeal directly, it would seem more productive to take advantage of the opportunities at the street-scale for design of imaginative and safe play areas. Bannon (1976) uses the example of Central Harlem, New York, to illustrate the potential for transformation of small blocks of vacant land in built-up areas, into 'vest-pocket parks' and 'tot-lots' as a viable alternative to the streets for play, or for quiet relaxation by older residents. 'Adventure playgrounds', where children are allowed and encouraged to create their own play environment under non-restrictive supervision, provide an unorthodox, but potentially very important, setting for spontaneous enjoyment:

> In urban areas where space of any kind is at a premium... adventure playgrounds are the closest we have come to emulating some of the mysterious and exciting pleasures of childhood... The land is left in its original state, with building materials (such as wood, cardboard boxes, logs, planks, bricks and so forth) provided for the children to build almost anything they desire... Building a house, planting a garden, digging a tunnel, cooking a meal, swinging on ropes from trees, creating a mysterious artefact, anything children enjoy which does not endanger them or others is permitted.
>
> (Bannon 1976: 205–6)

By way of example, the Lenox-Camden Playground, Boston, was run from April to October, 1966, and carefully studied by Robin C. Moore (Bengtsson 1972).

147

From Moore's assessment of his experience, a number of observations warrant mention, and are listed in Table 7.2. Whereas many playgrounds are now developed along less creative and flexible lines, the Lenox-Camden Playground serves as an important reminder of the need for, and importance of, less 'structured' playgrounds

Street closures for an hour, or a day, or for longer periods, perhaps with the introduction of mobile recreation programmes, are another means of harnessing and redirecting the attraction of the streetscape as a neighbourhood recreation resource to provide *ad hoc* play-space for children.

Use and non-use of urban recreation space

Failure to recognise urban recreation opportunities is not always confined to city administrators. Potential users, too, seem reluctant at times to avail themselves of the facilities which are provided. Field observations suggest some surprisingly low levels of utilisation of recreation space, especially in the inner core of some cities. An Australian study in the inner suburbs of Melbourne found that a neighbourhood park was not 'a particularly vital part of most residents' perceived environment' (Cole 1977: 93). Although considerable diversity was discovered in user groups and activities, the dislike for the park displayed by children, in particular, was traced to constraints on natural patterns of active child behaviour; possibilities for creative play in the park were virtually non-existent.

The broader issues of non-use and under-use of urban parks were first high-lighted by Gold in 1972. Gold concluded that the major constraints could be grouped into three categories – behavioural, environmental and institutional (Table 7.3). Not all of these inhibiting factors are easily countered, but obviously convenience of access, site characteristics, location, level of facilities, safety considerations and management and maintenance are subject to manipulation. Gold's comments support the view that non-planned designation of open space in urban areas, with little thought to effective location, size and quality, will probably ensure that it remains open space – empty and ignored.

Patmore (1983) also pointed to several factors impinging on patterns of use and non-use of recreation facilities. As noted in previous chapters, effective access

Table 7.3 Major causes of non-use in neighbourhood parks

Behavioral	Environmental	Institutional
User orientation*	Convenient access*	Goal differences*
Social restraints*	Site characteristics*	Personal safety*
Previous conditioning	Weather and climate	Relevant program
Competing activities	Physical location	Management practices
User satisfaction	Facilities and development	Maintenance levels

*Most significant in each category relative to all factors

Source: Gold (1973: 103)

Table 7.2 Lenox-Camden playground, Boston

The playground was run from April to October 1966 and carefully studied by Robin C. Moore. From the assessment of his experience the following observations have been abstracted:

1 Creative play is an opportunity for children to manipulate their environment to achieve their own ends, and to sense that the world around them can be changed and need not be taken as given.

2 Activity was the initial attraction and reason for coming to the playground; but if the setting allowed people to sit around watching, they did so – talking, singing, joking, flirting, etc., while the more intensive activity of the playhouse, tower and basketball court provided a background interest.

3 It often appeared that activity on the playground was only one link in a chain of play activities occurring in and around the child's home.

4 The playground was always the first place to 'check out' when looking for friends and/or action. This leads to the conclusion that play-spaces should be incorporated intimately into housing areas.

5 The most important observation in terms of age was that it bore little relation to physical ability, to courage in particular – as well as spills. A six-year-old girl would, for example, climb up the tower without a second thought, while an eleven-year-old boy would be scared and unable to take the same route. This observation has many implications for design, such as the non-segregation of different age groups.

6 One aim was to discover the most popular moveable materials. These turned out to be mild crates, large timber cubes 1 foot to each side, two in thick timber up to 12 inches wide and 5 feet long, sheets of masonite and ship-board, 50 gallon barrels and many other kinds of robust junk.

7 Moveable materials did raise a number of practical problems. The less robust items tended to get smashed and lost their usefulness. They had to be cleaned up and disposed of. After a while, moveable materials became dispersed over the playground, tending to reduce their play potential. Stimulation was increased if they were reassembled frequently by the adults.

8 The greatest amount of creative activity, in terms of frequency and span, took place behind, and in the playhouse. It is suggested that one of the reasons for this was the sense of enclosure there: spaces of adequate size for constructive activities, cut off psychologically from the surroundings, even though other activities were going on in the immediately adjacent area.

9 A critical difficulty is trying to comprehend the very small-scale environment that children operate in.

10 Materials that would normally appear as 'junk' in other people's eyes, are very relevant to much creative and imaginative play. Useful junk consisted of objects that could be used for building construction, or objects that previously had a specific function and could still be used as such: the steering wheel of a car, for instance, became the steering wheel of a 'fire engine'. These materials were often used individually as props to the imagination, and many times functioned as the initial stimulus, setting the child's thinking along a particular line.

Source: Bengtsson (1972: 151)

149

is not related to convenience of location alone. Patmore categorised four types of barriers to access:

- *physical barriers*, which include personal limitations and the nature of intervening space;
- *financial barriers*, which impose a direct economic constraint through high levels of admission charges or equipment costs;
- *social barriers*, which arise from the association of the images of certain recreational pursuits with social status;
- *transport barriers*, which relate to lack of access to a vehicle and associated time/cost deterrents on participation.

Godbey (1985) translated many of these constraints on participation into a useful model for summarising the reasons why people do not participate in a specific recreational activity (Figure 7.2). The model, when applied to an urban park, can identify options for action by management. Such remedial measures need not be elaborate and can be as simple as relocation of an entrance or better maintenance of grounds. Some of the techniques used to identify and redress causes of, and responses to, underuse of recreation facilities at water storages in the US are set out in Table 7.4.

More generally, solutions to a lack of readily accessible recreation opportunities in cities, rest with more enlightened planning of the urban environment to provide

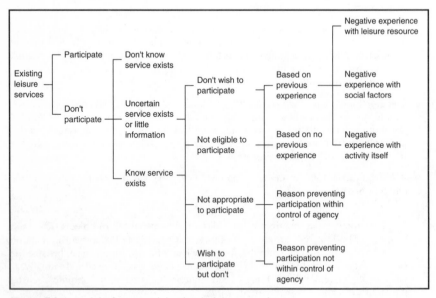

Figure 7.2 A model of non-participation in leisure services

Source: Godbey (1985)

Table 7.4 Identifying and solving underuse

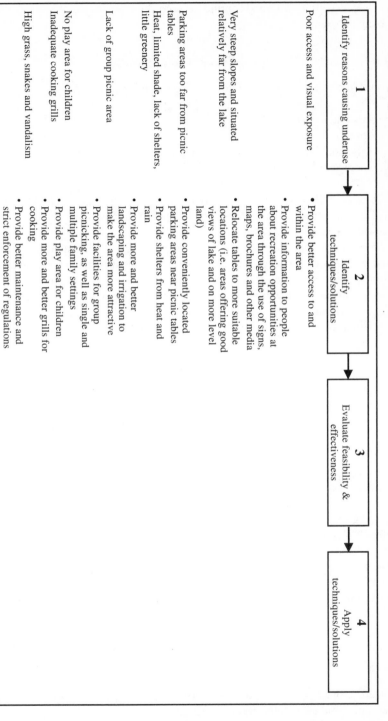

1 Identify reasons causing underuse	2 Identify techniques/solutions	3 Evaluate feasibility & effectiveness	4 Apply techniques/solutions
Poor access and visual exposure	• Provide better access to and within the area • Provide information to people about recreation opportunities at the area through the use of signs, maps, brochures and other media		
Very steep slopes and situated relatively far from the lake	• Relocate tables to more suitable locations (i.e. areas offering good views of lake and on more level land)		
Parking areas too far from picnic tables	• Provide conveniently located parking areas near picnic tables		
Heat, limited shade, lack of shelters, little greenery	• Provide shelters from heat and rain • Provide more and better landscaping and irrigation to make the area more attractive		
Lack of group picnic area	• Provide facilities for group picnicking, as well as single and multiple family settings		
No play area for children	• Provide play area for children		
Inadequate cooking grills	• Provide more and better grills for cooking		
High grass, snakes and vandalism	• Provide better maintenance and strict enforcement of regulations		

adequate recreation space and appropriate recreation facilities to meet the demands of their citizens.

Urban recreation planning

Several questions are fundamental with respect to urban recreation space:

* how much is needed?
* what form should it take?
* how should it be managed?
* where should it be located?

The first of these questions concerns measures of quantity, and this inevitably involves reference to space standards – specific numerical indicators of the adequacy of recreation provision.

From time to time, attempts have been made to arrive at desirable and practical standards for parks and open space, relative to user populations. In urban situations, the most frequently cited standards range from 7 to 10 acres (approx. 2.8 to 4.0 ha) per 1000 people, the total encompassing parks and playgrounds under various categories (Figure 7.3).

Whereas the standards approach specifies a total area of urban space set aside for outdoor recreation, it does little to ensure that such space will:

* be part of an overall scheme to ensure accessibility;
* be designed for specific purposes and community needs;
* complement the regional open space system;
* take into account natural features (Schomburgk 1985: 22).

The space standards for Canberra, Australia (see Figure 7.3), were considered by planners to be appropriate for the particular type and size of population of that city. However, it is obviously unrealistic to attempt to apply common standards across contrasting communities – standards which are inflexible and unrelated to changing socioeconomic profiles of potential users, or to varying space needs for different recreation activities. The fact that supposedly universal norms have not always been attained, reflects the many factors which should influence a more realistic definition of space standards.

Clearly, any set of recreation space standards should only be used as a guide, to be modified as required and applied sensibly in the context of the sociocultural characteristics of the community involved, and the resource attributes of the subject urban environment. In particular, rigid adherence to uniformity should not be allowed to obscure the many possibilities for innovative planning, management and design of leisure opportunities that are less demanding of space. In other words, a strict standards approach confuses recreation opportunity with area and recreation space *per se*. Standards, originally prescribed as minimums, become maximums and even optimums in some cases.

Again, the pace of modern city development quickly invalidates the setting of inflexible standards. It is not always a lack of conviction on the part of recreation planners regarding the desirability of departing from space standards advocated, so much as the unavailability or cost of land. Although application of standards might be marginally better than a completely *ad hoc* process, it cannot cope with the emergence of 'new' recreation resources, and makes no provision for community input or the involvement of private or commercial enterprises.

It is as well to remember, too, that in fully developed urban areas, it is not generally practicable to redistribute recreation space to match changing needs. However, given sufficient flexibility, the type of facility and the balance between sporting use and informal recreation pursuits, can be adjustable over time.

Moreover, mere figures have little to say about the form, quality and essential characteristics of the recreation space designated, under the idealised standards adopted. Too often, the urban recreation system has to make do with 'left-over' or

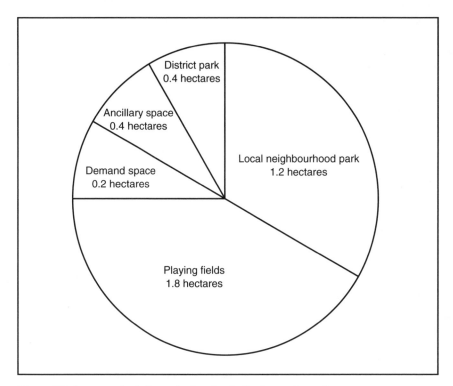

Figure 7.3 Area standards for park planning in Canberra, Australia

Notes:

1   Ancillary space refers to space for screen and shelter planting, sound-reducing planting, landscape development and easements for overhead powerlines and floodways.
2   Demand space refers to space for tennis courts, swimming pools, bowling greens, squash courts and 'concessional entertainment' for organised and social sport.

Source: National Capital Development Commission (1978: 10)

derelict areas, for which no other use can immediately be found. Minute, isolated parcels of low-grade land, devoid of vegetation or other natural features, that are unimaginatively designed and inadequately equipped, may meet the arbitrary space standards set, but do little to meet the recreation requirements of a neighbourhood.

Just (1987) provided a 'planner's checklist' (Table 7.5) to set out guidelines for planning decisions regarding urban recreation space. Size remains important, as research has shown that open space of less than 1 hectare (approx. 2.5 acres) is perceived as too small and is not well patronised. As the checklist suggests, the questions to be addressed are whether the space designated can be enlarged, or whether a monetary contribution would be the preferred option.

As noted above, the type of area, including the terrain and configuration, set aside as recreation space in new subdivisions is important for future use. Location is another obvious consideration, not only with respect to potential users, but also in relation to neighbouring land uses (e.g. residential buildings and roadways), and to pedestrian access and bikeways.

Accessibility, generally, is a fundamental concern when decisions are being made about the provision of recreation space (see Chapters 2–4). Access to a diversity of recreation opportunities within urban areas is generally assured for those with automobiles who are willing to travel reasonable distances. As noted earlier, such opportunities are severely limited for people without access to a car – the elderly, the young, the poor and the handicapped. These people, together with those who cannot drive or prefer not to use their cars, rely on public transportation, which is usually commuter-oriented, to work places and shopping centres rather than recreation outlets. Services are often reduced or eliminated during evenings and weekends when recreation demands are heaviest. This means that many city-dwellers are denied access to park and recreation facilities beyond walking distance. In these circumstances, the provision of close-to-home recreation opportunities is even more essential if equity in delivery and performance of recreation services is to be achieved.

According to Cushman and Hamilton-Smith (1980), a spatially equitable distribution of urban recreation facilities would ensure that no person was deprived of access by reason of distance, time, travel cost, or convenience. However, confusion can arise between efficiency and equity in location decisions. A recreation policy based on efficiency-related criteria of minimising costs and aggregate travel, and maximising attendance, would result in the location of a small number of large-scale facilities in high-density residential areas. At the same time, consumers living in lower density areas would be worse off.

Cushman and Hamilton-Smith (1980) advocate a compromise where efficiency is balanced against maximum equality of recreation opportunity. They believe that the degree of equity or inequity can be determined by reference to measurable elements of relative opportunity or relative deprivation (i.e. travel costs, constraints on recreation options arising from facility characteristics, and demographic variations in the population's ability to use services offered).

Table 7.5 Planners' checklist for assessing plans of proposed subdivision reserves

---

1. Reserve Function/Purpose
(a) State main function/purpose
(b) State secondary function(s)
(c) Is the main function of the reserve:

    (i)   Drainage or Screening (If "no" go to (ii))
        If "yes"   Could alternative sites be used for reserves?
                   Could alternative sites be acquired?
                   Would the monetary contribution be more useful?
        or   Are the proposed drainage/screening locations in accord
                   with a plan of proposed open spaces?
                   Can they contribute to a proposed regional train network?

    remember: Drainage and Screen Reserves area over-represented in
               Munno Para.

    (ii) Recreation
        Can the reserve provide for active pursuits?
        Can the reserve become a District Community Park?
        Can the reserve provide for two or more categories of
        recreational activity, e.g. walking, picnics/barbecues, informal
        active pursuits, formal sports, bicycle riding tracks, dog walking,
        jogging, play on playground, horse riding, etc.?

    remember: "Kick-around" areas on reserves and District Community Parks are
               under-represented in Munno Para.

2. Reserve Size
(a) Is the reserve greater than 0.5 hectares, 1.0 hectares, or between 5 to 10
    hectares (Yes)
(b) If the reserve is less than 0.5 hectares, can Council add to this area, or
    would the monetary contribution option be preferable?

3. Reserve Terrain
(a) Does the reserve have steep slopes or consist of a river valley/gully
    only? (No)
(b) Does the terrain provide an alternative to that which prevails in the
    area? (Yes)

4. Reserve Location
(a) Is the location :
    (i)   central for current and/or future land divisions? (Yes)
    (ii)   providing for through-access? (Yes)
    (iii) connected to schools? shops? other reserves? (Yes)
    (iv)   encircled with roads? (No)
(b) Are houses/units etc. oriented towards the reserve? (Yes)

5. Focal Points
(a) Does the proposed reserve incorporate:
    (i)   remnant vegetation? (Yes)
    (ii)   cultural artefacts (e.g. ruins, bridges, etc.)? (Yes)
    (iii) historical sites? (Yes)
    (iv)   encircling roads? (No)
    (v)   water features? (Yes)

Source: Just (1987)

Equity in location and access within an urban recreation space system must take account of these time/distance constraints and the circulation patterns of user groups. Studies of children's playgrounds, for example, indicate a highly localised service area of up to a quarter of a mile (approx. 0.4 km), and 75 per cent of all visitors to urban parks are said to come from less than a half-mile (approx. 0.8 km) radius. Distance, of course, is only one barrier standing in the way of individuals wishing to make use of a particular facility. Access to neighbourhood parks is often restricted by physical barriers such as highways, railroad tracks or industrial development. Chicago's lakefront parks, for example, have limited pedestrian access from surrounding neighbourhoods, due to the presence of Lake Shore Drive. Yet, these same parks can be easily reached by car. Similarly, a 'tot-lot', separated from its pre-school users by distance or busy streets, can have little role to play in meeting their need for recreation space.

Cushman and Hamilton-Smith (1980) suggest that the first step in reducing inequity is to identify, classify and map the spatial distribution of all recreation facilities in the city and the nature and level of services provided. Deprived residential sectors can then be determined, and deficiencies rectified. For urban parks, for example, the spatial patterns of playgrounds, neighbourhood parks, district parks and large urban parks, can be visually correlated and statistically analysed, according to the degree of dispersion and clustering of parks in each of the park types. In this way, the areas of the city being served and not served by parks in each of the park types may be determined (Cushman and Hamilton-Smith 1980: 171).

In a Canadian study, Smale (1990) went one step further in examining the issue of spatial equity in the provision of urban recreation opportunities, by taking into account variations in the demand for recreation resources, as well as the supply of them. An inventory of urban parks in one part of suburban Toronto was related to household demand indicators for recreation. The study revealed neighbourhoods which were 'supply-rich', in terms of recreation opportunities, and other areas which were 'supply-poor', indicating the need for remedial action.

A somewhat similar approach was used by Mitchell (1968, 1969) to evaluate spatial aspects of Christaller's (1963) central place theory in an urban recreation context. Part of Mitchell's purpose was to seek understanding of the interacting variables and processes which affect the distribution pattern of public recreation sites within the city of Columbia, South Carolina.

Such variables as relative location, distance, time and facilities, appear to be significant to consumers of recreational activities. On the other hand, public demand or pressure, available personnel, budgetary limitations and philosophical orientation, are also factors that seem to be important to producers of recreational services (Mitchell 1969: 104).

Mitchell discussed spacing of recreation facilities within a four-tier hierarchy of recreation units – playgrounds, play fields, parks and large parks – related to the criteria of function, size and service. He proposed a theoretical spatial distribution for each class within the hierarchy, based on uniform hexagonal patterns, equal spacing, regular size and shape of service areas, and standard threshold populations. When simplifying assumptions concerning the strict residential character of the city and its uniform population distribution were relaxed, a more complex distribution pattern emerged. This pattern reflected the overriding significance of population density as the key explanatory variable in understanding the location of public urban recreation sites.

The hierarchy approach, when coupled with the concept of the recreation opportunity spectrum (see Chapter 2), provides a suitable framework on which to base the planning of a functional system of urban recreation space. Figure 7.4a illustrates diagramatically how positions along the spectrum of recreation opportunities can be related to functional categories in an urban context. Figure 7.4b shows how different opportunity settings are linked with types of recreation activity.

As with Christaller's original work, the value and practical application of hierarchy techniques in the study of urban amenity provision, rest in discovering, explaining and correcting departures in an existing system from the idealised, theoretical framework. Such remedial action should not be necessary if sufficient regard is given to urban recreation requirements at the planning stage.

Summary

In many cases, the urban recreation planning process does not address the deeper behavioural needs of a leisure-oriented society. More often, it recognises and develops only conventional resources to accommodate present users and uses in stereotypical activities. By positioning a choice of urban recreation opportunities within a flexible hierarchy of recreation space, a functional recreation system can be created to provide for current and future community demands. However, any recreation system must have the capacity to cope with the inevitability of change.

In general, 'the temporal and spatial patterns and processes concerning outdoor recreation in urban areas have not been well conceptualised to date' (see Williams 1995: 20). Despite this neglect, and unless a dynamic element can be injected into the planning process, any recreation development initiative will lose impetus and be unable to respond to changing emphases in leisure behaviour, and associated pressures on resources and management policies. A flexible approach is the key to successful urban recreation planning, one in which priorities rather than rigid programmes are set down, and in which machinery exists for rapid review in the light of changing circumstances. Given this commitment, the recreation planner can make a useful contribution to generating a satisfying leisure environment for city dwellers in both established and emerging urban communities.

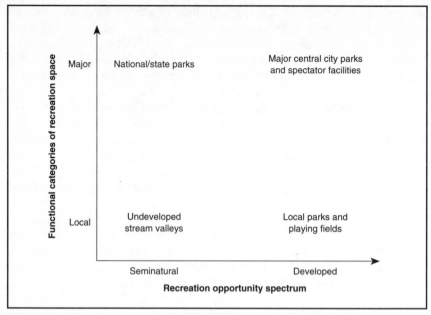

Figure 7.4a Functional hierarchy of recreation space and recreation opportunity

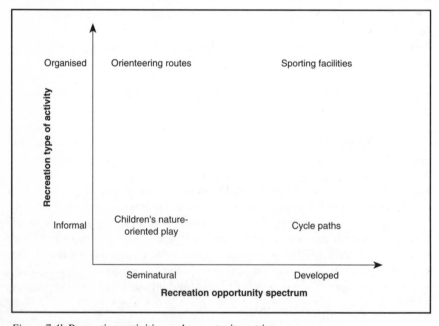

Figure 7.4b Recreation activities and opportunity setting

Source: Adapted from Ministry for Planning and Environment (1989: 7)

Guide to further reading

- *Outdoor recreation in urban areas*: Stanfield and Rickert (1970); Williams (1995).
- *Urban heritage and recreation and tourism*: Prentice (1993); Chang *et al.* (1996).
- *Urban parks*: Hamilton-Smith and Mercer (1991).
- *Urban tourism*: Ashworth (1989); Ashworth and Tunbridge (1990); Mullins (1991); Law (1992); Law (1993); Page (1995); Hall *et al.* (1997); Murphy (1997).
- *Waterfront development*: Craig-Smith and Fagence (1995); Williams (1995); Hall *et al.* (1997); Murphy (1997).
- *Planning for outdoor recreation and tourism*: Mercer and Hamilton-Smith (1980); Jansen-Verbeke (1992).

Review questions

1 Why is it difficult to devise a common strategy to address the changing recreation needs of cities?
2 What special problems arise in planning recreation opportunities in the inner core of older cities?
3 Identify the links between a hierarchical approach to provision of urban recreation space and the concept of the recreation opportunity spectrum.
4 How does the physical and human geography of a city affect its potential for outdoor recreation?
5 How might a compromise be reached between spatial equity and efficiency in urban recreation planning?
6 Suggest some specific reasons for underuse of urban parks and how these might be overcome.
7 Differentiate between urban open space and urban recreation space.
8 Discuss the contribution of urban waterfronts and urban bushland to urban recreation opportunities.
9 Examine the role of accessibility in the functioning of urban recreation resources.
10 What do you see as the respective (complementary) roles of the public and private sector, in creating an effective and satisfying spectrum of urban recreation opportunities?

8

OUTDOOR RECREATION IN RURAL AREAS

Explanation of the recreational appeal of extra-urban environments may be found partly in people's reaction to environmental stress (e.g. crowding and noise) associated with everyday urban living. Outdoor activities in a rural setting allow city residents to escape – to exchange the routine, the familiar, and boredom for the recreation opportunities perceived to exist in the surrounding countryside. Even knowledge or cognitive awareness of such outdoor opportunities is considered to act as a psychological safety valve for some in coping with environmentally induced stress (Iso-Ahola 1980).

Recent change in recreation and tourism activities

Rural areas in Western nations have long been used for recreation and tourism. However, since World War II, the nature of, and relationships between, the rural setting and the recreational activities engaged therein have changed significantly (Cloke 1993). Recreation and tourism in many areas are no longer regarded as simply passive, minor elements in the rural landscape. They are important agents of change and control of that landscape, and of associated rural communities (Butler *et al.* 1998).

Much recent change in rural areas has been linked to recreation and tourism. Until the 1960s and 1970s, rural recreation was mainly related to the rural character of the setting. Rural recreation comprised, primarily, activities which were different from those undertaken in urban centres, and which could be classified as relaxing, passive, nostalgic, traditional, low technological, and generally non-competitive (e.g. horse-riding, walking/rambling, picnicking, fishing, sightseeing, boating, visiting historical and cultural sites, attending festivals, viewing nature/ scenery, and farm-based visits) (Butler *et al.* 1998).

Whereas the above activities are still common, many other quite different activities are now engaged in, which bring new forms of conflict and impact, and require different planning and management responses. These new activities could be characterised as: active, competitive, prestigious or fashionable, highly technological, high-risk, modern, individual and fast. 'They include trail biking, off-road motor vehicle riding, orienteering, survival games, hang gliding, para-sailing, jet boating, wind surfing... adventure tourism, snow skiing, and fashionable shopping' (Butler *et al.* 1998: 10). In short, a far wider range of recreational activities are being pursued in rural areas, bringing a requirement for the establishment of specific facilities and settlements (e.g. resorts) to cater to the increasingly more sophisticated demands being placed on resources (e.g. see Sports Council 1991; Butler *et al.* 1998). 'Creation of an appropriate range of settings for rural tourism [and recreation] requires the deliberate selection and manipulation of features of the rural landscape to accommodate different types and styles of visitor use' (Pigram 1993: 163).

The role of rural landscapes in satisfying the recreational needs of a leisure-conscious society has long been recognised in Britain and other countries. For instance, since 1949, the Countryside Commission has been active in promoting the conservation of the natural beauty and amenity of the English countryside, within the framework of efficient agricultural use. A survey sponsored by the Commission in 1977, found that visiting the countryside was the most popular form of outdoor recreation for the people of England and Wales (Countryside Commission 1979). More recent figures in the UK and other countries provide further evidence of the popularity of the countryside. In the UK, more than 900 million day visits were made to the countryside in 1993 (CRN 1994), while in the US, more than 70 per cent of people participate in rural recreation (OECD 1993).

'For many urban dwellers, it is the rural ambience and the countryside experience which are the main considerations' (Pigram 1993: 161). Recognition of the strong correlation between recreational (and tourist) satisfaction and scenic quality of the recreation environment is an important step towards realisation of the contribution which rural landscapes, in both public and private hands, can make to the leisure opportunities of the city dweller. 'The success of rural tourism [and recreation]... is reliant upon the maintenance of a healthy and attractive rural environment. Implicitly, therefore, there is a need to effectively manage and balance all the various demands on the countryside...' (Sharpley and Sharpley 1997: 44).

Unfortunately, public resource-based recreation areas such as national parks and forests are in limited supply, and are not always close to centres of population. On those public lands which are accessible, visitation rates at peak periods are often pushed beyond carrying capacity so that fees, permits and other strategies for rationing use become necessary. At the same time, attempts to expand the resource base are frustrated by lack of land of suitable location and quality, and by budgetary constraints on park management services wishing to undertake further land acquisition programmes. Therefore, increasing attention has been given to the potential of private land for the provision of recreation opportunities within reasonable proximity of cities.

Rural recreation space: conflict and multipurpose use

Recreation is just one competitor for the use of rural land and water (see Green 1977). Many groups have an interest in rural areas, but for different, often competing, reasons. Other uses or interests include primary production (e.g. agricultural, aquatic, horticultural, pastoral and timber production), resource extraction (e.g. uranium and sand mining), conservation or preservation of the natural, cultural and built environments (e.g. national parks, wilderness areas and nature reserves), and transport and communication networks. According to Sharpley and Sharpley (1997: 23):

> In most industrialised nations, up to 80 per cent of rural land is still farmed or forested... although, significantly, the contribution of agriculture and forestry to income and employment in rural areas has gradually diminished during the twentieth century, as has their relative contribution to GDP in most countries.

The multipurpose character of the countryside represents both an opportunity and a constraint to recreation and tourism development.

> On the one hand, tourism and recreation can be viewed as a valid and valuable form of land use which, if carefully planned and managed, complements other uses and contributes to the economic and social well-being of rural areas; on the other hand, it may be considered that other, more traditional forms of economic exploitation of the land, including farming, mineral extraction and housing, should take precedence over its recreational potential. Thus, it is perhaps inevitable that, given the finite supply of the countryside, conflicts occur between different demands on the rural resource base.
>
> (Pigram 1993: 161)

Australia's large size and low population may suggest fewer constraints on recreation, and lower pressures on rural recreation resources, than in Britain or parts of Europe. However, this is not the case. Many groups have an interest in rural Australia, as they do in rural areas in other parts of the world. The perception that many areas are environmentally fragile or unique has encouraged a strong conservation ethic. Parts of the coastal strip are highly urbanised (see Chapter 7), and this places strong pressure on neighbouring rural areas that offer diverse and attractive recreation and lifestyle opportunities. Farming and mining activities predominate over much of the rest. These activities or outlooks are often in conflict, especially as knowledge, perceptions, attitudes and technologies change. People are even beginning to question:

> whether a socially-beneficial agriculture can be one which pollutes the land, poisons animal, bird and fish life, and leads to the destruction of the environment. It is becoming clear that ecological diversity and the aesthetic and amenity value of the countryside have become incompatible with modern agriculture and its modern practices.
>
> (Lawrence 1987: 66)

Recreation and tourism, particularly in ecologically sensitive areas, threaten environmental and cultural conservation, while both recreation and conservation pose threats to traditional views (i.e. agricultural, pastoral and mining activities) of rural Australia. The ensuing conflict undoubtedly serves to limit recreational access and tourism development.

The relationships between rural tourism and recreation, and other land and water uses, are largely influenced by landholder attitudes. Landholders include individuals (e.g. farmers), businesses (e.g. agribusiness and tourist resorts) and groups (e.g. recreation clubs) with private ownership rights; leaseholders and licensees (whose land use may be regulated by public agencies), and resource management agencies (e.g. national parks, forestry, nature/wildlife reserves, publicly owned recreational facilities and water reservoirs). Clearly, there is a wide variety of individuals and agencies with different value sets and interests with respect to the rural environment, and with different rights as landholders, according to land tenure and other institutional, legislative or contractual arrangements.

The complexity of institutional and ownership arrangements, and the multi-functional character of rural areas, have led to conflict between competing uses and between land managers. Land ownership and the exercise of landownership rights are thus critical elements in the supply of tourism and recreation opportunities. Access to land and water in this context is generally contingent upon legislation, public policy interpretations and landholder/management attitudes (Pigram 1981; Jenkins and Prin 1998). Effective recreation and tourism planning is hampered by numerous stakeholders: government agencies at different levels; conservation groups; developers; recreational groups and local communities generally.

Agreements and compromise between recreationists, responsible agencies and landholders are often difficult to achieve. On the one hand, as farmers seek to improve productivity (e.g. through more intensive land use practices, including the development of intensive feed lots), there are aesthetic and functional changes to the landscape, as well as impacts on recreational supply, and visitor experiences and satisfaction. On the other hand, rural recreation activities can become more contentious as their environmental impacts increase (e.g. large numbers of people visiting sensitive sites and the use of potentially destructive recreational technologies, including off-road vehicles). Recreational activities such as hiking, camping, fishing and nature observation may be passive and, therefore, depending on the resilience of the environment and its ability to resist impacts, have less inherent and actual potential to cause conflict between participants and land managers. However, landholder attitudes, perceptions and experiences may be influenced by small numbers of people who fail to consider the relationship between the type and intensity of their activities and the resulting impacts on the environment, including other people. For instance, there are those whose intentions and activities are deliberately environmentally destructive and illegal (e.g. indiscriminate shooting of stock, and stealing). Thus, any understanding of land use involves both an understanding of the values of the physical, biological, productive, spatial and visual/aesthetic attributes of land, and 'an awareness of the different standpoints from which land use may be considered' (Mather 1986: 6).

Conflicts

'Of all the resource-management conflicts in the countryside, recreation offers perhaps the greatest opportunities for multi-functional land use. Whilst inevitably many forms of recreational activity are compatible with others, some are not, and they have to be specially sited' (Cloke and Park 1985: 187).

Two features of rural areas in the last half century have been the difficulty of accommodating the structural changes which have occurred, and the much greater range of uses to which the areas have been subjected. Conflicts have arisen between recreation and tourist uses and other forms of land use, and between various forms of recreation and tourism. Conflicts between motorised and non-motorised recreational users of the same area can be severe, and often agreement and compromise is difficult to achieve, as can be seen in the case of disagreements between cross country skiers and snowmobilers, between non-mechanised trail users and off-road vehicle drivers, and between wind surfers and water skiers. Conflicts also exist between non-mechanised users of the same facilities, as witnessed by conflict between pedestrian trail users and mountain bike riders, between canoeists and anglers, and between hikers and hunters.

Such conflicts will probably become more severe as the overall demand for recreational and tourist use of rural areas increases, and as the range and types of uses widen. Compounding this problem in many countries, is decreasing public access to parts of rural areas because of changing patterns of ownership, and/or the reluctance of many landowners to accept public recreational access to private property. As increasing numbers of people acquire leisure or retirement properties in rural areas, they regard and treat those properties as private preserves; just as the landed élite zealously guarded their own leisure estates in past times (Butler *et al.* 1998). In Australia, such an attitude has a lengthy history dating back to early settlement (see Jenkins 1998).

Recreational use of private land represents multiple resource use and, as such, can generate conflict between recreationists and landholders. The basis for conflict lies in the various functions seen for rural land and the contrasting attitudes associated with these roles. Davidson and Wibberley (1977) suggest a strong polarisation between those whose dominant concern is the efficient production of food and fibre or other economic uses, and those who value more highly the intrinsic character of rural landscapes and wish to preserve this heritage unchanged. Between these two are other groups, for whom different attributes of the countryside are significant. City planners, for example, often view land, especially in the rural fringe, merely as a space and development reserve for urban expansion. For others, the primary role envisaged may be for communications facilities or specific resource uses such as extractive industries or water conservation. Transcending all of these in numbers are those who link the resource function of the countryside with leisure and outdoor recreation.

To some observers, this multiplicity of roles makes conflict almost inevitable if recreationists press their claims to private rural land (Green 1977). Whether conflict and confrontation are avoided depends essentially upon the goodwill and cooperation of the landholders: their attitudes are of fundamental importance in determining the amount of land available to the public for outdoor recreation. These attitudes, in turn, are a function of the landholder's personal beliefs and experiences, together with legal, economic, social and ecological considerations, national traditions and government policies, and the type and volume of the recreation activity involved (Cullington 1981).

The relationships between these factors are depicted in Figure 8.1. Essentially, the issue is one of balance between incentives and disincentives. Put simply: 'To increase the supply of private land for recreation, it is necessary either to increase the incentives, or to reduce the disincentives, or preferably both' (Cullington 1981: 8).

Incentives may be provided by governments in an effort to encourage wider recreational use of private land, and can include such measures as direct financial

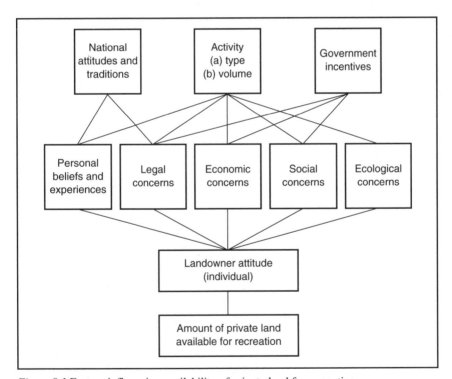

Figure 8.1 Factors influencing availability of private land for recreation

Source: Cullington (1981: 9)

support, compensation payments or the sponsoring of access agreements. Disincentives arise from landholders' concerns over such matters as:

- *Legal liability* for injury or damage to recreationists. The degree of the liability and the landholder's responsibility vary between jurisdictions and whether the visitor is an invitee, a licensee, or a trespasser. Several countries have enacted laws which attempt to allay landholders' fears of the liability problems should accidents occur.
- *Economic or financial implications* associated with the costs of providing access and possible dangers of changes in farming practice, set against any compensation payments or opportunities to derive income from recreational activities.
- *Social considerations* such as loss of privacy and problems with trespassing, which may be mitigated to some extent by the landowner's personal satisfaction from providing a community service.
- *Ecological impacts* on the farm environment as a result of recreational activities. These will depend to some extent upon how the farmer's perception of the problem is affected by the particular nature of the recreation activity and its incidence in space and time (Harrington 1975; Cullington 1981; Pigram 1981; Jenkins and Prin 1998).

All of these concerns are interrelated, and the degree to which they influence the landholder's decision to make available or withhold recreational access, is closely linked with the type and volume of recreation activity undertaken and government support and encouragement. The outcome rests very much with the individual landholders and how they perceive the balance between the incentives and disincentives. Undoubtedly, there will remain many who value the economic functions of rural land more highly than any amenity functions it may be deemed to possess. Moreover, conflict between these primary functions would seem most probable in the urban-rural fringe, where the economic value of the countryside is highest, and pressure for amenity and recreation space is greatest. It is here, too, that most problems and disputes over accessibility can be expected to arise.

Accessibility to recreation space

Accessibility has several dimensions, among them, technical, behavioural and sociocultural aspects. According to Moseley (1979), the concept cannot be divorced from the nature of the desired destination or experience. Certainly, much more than mere mobility is implied. Mobility, or the capacity to overcome space, is a technical and mechanistic condition, derived from such factors as vehicle ownership, travel time and costs, and individual physical attributes. Accessibility, on the other hand, is a broader concept, reflecting the opportunities perceived as available for travel. It is related to the behavioural notion of 'psychic space' or 'movement space'; that restricted area in which potential trip makers react to stimuli within the constraints of their value systems, experience and perceived environmental opportunities (Eliot-Hurst 1972). Moreover, the dimensions of a person's movement space are also, in part, a function of social and legal convention – the institutional 'rules of the game' to which an individual is exposed.

Thus, accessibility has many facets, and use of recreation space can effectively be denied in a variety of circumstances. Examples include: cost of travel, equipment and licence fees; lack of time, especially blocks of suitable time; inadequate information on recreation opportunities; ineligibility to participate on the basis of age, sex, qualifications, membership of group or social class; lack of transport; and special problems for people with disabilities (see Chapter 3). In the recreational use of the countryside, these circumstances may be compounded by the sheer difficulty of physical access; many sites are effectively closed off because of a lack of appropriate vehicles, equipment, stamina or expertise. Consideration of accessibility, too, can be complicated by disputes over property rights and by institutional and legal constraints on movement into and through recreation space.

Property rights

Central to the question of access and availability of private rural land for recreation is the issue of rights to property, and the privileges and responsibilities which ownership and control over land bestow. To some, property ownership, in a legal or economic sense, is the proprietorship of a bundle of rights (Wunderlich 1979). Others go further and question the concept of private property altogether, stressing that property should not be thought of as *things*, but as *rights*, the ownership of which is circumscribed (Dales 1972). In this view, ownership consists of a set of legally defined rights to use property in certain ways and a set of negative rights or prohibitions which prevent its use in other ways; a proprietor never *owns* physical assets, but only has the rights to their *use*.

In the context of recreational access to the countryside, ownership of the land itself is of no particular relevance. The crucial issues are ownership and exercising of the right to exclude others from use (Thomson and Whitby 1976). Difficulties arise because landholders are only one among several groups with an interest in how the resource is to be utilised and managed. The multiplicity of functions referred to above suggests a number of potential beneficiaries who may value the land for specific purposes. This would include the occupiers and would-be recreationists, but may also cover neighbours, passers-by and conservationists at large (Phillips and Roberts 1973). In economic and legalistic terms, the access issue can be seen as one of allocating among these interested parties the various rights over land in such a way as to maximise social welfare (Thomson and Whitby 1976). It could be that, where a landholder wished to retain exclusive rights to recreational resources (e.g. a stream, a beachfront, or a spectacular scenic view), purchasing such rights, over and above the price of the land, should be mandatory (i.e. the privilege of excluding the public would become taxable). However, others would claim that it is never equitable to permit the holders of land to alienate recreation space to themselves.

167

A finer definition of rights to property would certainly seem desirable in order to identify those which accrue to the property holder, to the state and to society. It could be held that private ownership rights become merely the residue, after public or communal rights to property are exhausted (Morris 1975). It could further be argued that ownership rights should not apply to the aesthetic component of the resource base, or extend to exclusive access to assets such as wildlife or fish that are found within a property. The landholder, when taking up occupation, also takes up effective control of countryside resources which may be valued by the wider community for recreation. This privilege, in turn, should imply a responsibility for making those resources available to society. It seems that few landholders are prepared to acknowledge this responsibility.

Recreational use of private land

While there is no universally applicable measure of access (Sharpley and Sharpley 1997: 78), and despite the constraints and disincentives noted, much private land is available in the 'old world' for outdoor recreation, either with the tacit or explicit consent of the landholder. In England and Wales, for example, individuals and groups have long enjoyed access to designated parts of the rural environment, although, even there, recreationists in large numbers are unwelcome. In parts of Scandinavia, recreation on private land is accepted and expected under a law known as 'Every Man's Right' (Cullington 1981).

In many parts of Scotland, visitors are allowed the freedom to roam by custom, but not by legal right (Sharpley and Sharpley 1997: 79). In Sweden, subject to certain conditions, visitors enjoy legal right of access to all public and private land in the countryside (*allemansrett*) (Colby 1988).

> A similar situation exists in Norway, where the great majority of the landscape is undeveloped and available for recreation. Visitors to the Norwegian countryside enjoy the right of Allemansretten, the right of access to most private land, although subject to certain conditions laid down in legislation. Germany, too, has a right of public access (*Betretungs-recht*) enshrined in law, although this is restricted mainly to forests, unenclosed land and along roads and paths.
>
> (Sharpley and Sharpley 1997: 78–9)

The establishment of trails and other means of facilitating recreational use of private rural land, is an indication of some relaxation of the access situation in countries of the 'new world', such as Australia, New Zealand and Canada. However, in many cases, countryside recreation remains inhibited by the prevailing attitude of particular landholders, who fear, with some justification, the consequences of thoughtless negligence or deliberate vandalism by visitors. Their experience suggests that, in many circumstances, recreation is simply incompatible with other uses of countryside, by virtue of its concentration in time and space, as well as problems of trespass, litter, property damage and general nuisance.

Conflict is most likely to occur closer to towns, where fringe landholders face higher levels of trespass damage, to the extent that some form of boundary protection may become necessary. In extreme cases, the actions of visitors may lead to drastic modification of farming practices or the abandonment of arable farming altogether.

Taking into account impacts of this scale, the negative attitude of rural communities to recreational use of private land can be better understood. Twenty five years ago, Phillips and Roberts (1973) reported that the continuing invasion of the countryside by urban dwellers seeking diversion, set against a background of rapid changes in farming, was leading to a situation in Britain where there was perhaps a greater degree of antipathy between farmers and visitors than ever before.

In Australia, too, where the concept of inviolate rights of property ownership is widespread and generally accepted, the lines were fairly clearly drawn between town and country. The attitudes of landholders to public recreational access were typified by the following statement:

Access to private land for sport or recreation is a privilege and privilege is not a birthright but something that can be earned by good behaviour and responsibility. This Association will not consent to accept the entry upon private land, without the permission of owners or occupiers, of any persons who are not performing a statutory function, as other than trespass.
(Graziers' Association of New South Wales 1975: 3)

This attitude has not diminished (see below). It was translated into real terms by means of a proliferation of cautionary signs at property boundaries, and warning notices in the rural press advising that all permits to enter land have been cancelled, and that trespassers will be prosecuted or face other dire consequences.

Thus, for many Australians, recreational contact with the countryside remains restricted, and is often confined to illicit and fleeting entry of private land, or viewing from a moving vehicle. Moreover, there seems little prospect of landholders being willing or able to divert resources voluntarily from what are seen as the land's primary functions – agriculture and the like – to providing recreation space for city dwellers.

In the UK, private land has a long history of use for recreation, and 'A central theme to the development of countryside recreation policies has been that of access' (Groome 1993: 5). However, recreational use of the countryside is not 'a public prerogative. It constitutes use of a domain, owned mainly (87 per cent) by private individuals, and with public access dependent on certain legal rights or lenient attitudes on the part of landowners. Even the national parks, areas designated specifically for landscape protection and public amenity, are largely in private ownership' (Shoard 1987, in Glyptis 1992: 156). British landowners have a long developed tradition of exclusive control of the countryside (Shoard 1996: 13). To make the matter more complex still, there are signs that land ownership is gradually becoming concentrated in the hands of fewer people rather than spreading more widely, while counterurbanisation is bringing new, affluent and mobile residents, who are accustomed to urban standards of service provision (Glyptis 1992).

Nevertheless, compared with Australia and Canada, the UK has enjoyed ease of access to rural lands for recreation, where the broad aim has been to promote and market leisure to as wide a public as possible (Pigram and Jenkins 1994). For instance, recent policy development in rural recreation and tourism has 'resulted in a change from the perceived need to control the public, to their wholesale encouragement. This has happened quite swiftly and has taken place in tandem with a fundamental reappraisal of the primary role of agriculture in rural areas' (Robinson 1990: 132). Despite initiatives to enhance public recreational access to the countryside in the UK, there are still concerns about landholder attitudes, the ineffectiveness of such initiatives in many cases, and, ultimately, the lack of access (see Jenkins and Prin 1998, for a broader discussion of public recreational access to private rural lands in Australia, Britain and Canada).

In Britain, a recent policy initiative for access to the countryside is the Countryside Commission's 'Countryside Stewardship Scheme' (CSS). The CSS was launched in 1991, and was undertaken in partnership with English Nature, English Heritage and the Ministry of Agriculture, Fisheries and Food. The Scheme encouraged farmers and landowners to conserve and re-create the beauty of five traditional English landscapes (chalk and limestone grassland, lowland heath, waterside landscapes, coastal land, and uplands) and their wildlife habitats, and to give opportunities for informal public access. 'The long-term objective is to develop a basis for a comprehensive scheme to achieve environmental and recreational benefits as an integral part of agricultural support' (Countryside Commission 1991: 1). This initiative broke new ground in environmental management, integrating conservation management and commercial farming, with a view to demonstrating benefits for landscape, wildlife, history/archaeology, access, or some combination of these objectives. Farmers and landholders were granted annual payments of up to £300 a hectare; a commitment costing the government £13 million for the first three years. Concerns have been raised as to whether the access payments have secured access, with many open access areas difficult to locate. However, a review of the Scheme (Countryside Commission 1998: unpaginated) stated that:

> Countryside Stewardship delivered significant environmental benefits over and above what would have happened in the absence of the scheme... An additional survey of public access to land under Stewardship access agreements found that a high percentage were well located and were easily accessible by the public. Sites were used by a wide variety of people of all ages and from all walks of life... day trippers as well as local people... The sites met the expectations of two-thirds of visitors. 20 per cent found the sites to be better than they expected. Only 4 per cent were disappointed with the sites.

Many other examples of recent initiatives to secure access to the UK country-side, and in particular private rural land, abound (see Watkins 1996). However, there is much debate with respect to such initiatives. Shoard (1996: 21) argues:

> Farmers now get £247 per mile per year merely for allowing people to walk along access strips ten metres wide, along the sides of, or across fields in ESAs [Environmentally Sensitive Areas] and £145 under the NRSA [Non-Rotational Set-Aside] (MAFF 1994: 6–7). The landowners of the past who established the idea of a right of exclusion would be amazed to learn the size of the potential bounty they have created for their descendants.
>
> The right of exclusion is being increasingly used to turn access into a tradeable asset. The government's endorsement of the right of the landowner to charge others to set foot on his or her land puts the official seal of approval on the notion that access to the countryside is a commodity to be bought from landowners rather than a free public good. This is like bestowing on one group a whole new form of wealth.

Yet, in the same volume of works (Watkins 1996), Curry (1996: 34) argues that one means of 'improving opportunities for access to closed land... should be pursued through direct payments from the consumer to the farmer and landowner, since, in the absence of non-excludability, this is both more efficient and more equitable'.

There are no right or wrong answers or simple solutions to the access issue, because of the disparate views on landowners' rights and the complex nature and outcomes of government intervention. This situation is unlikely to change markedly in the near future. The restructuring of rural economies and the difficulties that have been faced by some rural producers and communities, in tandem with an increased demand for recreational and tourist use of the countryside, mean that public sector incentives or disincentives and user fees are likely to become even more attractive to landholders. However, if 'The only honest and effective way of opening the countryside to the people is to require landowners to relinquish some of the rights in their asset, which they are otherwise bound to defend and exploit' (Shoard 1996: 21), then, in the absence of blanket public policy establishing such access, a significant reordering of landholders' values will be needed. More to the point, general access rights are unlikely to eventuate in the UK in the short to medium term. Indeed, before such a turnaround in values can be contemplated, an acute understanding of landowner attitudes is needed. To some extent, such understanding is becoming increasingly evident and detailed in the expansive literature on recreational access in the UK (see Watkins 1996), but is sadly lacking across Canada and Australia.

In the UK, the concept of sharing the countryside is generally well-established in comparison to Australia, because use of the countryside for rural recreation has a long history. Legislative provisions in the UK explicitly recognise the importance of recreational access to private lands. Public and private cooperation with landowners is supported by education and formal and informal consultation mechanisms. Furthermore, despite the development of numerous, sometimes very lengthy walking tracks, little progress has been made with respect to public access to private lands in Australia and Canada.

A recent study of rural landholder attitudes to recreational access to private lands in central western NSW, Australia, was designed to assist in the identification of barriers to recreational access to private lands; in the reconciliation of conflicts in the recreational use of private lands; and in the enhancement of the spectrum of recreational opportunities in the countryside (Jenkins and Prin 1998). The study's objectives were:

- to identify rural landholder attitudes to recreational access;
- to identify the underlying dimensions which explain those attitudes, and therefore to determine whether attitudes differ because of ownership, land tenure, government incentives, income and other arrangements;
- to review the systems of incentives and disincentives under which landholders are operating;
- to identify means of ameliorating rural land use conflict by way of incentives and removal of disincentives identified by landholders.

The study indicated that much resistance, if not direct opposition, remains to the use of private rural lands for public recreation, at least in this study area. Landholders expressed little or no interest in entering into incentives or agreements to facilitate recreational access to private property, and very negative reactions to recreational access were made by some respondents (see Table 8.1). However, the study revealed some prospect of endorsing public access, providing specified conditions were met. These conditions related to control of the type of recreational activities permitted, their location and the timing and duration of access, and provisions for legal liability and payment for use.

Clearly, the ability of landholders to regulate recreational activities, and the absence of legal liability of farmers for people who use lands for recreational activities, appear to be aspects warranting further research. They also represent potential public policy avenues for increasing recreational access to private lands.

Water-based recreation

The presence of water is often regarded as a fundamental requirement for outdoor recreation, either as a medium for the activity itself, or to enhance the appeal of a recreational setting. Water provides for a diversity of recreation experiences, some requiring direct use of the water itself (with or without body contact), and others merely requiring the presence of water for passive appreciation and to add to the scenic quality of the surroundings. The more active types of water-based recreation range over boating (sailing, power-boating, rowing and canoeing), fishing in all

172

Table 8.1 Landholders' comments

- The bible says to forgive your trespasses. We don't. We shoot the bastards.
- Government should provide the land for public recreational activities, for example, State forests... the farmers and landholders have a big enough responsibility just surviving.
- A lot of people have no idea about farming. If they come onto your place, they are just as likely to drive over new crops etc. and do damage or get bogged and then you have to pull them out. Irresponsibility with fire is our major fear, dog control also worries us.
- Most people that visit the property are honest and attempt to do the right thing. We are concerned about the small number of less civic-minded souls, who do on occasion turn up. They ruin the property, the owner's confidence and trust and make one wary of strangers in general.
- If we were to adopt a socialist agenda, it must apply to all urban and rural lands, i.e. open house to all. This must be a decision by all people to affect all people, not just a minority, i.e. rural landholders.
- We feel that private property should be just that. Access on invitation only.
- Once access is given, it's very hard to stop people – they think that they have an inalienable right to your place.
- Would you like people camping in your garden without permission?
- Surely there is enough land controlled by Crown lands to satisfy the bush walkers etc. without the farmers of Australia having to put up with yet another intrusion into their privacy.
- Have you asked these questions of town or suburban land owners?
- Having been burgled... to the extent of $15 000, I am now very reluctant to draw any attention to our remoteness, as we are not in residence permanently.
- I am entitled to have quiet enjoyment of my own land. This is given by the Law of the Land and must be retained unless I choose to give up those rights.
- Would you like the general public camping and walking around your front yard... we like our privacy as much as city dwellers.

Source: Jenkins and Prin (1998)

its different forms, and swimming (including sub-aqua diving, water-skiing and surfing). Some of these are associated more directly with coastal waters, while others are concentrated on rivers and inland waterbodies. All have experienced a remarkable upsurge in participation during the past two or three decades. In some cases, this upsurge has strained the capacity of the resource base to meet the growth in demand, and, in turn, has generated conflict between users and uses of water resources.

Water figures prominently in at least three aspects of recreation and tourism development:

- The quantity and quality of available water can represent major constraints on the location, siting, design and operation of tourism facilities. As pressure grows on increasingly scarce water resources, the potential of areas, otherwise suitable for tourism development, may be compromised by inadequate water supplies.
- The presence of water serves as an additional dimension to a recreational or tourist facility, enhancing the scenic quality and appeal of the setting, and

contributing to the attraction and intrinsic satisfaction derived from the tourist experience. An environment that is rich in water often forms an aesthetically pleasing setting for tourism.

- Water is essential for recreation and tourism – for drinking purposes, for sanitation and waste disposal, for cooling purposes, for irrigation and landscaping, and for the function of particular forms of water-related activities (e.g. swimming and boating). Water for the making of artificial snow is an issue in alpine and cool climate areas (Pigram 1995: 211–2).

There is ample scope for conflict over use of water for outdoor recreation, and competition can become particularly intense where water resources are in short supply. Conflict can occur between:

- recreation and other resource uses, such as control structures within the river system or agricultural practices and other land uses within a drainage basin;
- incompatible recreation activities, amongst which power-boating and water-skiing probably arouse most opposition from less aggressive forms of recreation such as swimming and fishing;
- recreationists and the environment exposed to use (e.g. the water and shoreline, flora and fauna, and nearby human settlements and communities).

Conflicts are not confined merely to the water surface, but can occur at access points over ancillary facilities such as boat ramps, parking, campsites, access roads and the like. Even within the one specific recreation activity, excess usage can generate conflict over space at peak periods. Part of the problem is the inability of all waterbodies to satisfy the requirements for particular forms of water-based recreation. At least two aspects are critical (Mattyasovsky 1967).

First, the 'form' or nature of the water, and associated features, is fundamental. Certain wave conditions are an obvious pre-requisite for surfing; 'white' water is ideal for wild river-running; and relatively static waterbodies may be preferred for water-skiing, sailing and rowing. Features of the shoreline and the area beneath the water can be important, as are the quantity, permanency and seasonal distribution of the waterbody. Boating enthusiasts who have to carry or drag their craft some distance to the water line from a poorly sited boat ramp, can vouch for the problems caused by water level fluctuations and drawdown of reservoirs in dry weather, or after large releases of water.

Second, the quality of water (i.e. clarity, purity and temperature) that is appropriate for different recreational uses, needs consideration. Water quality often has to be a compromise, so that minimum criteria are stipulated rather than 'ideal' standards. For some types of recreation, even low levels of pollution can be tolerated, depending upon the pollutants and the activity in question.

Portugal's Algarve region demonstrates how a precarious water supply and fierce competition between actual and potential uses can threaten the viability of the tourist industry (Martin *et al.* 1985). In the Algarve, water resources have not kept pace with demands from the intensification of agriculture and rapid urbanisation. The tourist industry accounts for up to 40,000 users, with peak demand coinciding with the period when water supplies are at their lowest. Groundwater is the primary source of supply, and the major problem is proximity of demand to the coast and the risk of saline intrusion as water levels in wells are depleted. Firm planning control and management of the groundwater resource and alternative sources of supply will be necessary in order to avoid contraction of tourist activity and possible abandonment of irrigation agriculture in the region.

The incidence of such problems is unlikely to recede as increasing population pressure and growing sophistication in water demand generate conflict between users and uses. The availability of water, in sufficient quantity and quality to satisfy such uses, has emerged as an important concern in many parts of the world. As competition for water increases into the next century, tourism will be forced to justify its claims on the resource, against a range of more conventional uses and priorities. The problem can be clearly illustrated with reference to the water situation in North America and Britain (Pigram 1995).

The low priority given to instream uses of water for recreation is apparent in North America. In the US, recreation resource allocation, especially in rural areas, has tended to be *ad hoc*, and provision for tourist opportunities is often a byproduct of other major resource developments. The result is resistance to those tourist development initiatives seen to threaten established claims on the resource base. The negative reaction in Hawaii to proposals to develop or expand golf resorts, for example, is probably partly a reflection of anti-Japanese sentiment. However, it is also based on the consequences for agriculture from increased pressure on limited water resources on some of the islands (Pigram 1995).

In Canada, the value of rural water for tourism and recreation is explicitly recognised in the resource appraisal procedures of the Canada Land Inventory. However, public sector initiatives to develop this potential have been intermittent and generally reactive to perceived exploitation of the natural environment by private interests (Butler and Clark 1992). Again, in park development, the emphasis has been mainly on environmental protection. Attempts to implement an integrated approach to the provision of opportunities for water-related tourism have received less attention (Pigram 1995).

In Britain, people have long enjoyed comparative ease of recreational access to rural land and water. The coastline is generally within easy reach, and increments to the stock of recreation water space continue to occur from the construction of new reservoirs, restoration of canals and the flooding of disused gravel pits and mineral workings. Since 1974, regional waterbodies have had a statutory obligation to provide for recreation in all new water projects. Yet, few authorities have the personnel or necessary skills to plan and manage facilities in order to satisfy an increasing demand for water-related recreation and tourism (Blenkhorn 1979). Some concern has also been expressed about recreation opportunities for domestic and international tourists at water supply projects, following privatisation. Although legislation provides for public access to water authority land, the requirements are vague and open to differing interpretations (Pigram 1995).

More generally, the value of water for leisure and recreation in rural Britain has been recognised by the Countryside Commission (1991). In its guide to sustainable tourism, the English Tourist Board acknowledged the role of clean waterbodies as an attraction for visitors, as well as the need for adequate water, in quantity and quality, for human and operational needs at tourist destinations. Clearly, the emphasis is on management of water to cater for the many ways in which it can function as a resource for recreation and tourism (Pigram 1995).

With sport-fishing, water quantity and quality are both significant, and for some species, temperature can also be a critical aspect of the fishing environment. It is important to consider fishing conditions for anglers, as well as the fish habitat, in physical and ecological terms. Habitat requirements vary and will almost certainly deteriorate with increased use. Management of the resource may require attention to the form of streams, e.g. construction of fish ladders and remedying pollution and other deficiencies in the condition of waterbodies as well as control of undesirable species. The quality of water is a less important consideration for recreational boating; more important are the size of the waterbody, depth, subsurface features such as rocks, any aquatic vegetation present, and compatibility with other users and uses (Mattyasovsky 1967). Boating of any kind is space-demanding, and power boating, in particular, can cause interference and danger to others, as well as water pollution and bank erosion. In addition, marinas, service facilities and boat launching ramps are often necessary. Provision of sufficient on-water mooring space can be a particular problem in popular, crowded waterways.

Although the primary concern must be provision of an adequate quantity of clean water of suitable quality, modern treatment facilities make many forms of water recreation compatible with this aim. Where recreation is permitted, bank and shoreline activities, as well as fishing and non-powered boating, are usually accepted without question. However, even body-contact forms of recreation could be permitted where water treatment is of a high standard. In any case, often there are many other 'natural' sources of water pollution stemming from agriculture, native birds and animals, and contaminated precipitation, as a study in northern New South Wales demonstrated (Burton 1975). In inland Australia, water for any purpose is generally in short supply, and recreational water space is severely restricted away from perennial streams. In this context, opposition to recreational use of domestic water supply storages is coming under increasing scrutiny, and there are indications that a more reasonable attitude to the issue may eventually emerge.

Summary

The relationships between recreation, tourism, and rural regional development, as in urban areas, are now significant economic, social and political issues warranting attention. Recreation and tourism can contribute greatly to rural development and prosperity, but this is not always the case. Tourism, in particular, has been utilised by governments, industry, regional authorities and other interests, as a means of diversifying and restructuring local economies in response to local and global forces, and more particularly to economic and population decline, and the need to create employment opportunities. Ultimately, the task in utilising tourism is to encourage economic and social development, to maintain or enhance the quality of life of residents and to maintain or enhance the quality of the physical environment.

Planning high-quality and sustainable recreation and tourism developments, which marry rural resources with local and tourist needs and preferences, has not been easy. Perhaps this is related to the lack of sound theories and concepts guiding the role and management of recreation and tourism in rural areas.

Guide to further reading

- *Rural recreation and tourism*: Simmons (1975); Middleton (1982); Patmore (1983); Cloke and Park (1985); Perdue *et al.* (1987); Wall (1989); Glyptis (1991); Groome (1993); Ibrahim and Cordes (1993); Watkins (1996); Page and Getz (1997); Sharpley and Sharpley (1997); Butler *et al.* (1998);
- *Recreational access to private rural lands*: Cullington (1981); Pigram (1981); Sanderson (1982); Butler (1984); Ravenscroft (1996); Watkins (1996); Jenkins and Prin (1998).
- *History of recreation in rural areas*: Towner (1996); Butler *et al.* (1998).
- *Rural change*: Bowler *et al.* (1992); Cherry (1993); Cloke (1993); Cloke and Goodwin (1992, 1993); Glyptis (1993 – in particular see chapter by Cherry and Cloke); Ilbery (1997); Butler *et al.* (1998).
- Readers should refer to the work of such agencies as the Countryside Commission for recent developments in rural recreation management.

Review questions

1 Compare and contrast definitions of 'rural'. To what extent is 'rural' a geographical term or an experiential concept?
2 What are the main forces influencing rural change? Discuss the relationships between economic change and restructuring, and rural recreation and tourism development.
3 What are 'private' lands?
4 Distinguish between public and private lands. What rights of ownership do public and private lands present for individuals and wider society?
5 Public recreational access to private rural lands is a contentious issue in many industrialised nations. Why is this so? What are your views on public recreational access to private rural lands? Why might there be strong attitudes towards ownership rights in countries like Australia?
6 How might the management of public and private rural lands be better integrated?
7 Discuss the importance of rural recreation as a means of escape for urbanites.

9

PROTECTED AREAS AND
OUTDOOR RECREATION

In an increasingly complex world, the need to set aside certain areas free of development, where conservation values can be protected, is seen as crucial and of growing importance. The International Union for the Conservation of Nature (FNNPE 1993) lists ten protected area categories (scientific reserve; strict nature reserve; national park; natural monument; managed nature reserve or wildlife sanctuary; protected landscape; natural biotic area or philanthropological reserve; multiple use management area/managed resource; biosphere reserve; and world heritage site). This chapter focuses on national parks, a critical factor in the supply of outdoor recreation and tourist opportunities. Systems of national parks and reserves can now be found in most countries of the world as nations and people recognise the important contribution of protected areas to society.

The urgent need to establish a comprehensive range of protected areas across the globe is demonstrable. Recent estimates indicate that, since 1970, the number of protected areas across the globe has increased 185 per cent to 9,932, while their geographical spread has increased by 515 per cent to 926,349,646 square kilometres (IUCN 1994 and WRI 1992, in Eagles 1996: 29). However, with this comes the realisation that both existing and new reserves must be managed effectively in the interests of conserving biological diversity. Less than 5 per cent of the planet's land surface is subject to a protective regime, and, in a world marked by rapid change, economic imbalance and variable access to resources, even that small fraction faces considerable pressures on its integrity and viability.

These concerns are reflected in the Caracas Declaration which emerged from the Fourth World Congress on National Parks and Protected Areas in 1992. The focus of the Congress was on the theme 'Parks for Life', and on the challenges threatening protected areas of the world in making a practical contribution to the health and well-being of humanity. The Caracas Declaration emphasised a number of fundamental principles:

- that nature has intrinsic worth and warrants respect regardless of its usefulness to society;
- that parks protect areas of living richness, natural beauty and cultural significance;

- that such areas are a source of inspiration, as well as places of spiritual, scientific, educational, cultural and recreational value (Lucas 1992).

Similar principles have been embodied in moves to establish national parks and reserves worldwide, but there is by no means unanimity in the philosophy or practice of natural area protection. Even the concept of a national park – the most common type of protected area – has evolved independently over more than 100 years. As a result, and in spite of some common features, there are as many variations on the national park theme as there are park authorities. Add to this the various parts of the environment which come under the description of 'wilderness area', 'marine park', 'nature reserve', 'state/provincial park', 'regional park', 'country park' or 'landscape park', and the picture becomes even less clear-cut.

This is not surprising when the wide-ranging perception of national parks and their role in society is appreciated. Right from their beginnings in the US last century, national parks that were established, were justified, in part, in terms of their potential to generate economic benefits. Much more recently, the same argument has been used to support efforts to expand the US national parks system: '...national parks are good business. They attract tourists and boost economies wherever they are situated. In part, that's why most Californians, including both senators from that state, favour the desert parks... In California, as elsewhere, that's the smart investment' (*USA Today*, 11 April 1994: 7).

Such sentiments are widespread. A survey of visitors to Dorrigo National Park, southeastern Australia, documented the scale and diversity of expenditure associated with park visits. The economic worth of the national park was substantiated by a survey of businesses in the nearby town of Dorrigo, in which the perceived importance to the local economy of park-associated tourism was emphasised. Some respondents went on to suggest that Dorrigo National Park, through exploitation of its tourist potential, could become the engine of regional development in this declining rural area.

Whereas considerable efforts have been made to demonstrate the magnitude of the economic benefits of national parks (e.g. McDonald and Wilks 1986a, b; Lambley 1988), there is no doubt that such parks also involve costs. Such costs can be divided into at least two categories – *direct* expenditure on establishment and maintenance, and the *indirect* or opportunity costs of commercial exploitation of resources, usually forgone as a result of the creation of the park. The fact that governments and the community are prepared to accept these costs and support the public funding of national parks, suggests that many people continue to regard them as worthwhile.

Among the range of values claimed for national parks, it would seem that one of the primary justifications lies in the inherent nature of society, and the demands which expanding populations and technological progress place on the natural environment. Without this pressure, large tracts of country would remain under-developed, and there would be less of a reason for national parks and nature reserves. Thus, for many people, the greatest value of the parks lies in their ecological role, in protecting areas and features of outstanding scenic and historical worth, and in preserving distinctive ecosystems, essentially unimpaired, for future generations. For others, provision of recreational opportunities is pre-eminent; the parks being seen as the means of physical and spiritual refreshment in a natural outdoor setting. A more limited segment of the population regards national parks as the vehicle for scientific research, retention of genetic diversity, and the study of natural phenomena in undisturbed surroundings.

Whatever the point of view, there is obviously widespread appreciation of national parks, and considerable support for the development and expansion of parks systems. The concept of a national park, however, is a relatively recent phenomenon.

In Australia, Canada, New Zealand and the USA, the establishment of national and state/provincial parks for recreational and tourist opportunities, and the protection and maintentance of representative environments, came about in the nineteenth century. Interestingly, the first national parks in all four countries were set aside for conservation and recreation purposes because the land was considered worthless for such rural activities as intensive agriculture, lumber, mining and grazing (Hall 1992).

Yellowstone National Park, established in the US in 1872, is claimed to be the first national park. It was followed, in 1879, by Royal National Park, on the southern outskirts of the City of Sydney, Australia. This claim has recently been challenged on the grounds that Yellowstone was initially set aside 'as a public park or pleasuring ground', not as a national park. The first time the term 'national park' was used, was in the legislation to create Australia's Royal National Park. The first legislative reference to Yellowstone as a 'national park' did not come until 1883.

The merits of the conflicting claims, however, are not the crucial issue here. Rather, since these beginnings and in the space of little more than a century, the modern parks movement has grown to worldwide dimensions. National parks can now be found in all continents, under a variety of economic and political systems.

National park concepts

The evolution of present-day national parks owes much to the American park movement of the nineteenth century, and to the efforts of conservationists such as Olmstead and Muir. The American park movement was motivated by regard for nature, and the revitalising powers of wild landscapes in an increasingly complex society. The dominant themes were the preservation and protection of the resources of nature, and the opening-up of these resources for the recreational needs of the nation. This movement culminated in the reservation of the first extensive area of wild land, primarily for public recreation, in the United States: the Yosemite Grant (in 1864). This was followed by Yellowstone, eight years later, and the Niagara Falls Reservation in 1885.

The International Union for the Conservation of Nature (IUCN) has attempted to clarify the concept of a national park by proposing a standardised definition. For management and planning purposes, a national park is defined as:

> a natural area of land and/or sea, designated to (a) protect the ecological integrity of one or more ecosystems for present and future generations, (b) exclude exploitation or occupation inimical to the purpose of designation of the area, and (c) provide a foundation for spiritual, scientific, educational, recreational and visitor opportunities, all of which must be environmentally and culturally compatible.
>
> (IUCN 1994)

As might be expected, the rather restrictive tone of the definition provoked some reaction. The clear bias towards preservation of ecosystems, and the implicit limitations on human use, meant that many so-called national parks in some countries would not qualify as such. Any exploitation of natural resources (including hunting and fishing), all construction (including water impoundments, roads and amenities) and, strictly speaking, all means of transport and communication, could be excluded. In practice, of course, many of these land uses and facilities are permitted, if only to provide the necessary infrastructure to allow the park to function. In most cases, consumptive recreational pursuits (e.g. sportfishing and even hunting, under certain conditions) are accepted, along with non-consumptive resource uses (e.g. hiking, boating, viewing, mountain climbing and scientific research). Active recreation is provided for and encouraged in many North American parks. Tourist amenities (often concessionaires) are accepted (but controlled), even within the park boundaries, under the operation of the management authority. Many observers would argue, too, that created bodies of water can enhance a park landscape.

Modifications of the IUCN definition have expanded the function of national parks to include protection of cultural heritage, as well as the conservation of nature. Nonetheless, there is a popular view that at least some facilities for visitors and administration are necessary for the management and enjoyment of a national park. However, the definition probably still applies to parks in Africa, North America, New Zealand and Australia. Few parks would qualify in the Old World, where very little unaltered natural landscape remains. Even this qualification could be challenged, depending on how one interprets and understands physical and other kinds of change. The problems of making generalisations about the concept of national parks, can best be illustrated by reference to representative park systems across the world.

National parks in the United States

The US national parks system encompasses nearly 300 different protected areas of diverse sizes and types, totalling some 32 million acres (approx. 13 million ha).

National parks are the best-known units within the system, but the Parks Service is also responsible for several other areas, with designations such as national monuments and national memorials (only a few of which are actually statuary or historic buildings); national historic sites (especially those associated with American military history); national lakeshores; seashores; parkways; and wild and scenic rivers. In addition, the Service administers a large number of lands and buildings in and around the national capital – Washington DC. The size and complexity of the American parks service make comparisons with other national systems difficult. Many of its features, however, in particular the approach to national park management, have been adopted by, or have at least influenced, other newly settled countries such as Australia.

National parks within the US system are predominantly large, natural areas, containing a variety of resources, and one or more distinctive attributes or features of such scenic quality and scientific value as to be worthy of special efforts at preservation and protection. In a sense, the American national parks are regarded as 'outdoor museums', displaying geological history and imposing landforms and habitats of interesting and rare fauna and flora. In 1979, there were thirty seven national parks in the US; most of them in the western states, with a total area of nearly 16 million acres (approx. 6.5 million ha). This figure had grown to fifty three in 1998. The better-known national parks (e.g. Yellowstone, Yosemite, and Grand Canyon) contain some of the most spectacular scenery in the world, attracting vast numbers of visitors both from North America and foreign countries. Indeed, the sheer numbers of people wishing to visit the parks in peak periods has led to concern for the natural resource base, and has prompted a review of park philosophy and management principles in the US parks system.

Reservation of parkland for recreational purposes was a potent force, if not the primary one, in the early days of the US national parks. This initial viewpoint is interesting in view of the later change in emphasis towards the conservation of nature. In the early decision-making years, the attitude of park authorities was one of active encouragement of visitation by the public (Fitzsimmons 1976). Part of the rationale for these efforts was that exposure to nature would prompt visitors to appreciate and support the parks. Broad popular support was also seen as a means of counteracting political and economic interests hostile to the national park concept. If enough visitors could be attracted, parks would become self-supporting and would provide the income needed for their role in the preservation of natural species and landscapes.

These efforts at 'popularisation' of the parks read a little strangely in view of latter-day problems in North America, stemming from visitor pressure, congestion and fears of deterioration of park landscapes. However, the historical context should be borne in mind. The first parks were remote and difficult to access; transport was relatively slow and primitive, and public funds for park development were very limited. While patronage remained low, there must have seemed little contradiction between use and preservation, nor any need for management plans to maintain ecological values. The major problem, presumably, was how to boost attendance, and justify the viability and continued existence of the parks. Funds generated by publicity programmes 'were in turn used to provide more recreational attractions and visitor services in a spiralling development cycle' (Forster 1973: 17).

Following the end of World War II, all the features of the modern outdoor recreation phenomenon emerged and brought unprecedented pressure on national parks and similar resource-based areas. Rapid rises in population, coupled with economic expansion, increased affluence, leisure and mobility, brought new waves of visitors to the parks, seeking more diverse and sophisticated forms of amusement, not all of which were compatible with park values.

During the 1960s, increasing public concern over the impact of rapidly accelerating use and modern technology led to greater awareness and acceptance of the need for positive steps to contain visitor activity and restore park environments. The balance in park management philosophy and practice tipped in favour of restoration and preservation of the resource base. According to Leopold *et al.* (1963), the goal of park management became to preserve and, where necessary, recreate the ecological scene as viewed by the first European visitor. Clearly, the protective function of national parks, and the obligation to maintain the natural heritage 'unimpaired for the enjoyment of future generations', was now to receive priority. Although provision for public enjoyment and recreation remained an objective, it became subservient to preservation of natural features and ecological values.

More recent management decisions by the US National Parks Service reinforce support for nature conservation as a primary objective. Restrictions on access to national parks are commonplace because of environmental damage and use conflicts. Motor vehicles have been excluded from some park areas, to be replaced by shuttle buses and mini-trains. Speed restrictions, one-way traffic systems and limited parking facilities have been introduced to dampen visitor use. Although the regulation programme has apparently received general public acceptance, there are some who believe it does not go far enough, and others who oppose the restrictions imposed. Some blame the tour operators and other concession holders, and advocate an increase in fees as part of the answer. A letter to the *National Parks Magazine* targeted tour buses which:

> ... dump 50 or more visitors at a time in one area. The visitor centers, bathrooms, and viewing areas become chaos. The buses are also noisy, spew stinking black exhaust, and take up parking space... Commercial interests have taken over the parks but very little money spent by park visitors is returned to the Parks Service... Park fees are ridiculously low... less than the price of a movie ticket... it's time to return our parks to the taxpayers.
>
> (Linz and Linz 1996: 10)

On the other hand, a proposal to restrict vehicle access in Yosemite National Park, and to require visitors to leave their cars in a parking lot and use a shuttle service, encountered strenuous opposition. Whereas some considered the proposal an infringement on their rights, others thought it did not go far enough, given that the ultimate goal of the Parks Service Management Plan for Yosemite is to remove all private vehicles from the valley (Nolte 1995: A21).

Despite these initiatives, the strict goal of preservation is obviously unattainable in the absolute sense while *any* level of use is permitted. The implications of this use/preservation dilemma are discussed in Chapter 10. However, it seems that for the present, at least in developed countries, perpetuation of natural and cultural heritage is now recognised as the prime function of national parks. How long the notion of national parks as predominantly nature reserves can be maintained is open to question if community support is alienated in the process and puplic funding continually reduced. The expectation that national parks will be generally accessible to the community is widely held. The further assumption that they will be, to a degree, self-supporting, is also important in the development and expansion of the national parks system. This situation may prompt renewed support for the involvement of the private sector in national park management.

Canadian national parks

Not surprisingly, the parks system in Canada has features in common with the US; in fact, there is some shared responsibility for certain natural and historic sites along their common border.

Parks Canada manages a system of protected areas consisting of 36 terrestrial national parks and 4 national marine conservation areas. Among the 36 national parks, some, such as Banff and Yoho, date from the late 1880s; others, such as Ivvavik and Vuntut (formerly, Northern Yukon), Grasslands and Bruce Peninsula, are recent additions, becoming national parks in the 1980s. The programme to establish national marine conservation areas (and its predecessor, the national marine parks programme) came into being in 1986, with the adoption of the National Marine Parks Policy.

Both programmes depend upon a system of regionalisation, which divides the country into thirty nine terrestrial and twenty nine marine natural regions (including natural regions in the Great Lakes) (Canadian Environmental Advisory Council 1991). Parks Canada has a mandate to establish representative protected areas in each of these natural regions. To date, sixteen terrestrial and twenty six marine natural regions contain no national parks or national marine conservation areas, respectively. Therefore, the protected area system at the federal level in Canada is not yet complete. Parks Canada continues to work to establish protected areas in the natural regions where currently there is no representation. On the other hand, some natural regions in western Canada are over-represented by the establishment of several national parks as tourism destinations in earlier times (Payne and Nilsen 1997).

As with the parks of the western US, recreational opportunities, as well as commercial considerations, were of prime concern in the early years. Interest in the first Canadian national park at Banff dates from the discovery of hot mineral springs in the 1880s. Curious as it may seem, in the light of the magnificent Rocky Mountain scenery in the area, the original reason given for the reservation of land at this site was the 'sanitary advantage' of these waters, and the need to protect them from commercial exploitation and control them for the benefit of the public (Scharff 1972). In 1887, Banff Hot Springs Reserve, when enlarged to an area of 260 square miles (approx. 670 km^2), officially became Rocky Mountains National Park. The name was later changed to Banff National Park, and the Canadian government and the railroads combined to develop hotels and facilities for visitors to the area.

It is worth noting that the *Rocky Mountains Park Act* of 1887, specifically reserved the area as 'a public park and pleasure ground for the benefit, advantage and enjoyment of the people of Canada'. This wording is almost identical with that proclaiming Yellowstone National Park. In addition, the Act went on to spell out the protective aspect, emphasising that no development was to be permitted that could impair the usefulness of the park for the purposes of public enjoyment and recreation.

As more and more parks were added to the Canadian system, transport networks were developed, all manner of visitor facilities were provided, and entrepreneurs were encouraged to maintain a high level of service to promote patronage. In some cases, the recreation facilities at sites like Banff and Lake Louise themselves became major tourist attractions to complement the scenic grandeur in the surrounding park landscape. As with the American parks, concern for nature preservation was to come later as visitor pressure mounted on the park environments, and the depredations brought about by indiscriminate hunting, mining and timber-getting became obvious. According to Nelson and Butler (1974), it was only in the period after World War II, that a strong preservationist movement emerged in Canada. The traditional view of tourism and recreation as fundamental underpinnings for parks was increasingly brought into question and, ultimately, the preservation and protection of park landscapes came to be regarded as first priority.

In 1994, Parks Canada revised its national parks programme policy to designate ecological integrity as a prime agency goal, and ecosystem management as the prime means of achieving it. Central to this new policy direction is the acceptance that ecosystem management must address the full range of human issues in establishing and managing parks, including the impacts of human use on natural systems and the impacts of park establishment and operation on human use systems. Although Parks Canada is directed to consider human use issues such as opportunities for public understanding, appreciation and enjoyment in establishing national parks, and while the agency possesses a range of tools (such as the Visitor Activity Management Process – see Chapter 6, as well as the associated Appropriate Visitor Activity Assessment and Risk Management processes), such human considerations have not yet figured in new park establishment in any major way (Payne and Nilsen 1997).

A particular concern is the question of prior human habitation in areas designated as national parks, and the problem of accommodating traditional resource uses within park management programmes. Proposals for new national parks and reserves in the more remote regions of Canada, such as the Yukon and the Northwest Territories, are examples. Special attention is being paid to protecting wilderness values, while maintaining the rights of native peoples to continue traditional extractive activities, such as hunting, fishing and trapping, in areas like Baffin Island. In many respects, this problem resembles that encountered in tribal territory in developing countries (see below). Prior human habitation also represents a problem, but of a different kind, in the older, more densely settled countries of Europe.

National parks in Britain, Ireland and Western Europe

It is clear that the IUCN definition of national parks is inappropriate, and largely irrelevant, for a country like Britain, with a long history of human settlement and no great reserves of unoccupied lands in which to create national parks in the North American mould. Moreover, by the time the first moves were made to establish national parks in Britain at the end of World War II, widespread acquisition of private land was prohibitively expensive and politically unacceptable.

The result is that areas designated as 'national parks' remain almost entirely in private ownership and productive use. Agricultural holdings, fenced pastures, forestry plantations, quarries, farm structures, transport routeways, and even villages and towns are all found inside the park boundaries. Management plans endeavour to reconcile conflicting interests between landholders and park visitors. At the same time, attempts are made to maintain and enhance the scenic quality and appearance of the landscape by controls over the location and nature of new facilities and proposals to alter existing structures.

A National Parks Commission (later Countryside Commission) was set up in Britain in 1949, and the first park, the Peak District National Park, became a reality in 1951. Since then, another eleven national parks have been created. The twelve national parks in England and Wales are designated:

- to conserve and enhance outstanding landscapes; and
- to make provision for people's enjoyment of the countryside.

The parks are run by national park authorities, which have a combined national and local membership, and shared responsibility for funding. Tourism is a major feature of the national parks of England and Wales, and in some of the national scenic areas in Scotland. In 1991, it was estimated that the national parks of England and Wales attracted 103 million visitor days a year, with the greatest number visiting the Lake District and Peak District national parks, some 20 million visitor days a year, each (National Parks Review Panel 1991). More recently, it was reported that in 1994 a minimum of 76 million visitor days were spent in the parks as a whole, but that this figure was likely to be a substantial underestimate of the actual totals, because survey methods did not cover all categories of visitor. Average daily expenditure, excluding accommodation, was estimated at £9.78 per person. These figures, though underestimates, clearly demonstrate the economic contribution of parks to local economies (Countryside Commission 1998).

The national parks were joined in 1989 by the creation of the Norfolk Broads Authority, a national park in all but name. The Authority is similarly constituted to those for existing parks, with additional powers over navigation on the ancient waterways. Restoration programmes for this unique water-dominant landscape include removal of sludge to open up new areas for boating, and the preservation of the remaining historic drainage windmills.

Interestingly, there are no national parks in Scotland. Although proposals were made as early as 1945 for five national parks, pressures on the countryside were much less than those in England and the idea lapsed. The Countryside Commission for Scotland, set up in 1967, established some quite small country parks for intensive recreational use, and proposed a new parks system to encompass urban parks, country parks, regional parks, special parks and national scenic areas. National parks were felt inappropriate in a Scottish context because, under internationally accepted standards, 'conservation must always take precedence over recreation and other land uses' (Foster 1979: 4). Such an approach was seen as lacking flexibility and inhibiting retention of desired characteristics in 'a living, in-use way rather than in a museum sense'.

In 1990, the Countryside Commission for Scotland recommended the establishment of four national parks, the first parks north of the border. Heritage landscapes such as the Cairngorms, Ben Nevis and Loch Lomond were proposed for protection, using a system of zoning for core areas, surrounded by management buffer zones and a transitional community development zone. Scotland still has no national parks, but the recent change of government may mean that the unique landscapes, remote qualities and cultural values are finally recognised.

Reviews of the concept and purpose of national parks in Britain have led to more emphasis being given to management procedures to ensure that recreational use does not threaten the scenic beauty and wildlife, and that forestry and agriculture within the parks does not detract from the appearance of the landscape. Concern has also been expressed about quarrying, the design and construction of reservoirs, housing and recreational facilities, and visitor pressure on roads not designed for heavy traffic.

One of the most contentious issues surrounding the management of Britain's national parks is their use for military purposes. Table 9.1 shows the widespread nature of military activities in the parks. Not surprisingly, frequent protests occur, as new proposals for development of training facilities are put forward.

The Otterburn Training Area occupies 58,000 acres (approx 23,500 hectares) in Northumberland National Park. This represents 22 per cent of the park's area, with further expansion being planned. A new army training camp has recently been built in Dartmoor National Park, where the Ministry of Defence refuses to abandon its training programmes, which include live firing.

Table 9.1 Military use of national parks

	Live firing	Dry/adventure training	Army camps and bases	Low flying aircraft	Restricted public access
Dartmoor	•	•	•	•	•
Exmoor		•		•	
Brecon Beacons		•	•	•	
Pembrokeshire Coast	•	•	•	•	•
Snowdonia	•	•		•	
Peak District	•	•		•	•
Yorkshire Dales		•		•	
North York Moors		•	•	•	
Lake District		•	•	•	
Northumberland	•	•	•	•	•

Source: Adapted from Lunn (1986: 6)

The military vigorously defends its need for training facilities, and points to the success of its conservation and restoration programmes. However, the Countryside Commission has long argued that military use of national parks is inconsistent with national park purposes, and opposes any extension or intensfication of military activity in the parks.

Clearly, problems will always exist where privately owned resources play the major role in providing recreational opportunities for park users, and where private interests may conflict with national priorities in conserving the natural beauty and amenity of the countryside.

The national parks of England and Wales were a product of the time and circumstances prevalent at that time. As these circumstances change, management has to adjust. Few would argue for the abolition of the parks, but their character may change and different solutions may have to be found in order to attain the objectives for which they were established. The designation of Country Parks is a move in this direction. The main purpose of Country Parks is the provision of recreational facilities in an outdoor setting, and in many ways they are the antithesis of national parks. More than 130 Country Parks have been recognised: these act as 'honeypots', providing readily accessible recreation outlets for large numbers of rural users, where existing, more natural areas are under threat from overuse. In this way, pressure on the national parks might well be relieved by provision of a greater range of alternative rural recreation opportunities, accessible to large centres of population. Park boundaries and features also need to be reassessed in order to identify areas and sites where management controls may be eased, or in other cases, tightened.

Whenever people are intimately involved, as they are in the British national parks, concern must be shown for their attitudes and welfare. The continued support and endorsement of the park concept by the inhabitants are vital for their continued success.

National Parks in Ireland

Few people appreciate that Ireland had a national park long before
Britain, or that Irish national parks conform to the strict guidelines
laid down by the IUCN.

(Dillon 1993: 5)

Nature conservation in the Republic of Ireland is the responsibility of the National
Parks and Wildlife Service, part of the Office of Public Works. Fauna and flora
are protected by refuges and nature reserves, and by 1200 Natural Heritage Areas.
Size apart, Irish national parks are similar to those in North America and Australia,
in contrast to those in Britain. Ireland's first national park was established in 1932,
near Killarney in the southeast, from the gift of a 4,000 hectare estate. It has since
been extended to 10,000 hectares (24,700 acres), and includes the lakes of Killarney
and surrounding mountains. The small Connemara National Park in Galway was
opened in 1980, followed in 1986 by Glenveagh National Park in Donegal, and
Wicklow Mountains National Park, near Dublin, in 1990. There are now five
national parks in the Republic (Figure 9.1), and more in Northern Ireland.

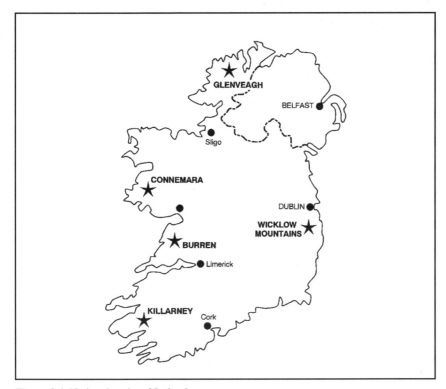

Figure 9.1 National parks of Ireland

Source: Adapted from Dillon (1993: 5)

While there appears to be general support for national parks in Ireland, controversy has arisen over a number of localised management issues. The stark beauty of the limestone pavements in the Burren National Park, in the central west, has been threatened by a proposal to build a modern visitor centre at Mullaghmore within the park. Concern was expressed over increased numbers of vehicles, pressure on roads and facilities, and degradation of the park's wilderness character (Don 1997). Money and jobs were also a consideration due to the belief that if roads were widened to take coachloads of visitors '... the cash rather tends to flow (out) with them' (Dillon 1993: 5). The work on this visitor centre and centres in other national parks is now on hold, pending design and location of more environmentally compatible facilities.

As in many national parks in the Western world, conservation of endangered species, and eradication of noxious exotic species and feral animals are ongoing concerns in Irish national parks. An example from Killarney is control of infestations of rhododendrons, an attractive flowering shrub, but one that represents a significant management problem for regeneration of native species.

National parks in Europe

The British Isles share, with much of the rest of Europe, the problem of developing a functioning park system within a landscape which has evolved over centuries of human use. In a country like the Netherlands, the task is made even more difficult; it has one of the highest population densities in the world, and a good proportion of the countryside is the direct product of human efforts to reclaim land from the sea. Yet, even there, 13 per cent of the country is still said to be in a more or less natural state – dunes, wetlands, woods or uncultivated areas – and a number of national parks have been created.

A national park in the Netherlands has been defined as:

> An uninterrupted stretch of land of at least 1,000 hectares (approx. 2,500 acres) consisting of areas of natural beauty, lakes, ponds and water courses and/or forest, with a special character as regards nature and landscape, and a special plant and animal life.
>
> (Netherlands Ministry for Cultural Affairs 1976: 2)

The Dutch Government is endeavouring to meet the requirements laid down in the IUCN definition, and to date twenty two areas have been selected which meet the criteria of size, quality and integrity of area and management. It is worth noting that in the Netherlands, 1,000 hectares is considered sufficiently large for a national park. This hardly compares favourably with around 4.5 million hectares (approx. 173,000 square miles) in the world's largest national park, Wood Buffalo, in Canada!

In addition to national parks, the Netherlands is developing an experimental system of National Landscape Parks, which are closely related to the National

Reserves in USA, the National Parks of Britain, the Regional Parks of France and the Naturparken in West Germany. The concept has much in common, too, with the idea of 'countryside parks' proposed for Australia (see below).

With this type of park, the concern is not with purely natural areas, but with areas shaped by humans and nature in combination over the course of many centuries. National Landscape Parks include villages and towns, agriculture, typical architecture, and other features of human activity characteristic of the Netherlands landscape. The concept envisages that landholders, in addition to working their land, should assist in the management of the landscape park and receive payment for activities concerned with its care, as well as compensation for loss of income as a result of any limitations on farming practice. Thus, farmers will no longer supply only grain, potatoes, dairy produce and meat, but will also provide the community with an attractive landscape in a healthy living environment. Moreover, they will get paid for it. In a country like the Netherlands, in particular, National Landscape Parks are seen as complementary to the national parks system, and as an appropriate way of encouraging people living and working in settled rural areas to maintain the natural and cultural values of the countryside.

The reunification of Germany and the changing geopolitical scene have led to the opening-up of former well-known national parks in central and eastern Europe. These include the Hochhanz National Park in the former East Germany, the Okjow National Park in Poland and the Tatra National Park near Zakopane, in the south of that country.

It has been predicted that the emergence of the European Union and the operation of the Channel Tunnel will lead to a new era of partnerships and linkages between Europe's protected areas (Simpson 1995: 5). Twinning arrangements and staff exchanges are already helping to disseminate best management practice, and to promote common policies and strategies between national parks in Britain and Europe.

National parks in Australia

'National' parks in Australia, have, until relatively recently, been the sole respon- sibility of the six State and two Territory governments. Technically, therefore, they did not qualify under the strict requirements of the earlier 1969 IUCN definition, that national parks be under the jurisdiction of the nation's 'highest competent authority'. However, from most other standpoints, they do meet the international guidelines – national parks typically consisting of sizable areas of predominantly unspoiled landscape, with the emphasis on nature conservation. Only since 1975, has an Australian National Parks and Wildlife Service functio- ned, with specifically 'national' parks being established alongside the State systems.

Originally, the provision of public recreational opportunities (as in North America) was the primary objective of national parks in Australia. The first park established, Royal National Park, near Sydney, provided holiday accommodation, sporting facilities and picnic areas, with the emphasis clearly on human pleasure and amusement. Since the early years, the concept of a national park has broadened beyond this recreational theme.

A gradual increase in the number and area of national parks and reserves followed the growth in environmental awareness which occurred in the 1970s and 1980s. New South Wales (NSW), the most populous State, has 103 national parks, 217 nature reserves, 13 historic sites, 10 aboriginal areas and 18 State Recreation Areas (SRAs) and 6 Regional Parks administered by the State National Parks and Wildlife Service. The total area is approximately 4.5 million ha (approx. 11 million acres), or about 5.69 per cent of the area of the State. The largest unit is Kosciuszko National Park, which occupies 640,000 ha (approx. 1.6 million acres) southwest of Canberra in the Australian Alps.

The Great Barrier Reef off the coast of Queensland also has a strong attraction for visitors. Following extensive lobbying by preservation-minded pressure groups, the Great Barrier Reef Marine Park Authority (GBRMPA) was set up, and sections of the reef region are being successively incorporated into the marine park. Such parks, by their very nature, present unusual problems in park management. Contentious issues in this case were the question of oil exploration on the reef, and the clash between the Queensland State Government and Australian Federal Government over administration of the resources of the region.

The federal government has become increasingly involved in park management since the formation of the Australian National Parks and Wildlife Service in 1975 (now part of Environment Australia). The Service works in collaboration with the States, and has sole responsibility for certain areas of nature conservation interest such as Norfolk Island and Christmas Island, as well as national parks such as Uluru and Kakadu, near Darwin, in the Northern Territory. The Service believes that the plans of management drawn up for Kakadu may well prove a model for the development of similar parks in 'frontier' areas. Certainly, the park has had to contend with some major problems. Apart from preservation of the park environment and providing for appropriate use by visitors to a remote area, protection of Aboriginal interests, regulation of mining (uranium) in and near the park, and control of feral animals all create difficulties for park management.

Australia's largest city, Sydney, with a population approaching four million people, is fortunate in that twelve national parks are located within a radius of 160 kilometres. To a great extent, this situation results from the reservation of land for parks in areas where the soil and terrain were considered unsuitable for agriculture and too rugged for housing. Royal National Park, south of the city, Kuringai Chase National Park, immediately to the north, and Blue Mountains National Park, to the west, are located on dissected sandstone plateaus, for which no economic use was perceived at the end of last century. Melbourne, on the other hand, in the State of Victoria, was ringed by good agricultural land, and there is now a serious dearth of parks and reserves close to the city.

The apparent contradiction and conflicts between outdoor recreation and nature preservation, and between different types and intensities of recreational use of national parks in Australia, suggest that the complexity of recreational demands should be matched by an appropriate array of recreational opportunities outside the parks. In the State of New South Wales this need has been met, in part, by the development of SRAs. These units, some of them quite large, comprise both natural areas and man-made features of scenic, historic and recreational importance. Several SRAs adjoin inland waterbodies, and others occupy coastal sites. SRAs were designed to take pressure off the national parks, and while haphazard destruction of their environment is obviously not allowed, they do cater for more intensive forms of outdoor recreation; even trail-bike riding and hunting may be permitted. In some parks, accommodation facilities have been provided, while others are designed for day-visitor needs only. Provided that the recreational emphasis continues, State Recreation Areas are a useful and popular complement to national parks, and an important additional unit in an integrated system of outdoor recreation opportunities.

The introduction of more recreation-orientated 'people's' parks, comes at an important stage in the development of the parks system in Australia. For many years, national parks have made a significant contribution to the recreation resource base. Now, in many parts of the country, and for a variety of reasons, opportunities for further expansion of the national park system are becoming limited. Indeed, it could be said that Australia has entered a 'mature' phase of park development, in which the initial stage of large-scale land acquisition is closing, to be replaced by careful appraisal, development and management of park resources already acquired. At the same time, consideration can now be given to alternative means of expanding the recreation opportunities of urban-based populations, by provision of a range of different park options in accessible rural settings.

Moreover, a good deal of questioning of the role of national parks has emerged in Australia in recent years. To some, parks and wilderness areas appear as enclaves of unproductive land, and havens for noxious plants and animals. Proposals to enlarge national parks on the north coast of New South Wales, and to establish additional national parks in the New England region, further inland, have generated significant local opposition. The continued relevance of the North American park model to Australia has been challenged, and the adoption of the British/European style park has been put forward as an alternative (Pigram 1981). The proposed new style of 'living' park environment, if developed along the lines of the National Reserves in the United States, would encompass farming communities and rural settlements within distinctive scenic landscapes. The creation of 'countryside parks' in this way should be seen as a complement for national parks (rather than as a replacement), and as an extension of the land management systems, providing for a dual network of inhabited parks developed in tandem with the traditional parks.

Some progress has been made in the conservation of biodiversity beyond the formal system of parks and protected areas. The Voluntary Conservation Agreements in place in New South Wales are an example, as are the Regional Parks of Perth, Western Australia (Moir 1995), some of which remain in private ownership.

Unfortunately, park authorities in Australia have generally shown a decided reluctance to depart from the existing national parks system. Brisbane Forest Park, near the capital city of Queensland, is a step in the right direction, although all land within that park is publicly owned. In the state of Victoria, the Regional Strategy Plan proposed for the Upper Yarra Valley and Dandenong Ranges on the northeastern outskirts of Melbourne, encompasses many of the features of the National Reserve concept. Non-urban land in private hands comprises 23 per cent (approx. 175,000 acres or 70,000 ha) of the total land area, with a further 3 per cent classified as urban. Approximately 105,000 people live within the region, and the Strategy Plan provides for protection of the special features and rural character of the area and the maintenance of recreation opportunities on both public and private land. In Sydney, eight areas have been designated as Metropolitan regional parks. These parks were conceived by the New South Wales Government as a way of providing the people of Sydney with green lungs; a similar justification as for the establishment of Royal National Park in 1879. Regional parks are areas of open space for recreation and for the conservation of fragile ecosystems. They vary in size (from 4,000 hectares to less than 50 hectares) and in the activities to which they cater, and represent a promising public sector initiative. (For details see the World Wide Web page at: http://www.npws.nsw.gov.au/parks/regprks.htm).

Apart from these initiatives, a generally negative attitude prevails at official levels towards the introduction of European-style national parks, or US National Reserves, in Australia. This reaction, coupled with growing resistance by rural landholders to any further acquisition of park lands, means that progress towards establishment of 'countryside parks' is likely to be slow. Yet, the need for innovations in park-planning should be obvious. In a developed country like Australia, with all the pressures on resources for greater output and more efficient production methods, the transition to large-scale stereotyped forms of land use can be very rapid. Therefore, there is an urgent need to adopt an alternative approach to allocating land for parks because the changing nature of agriculture and rural life acts as a disincentive for landholders to maintain the character and quality of the countryside in their keeping.

Despite scepticism and opposition from inflexible park bureaucracies, preservation-minded conservationists, and some feudalistic landholders, the countryside park concept could play a useful role in Australia, alongside national parks. If it can be shown by successful pilot projects that the economic and amenity functions of the countryside can be compatible, then a range of park types can be created, as and where appropriate. Given time and enlightened management, such parks have the potential to demonstrate the benefits of sharing the countryside, both for ongoing productive purposes and for outdoor recreation.

National parks in New Zealand

New Zealand was one of the first nations to establish a national park after the creation of Yellowstone. Tongariro National Park came into being in 1887, as the result of a gift from the Maori people of an area of volcanic peaks in the central north island. Since that time, twelve additional national parks, three maritime and

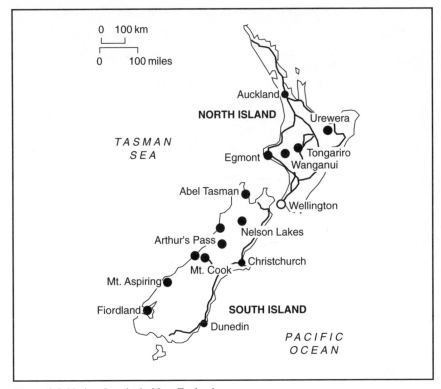

Figure 9.2 National parks in New Zealand

Source: Adapted from http://www.travelbank.net/nz/nzwalks/nz-walks.htm

two marine parks, twenty forest parks, and over 1,000 scenic and special reserves have been added. In all, more than one-fifth of New Zealand is under complete or partial protection; this is in a country which has a total landmass only of the US State of Colorado (Figure 9.2). (For further details see the World Wide Web page at: http://www.travelbank.net/nz/nzwalks/nz-walks.htm).

New Zealand has an environment that is unique in the world; its geographic isolation has resulted in the evolution of diverse fauna and flora. This, coupled with spectacular scenery, provides many opportunities for the creation of parks and protected areas. Many of the large reserves are focused on lakeshores and on the higher mountainous country and fiords of the South Island. Fiordland National Park, covering 1.25 million hectares (approx. 3.1 million acres), is one of the largest in the world. The park is a World Heritage Site and takes in areas of outstanding natural beauty such as Milford Sound. Another World Heritage Site, Te Waipounamu, covers 2.6 million hectares (approx. 6.4 million acres) and four national parks in the southwest, and is a focus for many forms of nature tourism. Some of the management problems concerning such protected areas are dealt with in the following chapter.

National parks in developing countries

Despite the existence of large areas suitable for designation as national parks, problems can arise with park establishment and management in developing countries. Although ecological considerations and the desirability of preserving unique ecosystems may certainly be recognised in the selection of environments and landscapes for inclusion in the parks system, park proposals are often assessed primarily against potential economic and social benefits. This means that, in negotiating land acquisition and planning the future operation and management of a park, it becomes critically important for the government authority to be able to demonstrate specific benefits, especially for the local people, by way of commercial opportunities and employment. Thus, economic factors may overshadow ecological considerations, to the detriment of the park environment.

National parks are now a reality in all corners of the developing world. Some of these parks and reserves reflect attempts to protect natural landscapes and wildlife for conservation and scientific purposes. In other situations, potentially large returns from tourism appear to have influenced their creation. Much of the stimulus for this tourist activity comes from worldwide interest in viewing nature, and the from diversity of animal and bird life to be found in the national parks. In the less developed countries of Africa, for example, most park visitors come from abroad. Whereas some newly-emerging nations may regard the parks as unwelcome vestiges of previous foreign dominance, parks are tolerated and even encouraged because of their role in providing local employment, and attracting tourists and foreign currency.

Large and varied species of wildlife can be found in national parks such as Tsavo and Nairobi in Kenya; Kilimanjaro and Serengeti in Tanzania; Matopos in Zimbabwe and Whangie Kruger in South Africa. Kruger National Park covers nearly 5 million acres (approx. 2 million ha), and is visited by almost 500,000 people annually, 25 per cent of them from overseas.

This also applies to Ras Mohammed National Park, Egypt's only national park (Egyptian Environmental Affairs Agency, undated). Established in 1983, the park covers an area of 480 square kilometres (approx. 120,000 acres) in the South Sinai peninsula. The park includes land and marine areas and shorelines along the eastern coast of the peninsula. Two other Managed Resource Protection Areas – Nabq and Abu Galum – have also been created further north on the Gulf of Aqaba. The national park and protected areas take in some of the world's best coral reef ecosystems and fossil coral platforms, as well as spectacular granite mountains and desert landscapes. Visitors attracted to those areas and to the rapidly developing tourist resorts at Sharm el Sheik and Dahab, are an essential feature of the economic development of South Sinai.

Although a number of national parks have been established in Southeast Asia and the Pacific Islands, considerable difficulties still have to be overcome. Countries like Indonesia, Papua New Guinea, the Philippines, Malaysia and Vietnam, have apparently endorsed the concept of national parks and appear convinced of the role they can play in nature conservation. However, such conviction cannot always lead to action in societies where wilderness is still considered an obstacle to progress and the value of conservation is not universally appreciated. There may well be difficulty in diverting money and manpower to the development of parks, and a reluctance to take land out of what is considered to be more productive use. Even in circumstances where the authorities do display enthusiasm and an awareness of the value of parks, obstacles may still surface in attempting to translate the concept into action.

Specific problems can occur in areas of prior human habitation, especially where land is in communal ownership and land use practices, such as shifting agriculture, timber getting and hunting, are destructive of the environment. Problems can be countered, in part, by raising standards of living above the subsistence levels that contribute to these rapacious forms of land use. Moreover, if the local population can receive some tangible benefit from the establishment of a national park, people may be more prepared to respect and maintain the integrity of the park environment. This calls for a fine balance between the creation of a strict nature reserve on the one hand, and a commercially orientated nature-based tourism enterprise on the other. If this is not achieved, there may be resentment and non-cooperation, where a more environmentally compatible, but less rewarding and beneficial type of park system, is imposed on local communities. In practice, as Cochrane (1996: 242) has argued, it is extremely difficult to achieve the aims of ecotourism and to improve the welfare of local people, simultaneously.

A major concern in these circumstances is the extent to which new or existing national parks and nature reserves may intrude upon the lives of local residents, leading to disruption of established patterns of land use and of the social fabric. A related issue for park establishment and management in developing countries, and one shared with the developed world, is the dilemma of promoting national parks as an engine of tourism, while maintaining the biophysical integrity of the park environment. In the absence of sound appreciation of park values, emphasis may be misguidedly placed on maximising visitor numbers in the interests of economic returns, to the detriment of the park itself (Pigram *et al.* 1997).

Wilderness

No discussion of protected areas would be complete without reference to what many regard as the ultimate in natural environments – wilderness. For much of history, wilderness has held a negative connotation; either as waste land, or some vast, hostile and dangerous place to be avoided if at all possible, or else to be tamed, controlled and exploited.

Today, both people and governments have come to think of wilderness in more positive terms, as something to be valued and preserved for a future world, in which it could become an increasingly rare phenomenon. Many people now perceive wilderness as a large natural area where animals and plants can live undisturbed, and where visitors can enjoy recreational activities of a primitive and unconfined nature. Hiking and canoeing are often given as examples of the types of recreation envisaged – those for which a minimum of mechanical aids is required.

The main benefits of a wilderness experience are often said to be the spiritual and psychological satisfactions gained. Other advantages of wilderness recreation are physical and mental stimulation, the aesthetic appreciation of beautiful scenery, and the experience of conditions similar to those encountered by the first settlers of a region. Wilderness serves as a sanctuary, either temporarily or permanently, for renewal of mind and spirit. In modern jargon, it has become a refuge for those who wish to 'drop out', momentarily, into a simpler, less complicated world; a place where self-confidence can be re-established through physical challenge and reliance on self-sufficiency and subsistence skills.

Wilderness areas are also valued because of their role in nature conservation and scientific research. The size, remoteness and variety of ecosystems represented in wilderness are important for wildlife preservation and the maintenance of ecological stability and genetic diversity. Apart from being a potential source of a wide variety of useful plants and insects, wilderness also provides a reference point against which to measure changes in settled areas, and in crops, forests and animal populations. Some proponents of wilderness argue that these areas also provide a buffer, or safety valve, against long-term disturbance of the global ecosystem, resulting from large-scale human interference. While this may be the case, it is a nebulous argument to use in trying to persuade decision-makers to close off public lands for exclusive use in scientific research. This argument has provoked a reaction in some quarters that wilderness is a selfish concept and the pursuit of a small and vocal élite. The restricted numbers and specialised forms of recreation associated with wilderness do little to destroy this impression (Sax 1980).

A further qualification concerns the degree to which conditions in wilderness areas can remain pristine. Conditions of total naturalness are impossible to find, even in Antarctica. Therefore, wilderness has to be a compromise, taking in areas where there remain no permanent traces of people (e.g. roads, buildings and modified vegetation).

The really large remaining areas of 'true' wilderness can be found, like the big national parks of the world, in North and South America, Australia, New Zealand, parts of Africa, and, of course, the Arctic and Antarctic. In general, these areas have not experienced heavy population pressure on their land and water resources. Wilderness is not a concept generally applicable in Europe; some limited examples of quasi-wilderness might be found, but as in Britain, Western Europe and Scandinavia, potential areas have, with few exceptions, been extensively used by humans.

Wilderness is land which retains its natural character and is without improvements or human habitation. Simple, non-mechanised forms of recreation are envisaged; to preserve wilderness values, it is necessary to protect the natural ecosystems present, and to maintain the topography and plant and animal populations in an undisturbed state. Thus, a prime purpose of wilderness management, is to keep the area as natural as possible by only allowing levels of use that are consistent with both ecological and perceptual carrying capacities. Wherever even minimal levels of recreational use are envisaged, their impact on natural systems has to be considered, as well as the impact of different user groups on each other. These objectives can be achieved by management techniques, such as zoning, access controls and imperceptible manipulations of the environment. Human influence, both direct and indirect, has to be reduced to a minimum so that both the land and the fauna and flora remain primarily the product of natural forces. In a sense, a conscious decision is taken to leave the land 'unmanaged', and to exercise control only over visitors and interference from outside. The degree of intervention depends upon the size of the wilderness and the effectiveness of any buffer zone in filtering out unwanted influences.

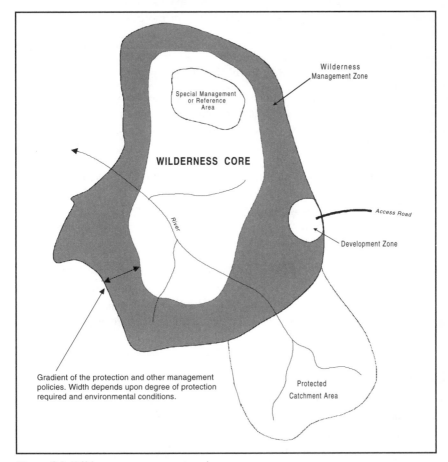

Figure 9.3 Wilderness management zoning

Source: Helman *et al.* (1976: 45)

Zoning is a common strategy employed in wilderness management. One approach to zoning is the core/buffer concept (Figure 9.3). Here, the wilderness core is surrounded by a wilderness management zone or protective buffer. The protective function is two-way – to protect the wilderness, and to protect adjacent land from disturbances such as wildfires which might originate in the wilderness. Subzones are determined for access and minor facilities, while separate scientific reference areas, with more restricted access, are set aside within the wilderness complex.

The designation of land as wilderness is a contentious issue. A long legacy of resource exploitation brands as strange and unacceptable the sterilisation of land with economic potential. Even among wilderness supporters, the formulation of management policies satisfying those advocating strict ecological preservation and those seeking a 'wilderness experience' is a difficult challenge to meet. Until wilderness is accepted as a legitimate form of land use, and its benefits, both for outdoor recreation and nature conservation, are more generally recognised, controversy will continue to surround the wilderness concept.

Guide to further reading

A number of useful references for further reading are cited below. However, readers are encouraged to access government documents, in particular those containing details concerning the management of specific protected areas. In addition, several other useful sources are cited in Chapter 10.

- General discussions about *the establishment and management of protected areas, particularly national parks*: Sax (1980); Hall (1992); Green (1992); IUCN (1994); Payne and Nilson (1997); Jenkins (1998).
- *Recreation and tourism in protected areas, including World Heritage Areas and national parks*: Forster (1973); FNNPE (1993); Eagles (1996); Shackley (1998).
- *Tourism in protected areas in developing countries*: Shackley's (1996) book makes frequent reference to protected areas in developing countries; Pigram *et al.* (1997) present a case study of Cat Ba Island, Vietnam.
- *Regional impacts of tourism and recreation in parks*: McDonald and Wilks (1986a, b).

Review questions

1 What is a national park? What should be the main goals and objectives of national parks authorities? To what extent should recreation be 'tolerated' in national parks?
2 Discuss important aspects of the early history of the establishment of national parks globally, nationally or regionally.

3 Select a national park or wilderness area. What are the main sources of recreational pressures on that park or area? What planning and management initiatives have been devised by that park's authorities to deal with such pressures? Have these measures been evaluated? How and to what extent have they been evaluated?
4 Protected places are not for people! Discuss.

10

NATIONAL PARKS MANAGEMENT

The previous chapter presented a broad overview of protected areas in various environments. This chapter develops a more focused and critical assessment of the management of national parks for nature conservation and outdoor recreation. Competing values, interests and priorities highlight the need for planning procedures that take account of environmental concerns, as well as recreational and tourism use of national parks, in an integrated and sensitive way.

The brief canvass of park systems in the previous chapter illustrated the many ways in which the national park concept has been interpreted. This diversity gives rise to an equally complex range of park problems and approaches to the management of national parks. Despite these variations, a recurring theme with all park environments is the need to strike a balance between conservation, or preservation, and use; that is, how to accommodate appropriate levels of human activity, while maintaining the quality of the natural environment for which the park was established. This is the 'dilemma of development' (Fitzsimmons 1976).

Resource use is primarily concerned with the present generation; conservation and preservation are linked more closely with generations to come. Expediency demands the satisfaction of current wants, whereas prudence suggests limitations on use in order to preserve park values for future generations. Conflict and compromise are inevitable since *any* use involves some disturbance to the park landscape and ecosystem. The aim of management should be, first, to exclude activities which are clearly inappropriate – power boating, organised sports and entertainment centres come to mind as ready examples. Second, care must be exercised to keep unavoidable disturbance to a minimum and to recognise and correct environmental deterioration before it becomes irreversible.

Given the popularity of national parks as a focus for outdoor recreation, it could be assumed that moves to expand the parks system would not be questioned. However, in countries like the US and Australia, proposals to establish new national parks frequently generate strong opposition. Part of this reaction stems from a conviction in some quarters that resources and personnel should be directed towards upgrading the facilities and management of existing parks before acquiring more land. Added to this is concern expressed by neighbouring landholders, who question the record of certain park management practices deemed incompatible with surrounding land uses. Part of the problem stems from inadequate attention to the setting of park boundaries in the first place.

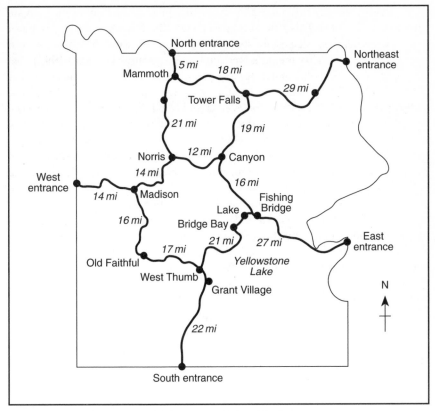

Figure 10.1 The boundaries of Yellowstone National Park

Source: Adapted from http://www.travelsmarter.com/mapynp.htm

National park boundaries

In the past, the process of establishing national parks might have been described as 'trying to put boundaries that don't exist around areas that do not matter' (Kimble 1951). Kimble was referring to regions rather than parks, but the point is well made. In many cases, park boundaries appear to have been determined with more regard for administrative and managerial convenience than for ecological and other relevant criteria. Examples include national parks which end abruptly at state borders or follow shire or county lines and similar cadastral features. The boundaries of Yellowstone National Park, for example, are mostly straight lines, and the park itself is contained almost within a square (see Figure 10.1). In Australia also, many parks and reserves, especially in isolated areas, are typically regular polygons with boundaries that rarely follow natural features.

The question of boundaries is of particular relevance to the management of Australia's national parks, whose boundaries often have been (1) determined qualitatively, and (2) influenced by political considerations and opportunism (Pressey *et al.* 1990). This situation gains added significance when it is realised that park boundaries, once established, are 'set in stone',with proposals for even minor changes to modify the park area requiring legislation (Howard 1997).

Yet, the allocation of land use, for any purpose, cannot be rigid. The process of allocation should be dynamic in keeping with new information and changes in technology and social preferences (Walker and Nix 1993). Certainly, experience in countries outside Australia does not suggest reluctance to review park boundaries. In Britain, a decision was taken in 1991 to reorganise the adminis- tration of national parks and review their boundaries. In Canada, delineation of park boundaries has also come under scrutiny (Theberge 1989: 1997). According to Theberge (1989: 22): '... it is ecologically indefensible to establish parks without provision for boundary or other management adjustments'.

In the US, following the 1980 State of the Parks Report to Congress, an increasing amount of research has been directed towards the formulation of credible and practical guidelines for establishing ecologically and managerially sound boundaries for national parks and reserves. This research culminated in the development of a comprehensive Park Resource Boundary Model for selecting critical park resource variables and determining their spatial extent (Sundell 1991).

Apart from establishing the spatial extent of a park, the determination of workable park boundaries is fundamental to maintaining the viability of the reserved area, its ecological integrity and biological diversity. An accepted part of the rationale for the establishment of parks and reserves is protection of areas and features of special natural or cultural interest. This task is often more acute along the edges of designated areas, where conservation values may conflict with neighbouring land uses. The need to integrate the area harmoniously into its surrounding environment is essential for the effective functioning of natural checks and balances with neighbouring ecosystems.

Some of the most significant problems in national park management occur in relation to park boundaries. Further, it can be argued that many of these problems could be avoided if more care was exercised in setting those boundaries. When inappropriate boundaries are established, the very resources which a reserve was intended to protect may become vulnerable to environmental stress from a number of internal and external threats. Thorough assessment of critical resource attributes is fundamental to the delineation of workable boundaries for a viable park or reserve.

Park relationships in Australia

In Australia, the National Parks and Wildlife Service takes seriously the challenge of neighbour–community relationships. In New South Wales, for example, there are more than 30,000 adjoining 'neighbours' to parks in urban areas, and some 8,000 in rural areas (Howard 1997). These interests cover a number of groups, including private and public sector landholders, and authorities managing utilities, services and transport. Among key issues of concern to these neighbours are fire management, control of noxious species and access.

Fire management

Growth in urban areas adjoining parks and reserves in NSW has increased dramatically in recent years, and in some high-risk areas, residential development has continued to the edge of park boundaries. Reflecting this trend, the Parks Service performance in fire management has remained at a high level (Table 10.1). Despite assertions to the contrary, the parks system has to cope with many more fires from outside protected areas than those which escape from the parks into adjoining land.

Whatever the origin, the disastrous bushfires of the 1993/94 summer, especially those close to the City of Sydney, tragically demonstrated the potential for loss of lives and property, and serious damage to the natural environment. There is wide diversity in the way Australian native vegetation responds to fire, and some species depend on fire for reproduction. A surprising number of native animals also survive fire by evading the flames or taking refuge in safe sites. Despite this resilience, long-term damage and decline can occur as a result of a sequence of fires. Strategies for fire prevention and control, therefore, are important elements in park management planning.

Control of noxious species

Noxious species of plants and animals can cause great damage to rural lands, affect livestock quality, and spread disease. They can also contribute to soil degradation and compete with native animals and plants. The New South Wales National Parks and Wildlife Service's (NSWNPWS) control measures, concentrate on introduced species and feral animals such as goats, foxes, rabbits, wild pigs, cats and wild dogs, including dingoes. In some parks, wild horses or brumbies, and introduced fish, in particular, European carp, are also a problem.

Table 10.1 Fire origin and progress – New South Wales

Year	Started on park, controlled on park	Started on park, moved off park	Started off park, moved on park	Total fires	Area burnt (hectares)
1989–90	142	8	99	249	66464
1990–91	303	29	93	425	125469
1991–92	266	21	109	396	66409
1992–93	167	6	40	213	21772
1993–94	216	25	50	300	382897
1994–95	173	15	62	250	89112

Source: Howard (1997: 399)

In the past, the Parks Service has been criticised for an apparently ineffective approach to control of unwanted animal species. However, a wide range of control methods are now in use, including physical, chemical and biological means. One controversial aspect of control of pest animals is the targeting of honey bees, which are said to compete with native bees, birds and small animals for nectar. For this reason, no new beehive sites are being allowed on park and reserve land, and current licences are being phased out.

Controversy also surrounds the Parks Service's attitude to wildlife management. Many landholders regard kangaroos, for example, as pests, whereas the Service has a responsibility to conserve and protect these native animals. Similarly, dingoes are native animals and the Parks Service aims to maintain existing dingo populations within park and reserve lands. Both these species, along with fruit bats or flying foxes, and wild ducks and other waterbirds, are considered 'bad news', particularly by adjoining landowners, who are critical of what they see as inadequate methods to control these 'pests' and contain them within park lands. Similar criticism is aimed at measures taken to eradicate weeds and noxious plant species.

The Parks Service is working hard to build bridges with neighbouring rural communities, and to avoid the parks and reserves being turned into 'islands' with little relationship to surrounding environments. An immediate risk when creating a park is the loss of opportunity to interact with adjoining ecosystems, and for natural checks and balances to regulate population expansion and changes. In ideal circumstances, parks are most viable when buffered by transition zones of extensive land use (e.g. forestry). Controlled zones, for hunting around parks with large wild animal populations, have also been advocated to regulate the growth and movement of herds.

Park management problems: external and internal

A different type of conflict can arise when incompatible resource uses are imposed *inside* a park from *external* sources. A park can be seen as relatively unused space, and, as public land, involves no resumption or compensation when claims are made on that space for public purposes. Near-urban parks, in particular, are regarded as 'fair game' for waste disposal facilities, motorways, airports, service corridors and water control structures.

Externally derived pollutants such as oil spills, chemicals, sewage, industrial effluents and biocides that are introduced into park drainage systems, can also affect park ecosystems and food chains. Pollution from pesticides is most likely in parks established near zones of intensive land use. Detergent pollution may occur in parks close to the urban fringe, and marine parks are obvious targets for oil pollution. Even when the source of pollution can be pinpointed, control may be difficult, and would probably only succeed as part of a wider programme to combat pollution generally.

Within a park, the principal recreation-related concerns are with access and the impact of recreation and associated human activities on the park environment. In part, the problem is a function of visitor numbers and the extent and nature of visitor management (e.g. marketing strategies; level and sophistication of facilities and services; educational programmes). Roads, parking, toilets, accommodation, food outlets, refuse and litter, and off-road vehicles, are just some of the ramifications of outdoor recreation that can place pressure on ecological quality and park resources. Much recreation activity is seasonal, and the problems are worsened during peak periods. In Australia's Kosciuszko National Park, for example, an ongoing debate continues over the place of snow sports in the park, and proposals for development of additional facilities and extension of access into more remote areas.

Park management: The question of access

While it is as well to recall that outdoor recreation is only one of the roles set down for national parks (Chapter 9), these areas provide a diverse range of opportunities for the enjoyment and appreciation of visitors. Access to parks and reserves for recreation purposes requires careful management to ensure that the activities of visitors co-exist harmoniously with the conservation of natural and cultural values, and with the educational and spiritual experiences parks have to offer. It is important also to recognise that national parks are one element in a system of protected areas and public lands. Recreational access to national parks should not be looked at in isolation, but from a regional perspective, taking into consideration the complementary access opportunities available in public land (e.g. State Forests and State Recreation Areas in Australia) as well as private land. Planning public access to national parks on a regional basis has the potential to increase economic, social and environmental benefits, and to maintain balance in the overall spectrum of recreation opportunities. It also mitigates against pressure to have national parks respond to unreal and inappropriate demands for outdoor recreation, and helps achieve compatibility between environmental protection and visitor use.

That said, community expectations remain high that access to national parks will remain open for a diversity of recreational pursuits. The New South Wales National Parks and Wildlife Service has produced a *Draft National Parks Public Access Strategy* (1997b), which provides a basis for offering a wide range of outdoor recreation opportunities, from pleasure driving, bushwalking, horse riding and biking, to various forms of boating, fishing and water-related activities. At the same time, a number of qualifications and limitations are indicated which constrain unfettered participation in these forms of recreation. Particular sources of friction are the restrictions placed on the use of four-wheel drive and other off-road vehicles; such vehicles permit access even to trackless areas of parks. The Parks Service is sufficiently concerned at their potential for degradation of the environment to declare the more remote parts of national parks and wilderness 'off-limits' to such forms of transport. Likewise, snow-mobiles are discouraged in alpine parks, and horses are banned from several near-urban parks, because of fears of erosion.

Pets, especially dogs, and firearms, are not permitted to be brought into parks; even fishing is subject to controls regarding which areas may be fished, the number and species of fish which may be caught, and the methods used. Although progress is being made in providing access for people with disabilities, many national parks remain effectively closed to these groups, as well as to seniors and young families. These limitations, along with the imposition of visitor fees, mean that there is still some way to go before opportunities for public access to national parks and reserves satisfy appropriate equity standards.

Whereas national parks in Australia have always attracted large numbers of visitors, and indeed, in earlier days, were justified on the basis of the recreation opportunities they provided (see Chapter 9), it is only now that the NSWNPWS is developing a policy and management strategy to promote the parks as a focus for nature tourism. The *1997 Draft Nature Tourism and Recreation Strategy* seeks to achieve ecologically sustainable visitor use of protected areas in the State. The strategy points to problems such as disturbance to wildlife, introduction of unwanted species, soil erosion, damage to vegetation, and escaped fires, associated with thoughtless or deliberate acts by visitors. The strategy aims to balance the protection of natural and cultural values with management of visitor use (Figure 10.2).

In putting forward the draft strategy for public comment, the Parks Service draws attention to the economic benefits flowing from tourism in the State's national parks and protected areas. Focusing first on the City of Sydney – 'City of National Parks' – the strategy identifies a select number of parks and nature reserves located in strategic locations within regional New South Wales, to be promoted as key destinations to international and domestic visitors (Figure 10.3). Themes ranging from local diversity of the natural environment to the cultural heritage of Aboriginal people, are among the varied experiences represented in the programme. The concept of the recreation opportunity spectrum will be used to develop an array of settings that appeal both to mainstream tourists and to niche markets. It is estimated that by the year 2005, some 28 million visitors will be attracted to the key regional park destinations identified.

Tourism and national parks

Promotion of tourism in national parks would seem to make good economic sense and, if planned as is envisaged in the strategy above, should be compatible with ecological objectives of park management. Without such planning, encouragement of visitor use can bring with it the risk of degradation of the park environment and, ultimately, loss of its appeal for nature tourism. The ongoing debate over the merits of tourism in national parks takes on added significance, due to the growing recognition of the economic contributions associated with tourist use of parks and protected areas, especially in developing countries. These benefits can be considerable and include:

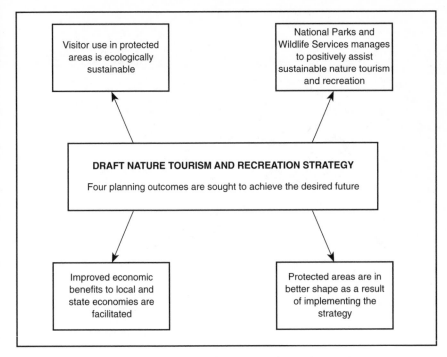

Figure 10.2 Planning outcomes of the draft nature tourism and recreation strategy

Source: NSW National Parks and Wildlife Service (1997a: 8)

- local employment, directly in the tourist sector and in support areas;
- stimulation of profitable domestic industries in accommodation, restaurants, transport, guide services and artefacts;
- foreign exchange;
- diversification of the local economy to meet new and growing demands for local products;
- improvements to local infrastructure and intercultural communication and understanding.

Against these positive aspects, must be considered a range of potential negative impacts from overuse of protected areas by visitors. Some of these are listed in Table 10.2, and the consequences are likely to be added to in fragile park eco-systems favoured by ecotourists (Chapter 11). Less obvious impacts can also ensue from resentment generated among local people, who may perceive the parks as being provided primarily for the benefit of foreign visitors or outsiders.

In an Australian study, Buckley and Pannell (1990) documented the kind of problems associated with visitor use of national parks for tourism and recreation. Table 10.3 is especially useful, in that it also locates parks where problems have occurred and includes references to support the incidence of impacts.

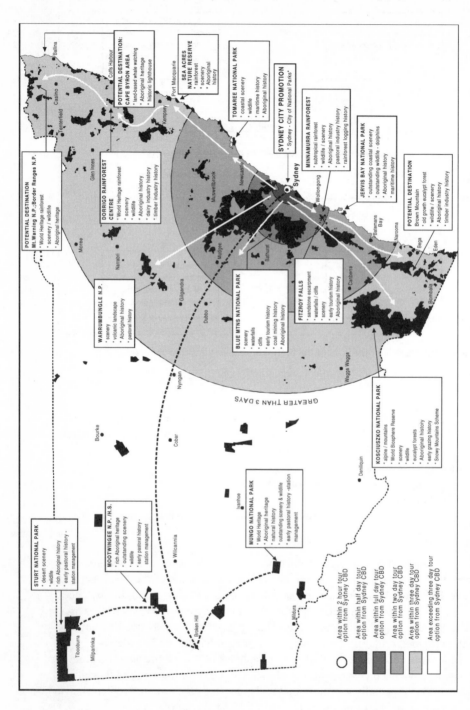

POTENTIAL DESTINATION:
Mt.Warning N.P.::Border Ranges N.P.
• World Heritage rainforest
• scenery / wildlife
• Aboriginal heritage

POTENTIAL DESTINATION:
CAPE BYRON AREA
• land-based whale watching
• Aboriginal heritage
• historic lighthouse

SEA ACRES
NATURE RESERVE
• rainforest
• scenery
• Aboriginal history

TOMAREE NATIONAL PARK
• coastal scenery
• wildlife
• Aboriginal history

SYDNEY CITY PROMOTION
'Sydney - City of National Parks'

DORRIGO RAINFOREST
CENTRE
• World Heritage rainforest
• scenery
• wildlife
• Aboriginal history
• dairy industry history
• timber industry history

MINNAMURRA RAINFOREST
• subtropical rainforest
• wildlife / scenery
• Aboriginal history
• pastoral industry history
• rainforest logging history

JERVIS BAY NATIONAL PARK
• outstanding coastal scenery
• outstanding wildlife - dolphins
• Aboriginal history
• maritime history

POTENTIAL DESTINATION
Brown Mountain
• old growth eucalypt forest
• wildlife / scenery
• Aboriginal history
• timber industry history

WARRUMBUNGLE N.P.
• scenery
• volcanic landscape
• Aboriginal history
• pastoral history

BLUE MTNS NATIONAL PARK
• scenery
• waterfalls
• cliffs
• early tourism history
• coal mining history
• Aboriginal history

FITZROY FALLS
• sandstone escarpment
• waterfalls / cliffs
• scenery
• early tourism history
• Aboriginal history

STURT NATIONAL PARK
• desert scenery
• wildlife
• rich Aboriginal history
• early pastoral history -
  station management

MOOTWINGEE N.P. /H.S.
• rich Aboriginal heritage
• outstanding scenery
• wildlife
• early pastoral history -
  station management

MUNGO NATIONAL PARK
• World Heritage
• Aboriginal heritage
• natural history
• outstanding scenery & wildlife
• early pastoral history -station
  management

KOSCIUSZKO NATIONAL PARK
• alpine / mountains
• World Biosphere Reserve
• scenery
• wildlife
• eucalypt forests
• Aboriginal history
• early grazing history
• Snowy Mountains Scheme

GREATER 3 DAYS

Area within 2 hour day tour
option from Sydney CBD

Area within half day tour
option from Sydney CBD

Area within full day tour
option from Sydney CBD

Area within two day tour
option from Sydney CBD

Area within three day tour
option from Sydney CBD

Area exceeding three day tour
option from Sydney CBD

Figure 10.3 Key regional destinations: key nature tourism destinations in regional NSW to be promoted nationally and internationally

Source: NSW National Parks and Wildlife Service (1997b)

Table 10.2 Potential environmental effects of tourism in protected areas: The types of negative visitor impacts that must be controlled

Factor involved	Impact on natural quality	Comment
Overcrowding	Environmental stress, animals show changes in behaviour	Irritation, reduction in quality, need for carrying-capacity limits or better regulation
Over-development	Development of rural slums, excessive man-made structures	Unsightly urban-like development
Recreation:		
Powerboats	Disturbance of wildlife	Vulnerability during nesting seasons, noise pollution
Fishing	None	Competition with natural predators
Foot safaris	Disturbance of wildlife	Overuse and train erosion
Pollution:		
Noise (radios, etc.)	Disturbance of natural sounds	Irritation to wildlife and other visitors
Litter	Impairment of natural scene, habituation of wildlife to garbage	Aesthetic and health hazard
Vandalism/ destruction	Mutilation and facility damage	Removal of natural features
Feeding of wildlife	Behavioural changes, danger to tourists	Removal of habituated animals
Vehicles:		
Speeding	Wildlife mortality	Ecological changes, dust
Off-road driving	Soil and vegetation damage	Disturbance to wildlife
Miscellaneous:		
Souvenir collection	Removal of natural attractions, disruption of natural processes	Shells, coral, horns, trophies, rare plants
Firewood	Small wildlife mortality, habitat destruction	Interference with natural energy flow
Roads and excavations	Habitat loss, drainage	Aesthetic scars
Power lines	Destruction of vegetation	Aesthetic impacts
Artificial water holes and salt provision	Unnatural wildlife concentrations, vegetation damage	Replacement of soil required
Introduction of exotic plants and animals	Competition with wild species	Public confusion

Source: WTO (1992: 14, adapted from Thorsell 1984)

Table 10.3 Environmental impacts identified in Australian national parks

National Park	Tracks and ORV's	Trampling (human or horse)	Weeds and fungi	Boats damage bank	Firewood collection	Human wastes	Camp sites	Water pollution
Western Australia								
Cape Range								
Stirling Range	*	*			*			
Northern Territory								
Kakadu	*			*				
Uluru	*	*						
South Australia								
Simpson	*	*	*		*			
Coongie	*				*		*	*
Flinders	*		*					
Coorong	*			*	*			
Queensland								
Carnarvon						*		*
New South Wales								
Mount Warning		*			*			
Kanangra Boyd	*						*	*
Blue Mountains	*	*				*	*	*
Ku-ring-gai								*
Kosciusko		*	*		*	*	*	*
Victoria								
Croajingolong	*							
Wilson's Promontory	*	*			*	*	*	
Tasmania								
Cradle Mt		*			*	*	*	*
St Clair								
Gordon River				*				
Southwest			*					

Source: Adapted from Buckley and Pannell (1990)

Changed water course	Water depletion	Disturbance to wildlife	Damage to archaeological sites	'Cultural vandalism'	Litter	Visual impacts: roads and buildings	Noise	Reference (see Buckley and Pannell 1990)
	*							Peerless n.d
						*		Brandis & Batini 1985
		*			*			ANPWS 1986b
*		*		*		*	*	Ovington et al. 1973
								ANPWS 86a
		*					*	SANPWS 1984
		*	*	*	*		*	Gillen 1988
	*				*	*	*	Williams et al. 1988
			*		*		*	SADEP 1984
			*	*				Pitts 1982
					*	*		NSWNPWS 1985
*					*			Brown 1988
					*	*		Brown 1988
*				*				Snelson n.d
					*	*		
								VNPWS 1985
	*				*	*	*	VNPWS 1987
					*			O'Loughlin 1988
								TDLPW 1985
								Cook 1985, Bayly-Stark 1985
								Neyland 1986

* = recorded in reference cited = observed by RB

A constructive approach towards a positive relationship between tourism and national parks calls for management of both the park resources and people. As noted in Chapter 6, resource management implies close monitoring of features of the environment in order to detect the rate, direction and character of change. In this way, remedial action can be taken before degradation reaches the point where the park environment becomes a source of dissatisfaction to visitors. Management, then, implies the maintenance of the park's resource base to enhance, or perhaps restore, satisfying recreation opportunity settings for visitors. These measures need to be complemented by efforts directed towards visitor management. A classification of visitor management strategies is provided in Table 10.4.

As noted in Chapter 5, some cynics have suggested that park management would be easy if it wasn't for the people. Certainly, the physical attributes of parks lend themselves to relatively straightforward procedures and technical and engineering-type techniques. With visitor management, a much more sensitive approach is required in coping with the many sources of conflict and manifestations of overuse. A good balance needs to be struck between regulation and modification of visitor behaviour, otherwise the benefits of tourism may be traded off through lost patronage resulting from regimentation of people in parks.

> The guiding principle for tourism development in national parks is to manage the natural and human resources so as to maximise visitor enjoyment while minimising negative aspects of tourism development.
> (World Tourism Organisation 1992: 12)

Manning (1979) distinguishes between *strategies* and *tactics* in the management of recreational use of national parks. Strategies are defined as basic conceptual approaches to management, setting out different paths to preservation of environmental and recreational quality. Tactics are defined as tools to carry out various management strategies. To clarify the distinction, a strategy, for example, might be a decision to limit recreational use. Within this basic strategy a number of tactics or tools might be pursued, including the imposition of fees, the requirement of permits or the erection of physical barriers. Manning classifies available strategies for park management into four basic approaches, each with a number of distinct sub-strategies. These are summarised in Figure 10.4; two of the strategies deal with supply and demand aspects of recreation and park lands, and two focus on modifying either the character of existing use to reduce adverse impacts, or the resource itself to increase its durability or resilience.

Typically, efforts are made to overcome 'people problems' by spreading the load in space and time. Manning's strategies offer four possibilities which are not meant to be mutually exclusive; nor are these approaches exclusive to parks management, but have application in other forms of recreation activity and tourism. Increasing the supply of park space may not be feasible in all situations, so that making better use of available space is the alternative. Redistribution of visitor pressure is an obvious option, as are measures to separate or limit uses at popular sites.

Table 10.4 Classification of visitor management strategies

Indirect strategies	Direct strategies
Physical alterations • Improve or neglect access • Improve or neglect campsites	**Enforcement** • Increase surveillance • Impose fines
Information dispersal • Advertise area attributes • Identify surrounding opportunities • Provide minimum impact education	**Zoning** • Separate users by experience level • Separate incompatible uses
	Rationing use intensity • Limit use via access point • Limit use via campsite • Rotate use • Require reservations
Economic constraints • Charge constant fees • Charge differential prices	
	Restricting activities • Restrict type of use • Limit size of group • Limit length of stay • Restrict camping practices • Prohibit use at certain times

Source: Adapted from Hendee *et al.* (1978, in Vaske *et al.* 1995)

Paradoxically, one way to approach the problem of excessive tourism pressure in national parks is to concentrate visitor use even more (e.g. see Hammitt and Cole 1991). Concentration of use can help control general site deterioration by attracting visitors to selected locations able to sustain high levels of use. Alternatively, additional opportunities for tourists can be created by diverting some visitors to underused sites, and by efforts to reduce seasonal or daily peaks in visitation through the use of incentives to extend operations into slack periods.

When essentially voluntary means of bringing about dispersal of use fail to achieve that objective, it becomes necessary to adopt a more direct approach to regulating visitor behaviour. Regulation of use implies some restriction over what tourists are permitted to do. Attempts at 'people control' come down to a choice between 'do' and 'don't' – the 'carrot or the stick'. Most park managers would be aware of the value of allowing the user to retain some sense of freedom of choice, and the role of interpretation in modifying of visitor behaviour is discussed presently. However, with certain management problems, such as vandalism, enforcement of rules, backed up by strenuous efforts at detection and punishment of offenders, may be the most effective means of control.

With regard to managerial directives, several of the tactics applied in resource management also require visitor regulation as a concomitant of site protection. The admonition to 'Keep off the Grass', for example, is clearly designed to bring

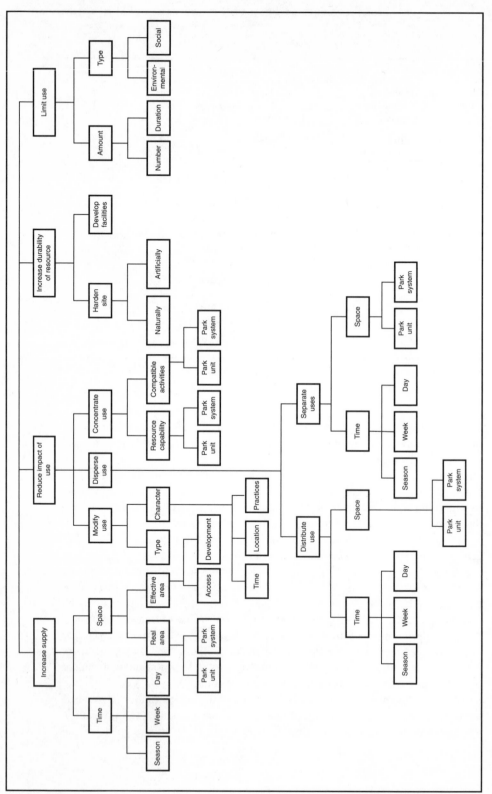

Figure 10.4 Strategies for national park management

"...but, Harold, there must be something we can do here...it's such a lovely spot..."

Figure 10.5 A question of interpretation?

about a particular response in use patterns, and thereby help maintain the condition of the site. Regimentation for its own sake, however, is indefensible and, in most instances, a positive approach is possible and preferable to an endless array of signs informing users of what they cannot do (Figure 10.5).

On-site control involves, among other things, site-hardening by way of trail establishment and maintenance and directions as to location, time and duration of park visits and activities, in order to attain the desired intensity of use for the area (Table 10.5). Among the most common procedures for visitor control are zoning

Table 10.5 Strategies and actions for reducing impact on campsite

Strategy	Possible actions
Reduce amount of use	Limit number of parties entering the area
Reduce per capita impact	
• Use dispersal	Persuade parties to avoid camping on highly impacted campsites
• Use concentration	Prohibit camping anywhere except on designated sites
• Type of use	Teach low-impact camping techniques
• Site location	Teach parties to use resistant sites for camping
• Site hardening/shielding	Build wooden tent pads on campsites
Rehabilitation	Close and revegetate damaged campsites

Source: Hammitt (1990: 21)

or scheduling, and rationing. These have the advantage, not only of limiting use, but of promoting dispersal of use, and reducing conflicts by separation of incompatible types of tourism activity, such as fishing and water-skiing.

Zoning involves the clustering of compatible uses in selected parts of a site. National parks are themselves a form of zoning, and certain areas within parks are designated for special purposes (for example, see Chapter 9). Different stretches of a river or lake can also be zoned for different uses, sometimes on the grounds of safety, or because the resource attributes do not lend themselves to all types of recreation, or simply to avoid mutual interference and maximise satisfaction between users. Spatial zoning is likely to be more successful where there is a logical and accepted basis for partitioning of the site. It is useful, too, if zone boundaries can be aligned to some natural or recognisable feature (e.g. different activities allocated to the opposite banks of a river).

Scheduling, or zoning by way of time limitations, is another useful procedure for visitor control in national parks. Recreation activities using the same site are allocated to specific time periods on an hourly, daily, weekly or seasonal basis in order to reduce conflicts and to ensure adequate rotation of use. The time-frame chosen depends upon the degree of conflict and the level of competing uses. A variation of time zoning, especially with linear resources, is the staggered scheduling of departure times of such activities as river tours. Ideally, schedules should be drawn up after consultation with user groups, and, if possible, tailored to fit normal recreational patterns.

Rationing refers to the mechanism through which opportunities to use designated recreation resources are distributed to users. Implementation of rationing assumes: that reasonable estimates of tolerance to use, or carrying capacities, can be established; that in the absence of rationing, use would exceed capacity at some sites; and that a reduction in use through rationing is the preferred management option (Grandage and Rodd 1981). Recreational use can be rationed by various means. Chubb and Chubb (1981) suggest three broad approaches:

1 Sharing the limited supply of recreational opportunities among potential participants by measures such as the issue of permits and licences; use of a reservation system; queuing and participation on a 'first-come, first-served' basis; or allocation of opportunities by lot.
2 Providing a limited experience to as many participants as possible, even though this may affect the quality of the experience (e.g. placing limits on the size of parties, or rotation or regulation of use to encourage high turnover).
3 Making it more difficult for people to participate. Use can be discouraged by restricting access; making reservations harder to obtain; insistence on merit standards (certain skills or knowledge) as the basis for entry; and the imposition of fees. Differential fee structures can also be used to redistribute recreational use patterns.

Any form of rationing discriminates to a certain extent against some participants and some forms of recreation. The use of fees or eligibility standards to apportion recreational opportunities has been criticised on the grounds of equity and effectiveness. Moreover, all rationing systems have shortcomings, so that, once again, monitoring of visitor behaviour and user preferences plays an important part in visitor management. Only through knowledge of what visitors do and want to do can appropriate management tactics be devised. Monitoring may be formal or informal, overt or covert, and it should be a two-way exercise concerned as much with the impact of the site on the visitor as how the visitor affects the site. The managerial process is complicated because it is not merely a question of ecological carrying capacity and over-use; social carrying capacity can also be involved, along with the various sources of conflict identified earlier.

To complement procedures for visitor regulation, management can seek to modify recreation behaviour by persuasion and the provision of information through interpretative services.

Information and interpretation

Interpretation communicates information and perception about a park, forest, structure, battlefield, or an entire region's distinctive features and influences. It helps visitors understand and appreciate the special natural and cultural resources, how these have influenced the way the region has evolved, and how its protection or disappearance will affect the area's future. The individual should thus enjoy a richer life through enhanced perception of landscapes and historical artefacts [or heritage].

(Knudson et al. 1995: xix)

Nearly 30 years ago, Wilcox (1969) wrote, in reference to America, that interpretation was a most challenging and provocative area of growth, with potential good for society in how it sparks imagination, and in how it can act as a tool to build a more meaningful life. Not much has changed.

The provision of an appropriate interpretation programme is an important supportive aspect of park planning and management, be it strategy or tactic. The main aim is to communicate to park users the objectives of management and the rationale for the various measures undertaken. In the long run, a sound interpretation policy may provide the key to resolving the dilemma between park preservation and use by developing in park users a deeper regard for national parks and a desire for a meaningful role in their care and management.

Interpretation has been described by Tilden (1977) as more than just instruction or communication of information. Tilden sees the chief function of interpretation as provoking and stimulating interest and awareness among visitors to a recreation site. This is to be achieved by revealing meanings and relationships in nature by reference to original objects, and by first-hand experience with common, easily understood examples and materials (for practical examples of interpretation plans and designs for trails, parks, historic and other sites, see Trapp et al. 1994; Veverka 1994; Knudson et al. 1995).

A second function of interpretation is to assist in accomplishing management objectives. It can do this by encouraging appreciation of the recreation environment and promoting public co-operation and responsibility in conserving recreational values. Much destructive behaviour results from ignorance rather than malicious intent, so that increasing the flow of information to the public is a preferable and probably cheaper means of reducing depreciative acts than prohibitions and censure (Lime and Stankey 1971). As Clark (1976) and Harrison (1977) explain, the key is often in pointing out 'why', when certain norms of behaviour are required:

- Why can't cars be driven off parking pads?
- Why can't tables be moved...?
- Why can't a tree be chopped down for firewood?
- Why can't initials be carved on benches or tables or trees?
              (Clark, 1976: 66; also see Beckman and Russell 1995)

Many people who visit museums, parks, forests and similar areas, welcome interpretation (Walsh 1991, in Knudson et al. 1995). In the dissemination of information about recreation opportunities, Jubenville (1978) makes the distinction between advice reaching the potential visitor before arrival at the site (Regional Information System) and information provided at the site (Area Information System). Prior information should reach the individual when choices concerning recreation participation are being considered. It is more often the task of government or regional organisations than the specific site manager, and can even involve zero, minimal or negative information, aimed at diverting attention or making heavily patronised sites less attractive. Thus, certain sites or facilities may be omitted from a map, or reference may be made to popularity and associated crowding; conditions which some will try to avoid.

Much more effort is directed positively into inducing desired patterns of behaviour on-site by increasing public awareness through publicity, education, interpretation and other less obtrusive methods or persuasion. Freedom of choice is seemingly not directly involved, yet the behavioural response sought is produced. Various means are available for transmitting information and communicating information between management and visitors. Typical approaches involve the use of maps and signposting, publications and brochures, electronic media and onsite contact by way of visitor information centres and guide services. These last methods for getting the message across are more often in the nature of interpretation than mere passive provision of information (see Trapp *et al.* 1994; Alderson and Low 1996).

On the basis that 'an informed public is a caring public', the Countryside Commission in Britain designed a number of self-guided trails around forests, farms, urban centres, ancient monuments and natural areas. The aim was to increase understanding and appreciation of these features, and thereby engender improved standards of behaviour and greater respect for the environment. Innovative interpretation schemes have been implemented in such countries as Australia (e.g. Beckman and Russell 1995; Hall and McArthur 1996), Canada (e.g. Graham and Lawrence 1990), and the US (e.g. Trapp *et al.* 1994).

A basic objective of recreation management is to provide a sustained flow of benefits for users. Concern for the quality of the visitor experience, then, is another justification for effective interpretation programmes.

> Increasing our contact with visitors can help them find out what the range of recreation opportunities and attractions is… recreational experiences may also be enhanced if visitors can be taught an understanding of the basic concepts of ecology and other outdoor values… By deepening their sense of appreciation and awareness of the natural environment, more recreationists could take better advantage of an area's recreation potential.
>
> (Lime and Stankey 1971: 181)

Care is needed in the selection and implementation of on-site interpretative methods. Mention has already been made of some of the means of communication available. Basically, the choice is between personal services (e.g. talks, demonstrations, guided tours and information services at a visitor centre) and self-directed services (e.g. printed materials, sign-posting, audio-visual media and the internet) (see Tilden 1977; Knudson *et al.* 1995). The advantages and disadvantages of such methods have been evaluated by several authors (e.g. Alderson and Low 1996). The choice of media will be influenced by philosophical and practical considerations pertinent to a particular site. The method selected should be capable of interpreting the recreation environment to the anticipated audience in an appropriate (exciting, interesting, comprehensible) fashion, as well as being

Table 10.6 Proposed protected area management guidelines

1 Involving local people in protected areas management
2 The preparation of protected area systems plans
3 Data management guidelines for protected area managers
4 Research tools for enhancing management of protected areas
5 Extending the benefits of protected areas to surrounding lands
6 Using protected areas to monitor global environmental change
7 Management of protected areas by private organisations
8 The application of an international review system for protected areas
9 Monitoring management effectiveness and threats in protected areas
10 Establishing and managing genetic resources
11 Integrating demographic variables in the planning and management of protected areas
12 Applying science to the establishment and management of protected areas
13 Effective management of marine protected areas
14 Expanding the world's network of protected areas
15 Effective management of transfrontier protected areas

Source: Harrison (1992: 23)

reliable, flexible, compatible with other media and reasonably vandal-proof (New Zealand National Parks Authority 1978).

Interpretation is more than mere mechanics, and a certain amount of caution is called for in the implementation of an interpretation programme. Too much interpretation can be counter-productive and destroy the sense of spontaneity and discovery in recreational activities. Participation can become 'over-programmed', and people may resent what they perceive as attempts at 'brain-washing' and efforts to force them into designated modes of use and enjoyment. Managerial attitudes can also intrude: elitist overtones and pre-conceived obsolete notions of what constitutes acceptable patterns of recreation behaviour (or of deviance), can distort the orientation of interpretation initiatives.

All procedures aimed at recreation visitor management, whether direct regulation or indirect modification of user behaviour, involve some loss of freedom. Some trade-off is required between freedom of choice and the adequacy of the resource base to meet the requirements of users and the objectives of management. However, positive manipulation of the physical and social environment to create and enhance opportunities for tourism and recreation, is surely preferable to reliance on negative forces or congestion, frustration, dissatisfaction and ultimately self-regulation, to produce their own solution.

Management guidelines

The International Union for the Conservation of Nature has proposed guidelines for the management of protected areas (Harrison 1992) (see Table 10.6). These guidelines arose out of the World Parks Congress in Caracas and will form the basis for a series of documents directed towards protected area managers. At least two of the guidelines call for further consideration:

- involving local communities in park management;
- managing park boundaries and transboundary zones (discussed earlier).

Community-related management of parks

The notion that local communities should be involved in park management follows logically from the view that parks should not exist in isolation from their surroundings, but should be considered and managed in relation to them. 'Parks and neighbouring communities are bound together by their individual actions into a collective future.' (Field 1997: 425). Unfortunately, parks and their neighbours are often in an adversary situation, as evidenced by the potential for conflict noted earlier in this chapter. It is far preferable for the relationship between parks and protected areas to be viewed as an asset, offering positive and productive benefits (Renard and Hudson 1992). The benefits of community-based approaches to park management include:

- building popular support for parks;
- addressing concerns of communities affected by parks;
- ensuring that benefits from parks reach local communities;
- supplementing public funds and personnel needs in park management;
- integrating community knowledge of park resources into management;
- being responsive to variations in social and environmental conditions; and
- providing training and opportunities for skill development for communities to participate in park management (Renard and Hudson 1992: 4).

For these benefits to be realised, meaningful partnerships between the park and the local community must be forged, reflecting the rights, aspirations, knowledge, skills and resources of that community. Co-management is based on:

- participatory planning, collaboration and shared responsibility;
- access to information;
- appropriate institutional arrangements; and
- sound legal, technological and financial support.

Advances along these lines are being made, with significant input to park management from local indigenous people. In Australia's Uluru and Kakadu National Parks, for example, local Aboriginal communities form a growing component of park management personnel, contributing important elements of their knowledge and culture to interpretation programmes, especially those of spiritual significance. Tours, trails and brochures reflect Aboriginal themes, and offer authentic aspects of the Aboriginal lifestyle by way of handicrafts and 'bush tucker' (native foods). Aboriginal participation in park management, in turn, gives the local communities a much greater sense of control and self-determination over environmental and conservation issues (Absher and Brake 1996).

Closer to the cities, partnerships can be built up between urban populations and neighbouring national parks. Once again, the inevitable interaction between residents, tourists and protected areas adjoining cities, is the rationale for colla-boration. In Britain, the Countryside Commission (1992) is active in promoting links between city communities and neighbouring parks. An experimental project run in conjunction with Birmingham City Council demonstrated the reciprocal benefits of the scheme – for the national parks in increasing awareness of what they had to offer, and for the city in opening up opportunities to experience the rural environment.

In Australia, guidelines have been drawn up for the development of viable, yet environmentally sensitive tourism facilities near major natural areas, including national parks (NSW Department of Planning 1989). The guidelines apply particularly to the NSW North Coast, an area of outstanding potential for natural-area tourism, but where population pressure and expanding urbanisation could easily intrude on those areas and erode their natural appeal. Raising awareness among urban dwellers and tourism developers of the special qualities of the region's natural areas should help build bridges between the management of protected areas and urban populations close by, as well as tourists. The outcome should be tourism developments which complement and enhance the values of the adjacent natural areas.

Whereas the guidelines relate specifically to tourist facilities, the attraction of which depends, in part, on their proximity to natural areas, similar thinking could apply to tourism development within parks, including development by the private sector. Natural-area protection has long been considered the sole responsibility of government and the public sector. Increasingly, financial stringency has raised the possibility of a role for private enterprise in natural-area management, if not in privatised national parks, at least in providing some of the facilities and activities of management required to service park visitors. Such a prospect raises serious questions about the respective roles of the public and private sector, and whether a balance can really be attained between nature protection and profit (Charters *et al.* 1996). Certainly, scepticism surrounded a recent proposal in the US, to allow a private developer to build and operate a new visitors' centre at the historic military park in Gettysburg, Pennsylvania, at no cost to the Park Service. However, the private tourist developments described in the Australian studies suggest that commercial operations can coexist compatibly with nature conservation, even in environmentally sensitive areas used for ecotourism. At the same time, an enhanced role for the private sector in park management will need a fundamental shift in the attitudes of government agencies and environmental interests before the opportunities for collaboration can be seriously addressed.

No longer is concern for the natural environment the preserve of public agencies. Increasingly, landholders in the private sector are taking up the challenge of setting aside portions of their land for nature conservation. Future management of parks and protected areas may well proceed in tandem with the private sector and community groups, in a collaborative effort to ensure sustainability and biological diversity of the natural environment.

Guide to further reading

- *Tourism and recreation management in national parks and reserves*: Myers (1972); Myers (1973); Black and Breckwoldt (1981); Boden and Baines (1981); Kelleher and Kenchington (1982); Edington and Edington (1986); Bateson *et al.* (1989); Olokesusi (1990); Eagles (1996).
- *Interpretation*: Two excellent and widely sourced works are Tilden (1977) and Knudson *et al.* (1995).
- *Economic, physical and social impacts in national parks*: Darling and Eichhorn (1967); Wall and Wright (1977); Jefferies (1982); McDonald and Wilks (1986a, b); Landals (1986); McNeely and Thorsell (1989); Buckley and Parnell (1990); McIntyre and Boag (1995); Liddle (1997).

Review questions

1 What are the important principles governing the declaration of national parks in a country of your choice? How is a national park defined in that country? Is the management of national parks in that country in line with IUCN definitions?

2 What are the major environmental impacts that threaten national parks? Have these impacts been addressed by parks management? If so, how? If not, why not? In your answers, make reference to case studies of one or more national parks.

3 Should national parks authorities be permitted to charge entrance/user fees? Explain your answer with reference to case studies.

4 Describe some of the main political issues arising out of recreational and tourist use of national parks both generally, and with specific reference to one or more case studies.

5 What are some of the fundamental differences in land ownership, and recreational use and management of national parks in the UK, the USA and Australia?

11

OUTDOOR RECREATION AND TOURISM

Tourism is an important human activity of physical, sociocultural and economic significance. As noted in Chapter 1, discussion of tourism in the context of outdoor recreation is logical. Much tourism is recreational, in that tourist activities are engaged in during leisure time, commonly outdoors, for the purpose of pleasure and personal/group satisfaction. Similarly, outdoor recreation overlaps with tourism in the distinctive characteristics and behaviour associated with each.

This chapter examines key tourism concepts, the significance of tourism, the factors influencing tourism patterns and processes, and tourism resources, landscapes and systems. The tourism opportunity spectrum (Butler and Waldbrook 1991) is presented and discussed in light of (1) its contribution to tourism planning, and (2) its explicit recognition of the close relationship between outdoor recreation and tourism planning and development.

Concepts and definitions

Definitions of 'tourist' and 'tourism' are many and varied, but most incorporate the notions of distance travelled, and duration and purpose of travel. Certainly, the term implies more than the French derivations – *tour* (a circular movement) and *tourner* (to go around).

Broadly speaking, anyone who visits an area other than the place of residence is a tourist. However, diversion, or the pleasure motive, is frequently seen as an essential element, and allowance is made for the time and distance involved in travel, and the duration of the visit. The picture is further confused by distinctions between domestic and international tourism, and the different definitions among countries and areas within countries.

According to the World Tourism Organisation (WTO 1993a), the definition of a 'tourist' has three dimensions. Clear distinctions are made between 'visitors', 'tourists', and 'same-day visitors':

A *visitor* is 'any person who travels to a country other than that in which s/he has his/her usual residence but outside his/her usual environment for a period not exceeding 12 months and whose main purpose of visit is

226

other than the exercise of an activity remunerated from within the country visited'.

A *tourist* is 'a visitor who stays least one night in a collective or private accommodation in the country visited'.

A *same-day visitor* is 'a visitor who does not spend the night in a collective or private accommodation in the country visited'.

A range of definitions have also been applied to the terms 'tourism' (see Chapter 1) and 'tourist industry', with some writers arguing that universal definitions for each of these terms will never be developed, and that there is no readily definable tourist industry. The WTO (1994) argued that 'Tourism comprises the activities of persons travelling to and staying in places outside their usual environment for not more than one consecutive year for leisure, business and other purposes'. However, tourism should be viewed in broader, more theoretical terms. Tourism is a form of human behaviour (Przeclawski 1986), and 'a category of leisure with special significance in individuals' total leisure patterns' (Simmons and Leiper 1993: 204). According to Leiper (1995: 20),

Tourism can be defined as the theories and practices of travelling and visiting places for leisure-related purposes.

Tourism comprises the ideas and opinions people hold which shape their decisions about going on trips, about where to go (and where not to go) and what to do or not to do, about how to relate to other tourists, locals and service personnel. And it is all the behavioural manifestations of those ideas and opinions.

Leiper's definition is, at least conceptually, somewhat closely aligned with Jafari (1977: 8), who stated that 'tourism is the study of man away from his usual habitat, of the industry which responds to his needs, and of the impacts that both he and the industry have on the host, sociocultural, economic and physical environment'. Pearce (1987: 1), too, presents a robust conceptualisation of tourism, stating '... tourism may be thought of as the relationships and phenomena arising out of journeys and temporary stays of people travelling primarily for leisure or recreational purposes'.

Clearly, defining tourism and tourists is problematical, while definitions themselves are often subject to debate and argument. This situation, however, should not be viewed in a negative light, because debate and argument do aid the development of theoretical and applied research. It is perhaps also worth noting that tourism, much like the study of leisure and outdoor recreation, has only recently received academic and wider social credibility (also see Chapter 1). One of the reasons for the turnaround in tourism's acceptance as a critical aspect of people's way of life, is its economic significance. The following discussion examines the growth in tourism on a global scale, and explains some of the forces underpinning that growth.

Global tourism

Tourism has become one of the largest (if not *the* largest) single items in world trade. From 1950 to 1972, annual tourist arrivals in all countries grew from 25 million to almost 200 million, an average growth rate of about 10 per cent per year. In the same period, total foreign exchange earnings from tourism rose from US$ 2.1 billion to US$ 24 billion, an average annual increase of about 11 per cent. By 1976, the number of global visitor arrivals was estimated at 220 million, an increase of more than 90 per cent in a decade, while travellers spent, in all, about US$ 40 billion. Arrivals had grown to 264 million by 1978, and expenditures to around US$ 63 billion in that same year (WTO 1979).

Tourism remains one of the highest industry growth areas of the 1980s and 1990s, in terms of both expenditure and foreign currency generation. Since 1950, international tourism activity has risen at a rate of about 7 per cent per annum in terms of international visitor arrivals, and by around 13 per cent per annum in terms of international receipts (WTO 1993b). In 1996, global tourist arrivals increased by 4.5 per cent to 592 million, and world tourism receipts, excluding air fares, increased by 7.6 per cent to US$ 423 billion (WTO 1997a). According to several estimates, and depending upon how it is defined, tourism has become the world's largest business enterprise, overtaking the defence, manufacturing, oil and agriculture industries. The worldwide gross output for tourism in 1992, was US$ 3.2 trillion (about 6 per cent of the world's gross national product), with employment encompassing 127 million people (Lundberg *et al.* 1995).

It is predicted that international tourist arrivals will increase at an average rate of 4.1 per cent, to reach 702 million by the year 2000 and 1.018 billion by the year 2010. At this rate of growth, the tourism industry will generate around 11.4 per cent of global Gross Domestic Product (GDP) in 2005, and employ some 338 million people (or account for one in every eight jobs) (World Travel and Tourism Council (WTTC) 1995).

Trends in world tourism since 1950, reveal a heavy geographical concentration of both tourist arrivals and tourism receipts. Approximately three-quarters of world tourism is intraregional travel, with Europe dominating international tourist arrivals (59 per cent) and world tourism receipts (more than 50 per cent since 1960). On a country-specific basis, 'The United States has been the largest recipient of tourist travel income for several years, and will probably continue to be, receiving more than twice as much international tourism income as its nearest competitor, France' (Lundberg *et al.* 1995: 8).

Distance and costs are constraints to travel, especially international travel. However, some countries of the Asia and Pacific regions, notably China, Japan and Korea, recently emerged as potential areas for tourist development, and as sources of international visitors. The rapid growth of tourism in the Asia and Pacific regions, particularly since the early 1980s, has been attributed to the increasing numbers of intraregional tourists (Mak and White 1992; Forsyth and Dwyer 1996). However, the recent financial and political turmoil in Asia and the Pacific has had an immediate impact on tourist travel to such countries as Australia.

For instance, in early 1998, Ansett Airlines was forced to cancel its twice-weekly flights to South Korea, a formerly very lucrative tourist market. More than 70 per cent of seats on the Seoul to Sydney route were vacant in late 1997 and early 1998. Ansett's management estimated that:

> Ansett would lose $16 million over the coming [1998] year if it continued flights... The announcement comes just four days after Qantas – the largest carrier in South-East Asia – also announced the suspension of its four weekly flights to Seoul...
>
> Tourism numbers from Asia are still expected to grow by 3 per cent, but this is much lower than the strong growth expected before the onset of the Asian currency crisis.
>
> The [Australian] Government's Tourism Forecasting Council has predicted that Australian tourism to Asia, excluding Japan, would take six years to return to 1996 levels.
>
> (Stott 1998: 2)

In brief, Western Europe and North America will continue to dominate world visitor arrivals and tourism receipts, and indeed there have been unexpected increases from these markets to such countries as Australia in 1997–98. This increase could be attributed to reductions in the value of the Australian dollar, intensified international marketing and promotion, and the staging of large-scale events. The East Asia and Pacific regions will be major tourist growth regions in terms of international travel and tourism development, but their recovery is unlikely to be speedy, and their subsequent growth is likely to be much slower and more cautious than mid/late 1997 forecasts.

Factors influencing tourism patterns and processes

Numerous factors have influenced the patterns and processes of international and domestic tourism growth, and, in some areas, decline, as well as more specific world and regional trends. These factors include industrialisation; freer trade (as policies of high and extensive protectionism are abandoned); widespread growth in wealth and leisure; increased environmental awareness; growing conflict among competing resource users; ageing populations; the ease and increased speed with which people can travel further; and changes in employment structures (e.g. decline in agricultural employment).

Production processes have become increasingly integrated across national boundaries (OECD 1990; Sorensen and Epps 1993; Fagan and Webber 1994; Lane 1994), the significance of multinational corporations is increasing, 'the pace of change in the direction and composition of world trade has quickened' (Fagan and Webber 1994: 26), and the international mobility of financial capital and people continues to escalate. In brief, the patterns of economic development for urban and rural areas have changed, and therefore, so too, have the ways in which communities operate in order to adapt and survive (e.g. for rural areas see Sharpley and Sharpley 1997; Butler *et al.* 1998; for urban areas see Law 1993; Page 1995; Williams 1995).

People's expectations of their visits to urban and rural areas are changing as greater emphasis is given to the conservation and maintenance of natural and cultural heritage, including the rights of indigenous people. At the macroeconomic level, the patterns of change in economies reflect global pressures towards the convergence of policies that are driven largely by the power of global financial markets and their policy preferences. Microeconomic reform has also been forced on countries and regions by changes in the world economy, with responses being largely domestic policy choices. The major forces of global and regional change, both generally and more specifically with respect to tourism, can be categorised as social, geopolitical, economic and technological (e.g. see Hall 1994; Wahab and Pigram 1997). The following discussion examines these forces in more detail. Emphasis is given to how socioeconomic and political conditions; technological developments; increased environmental awareness; altered tourist tastes and behaviour; and other forces, have served to alter the nature of tourism demand and supply.

Socioeconomic

Socioeconomic factors significantly influence recreational and tourist decision-making and behaviour, because of their effects on personal income and time, attitudes to, and perceptions of the world (i.e. world views), consumption patterns and, ultimately, tourist demand. World population has been estimated to increase at the rate of 170 persons per minute (Villeneuve 1991: 19, in Gartner 1996: 13), with the highest growth rates occurring in developing countries and accounting for around 95 per cent of that rate (Godbey 1995, in Gartner 1996: 13).

The economic recession of the 1990s, and the restructuring of many national and local economies, gave considerable light to tourism as an economically significant industry, providing a means of generating international trade links, foreign investment, industrial diversification, income and employment. Many economies are moving, or have moved, away from a dependence on primary industries such as agriculture, mining, gas or oil, towards a greater reliance on the industrial and service sectors. These moves have been significantly influenced by the globalisation of economies and the spread of multinational corporations seeking tax incentives, subsidies and lower wages to offset the costs of producing goods, and to increase their market penetrations and shares.

In wealthier countries, the period following World War II has been one of generally rising per-capita incomes and diminishing rates of population increase. Reductions in the size of the public sector, and concomitant reductions in expenditures with macroeconomic restructuring, have resulted from a public sector focus on market-led recovery and economic efficiency as a precursor to social welfare. Governments have been prompted to encourage private-sector investment in tourism projects such as casinos, waterfront development, coastal resorts and ecotourism, while much greater emphasis and resources have been given to international marketing and promotion to increase international visitor numbers and receipts.

Average life expectancy has risen substantially since the early 1900s, with increases of 20 years or more in developed countries. The ageing of Western populations, along with increased affluence and leisure time, and the desire of an increasing number of 'older persons' to remain active, have influenced recreation and travel patterns (also see Chapter 3). Towards the other end of the age spectrum, youth tourism, stimulated by education and other influences, is a growing market segment. Other developments include changing attitudes and patterns in marriage, child bearing in families, and less discrimination on the basis of sex, age and race. All of these developments have influenced travel and tourism patterns and processes, and, therefore, outdoor recreation demand and supply.

General and tourism-specific planning approaches, too, encourage more wide-spread public participation, which, in turn, affects awareness and perceptions of, and attitudes to, tourism. Relationships between hosts and guests, between the tourist industry (however defined) and local communities, and within local communities, are just three areas where this development has had significant implications in furthering the integration of tourism in local economic and social development, and in raising awareness of tourism's potential contributions (but, to a lesser extent, costs) to local communities.

An additional point is warranted here. The whole arena of tourism–community relationships requires much greater research. Pearce *et al.* (1996: 27–8) identified three major problems in the study of such relationships:

1 There are definitional and measurement problems with the concepts of tourists, tourism and community. In particular, there has been no attempt to investigate systematically the nature of the tourism phenomenon, and particularly the dimensions on which types of tourism and tourists differ. Nor has the concept of what constitutes a community been considered carefully by researchers.
2 The description and perceived impacts of tourism have provided a limited view of the nature and content of host perceptions of tourism.
3 There is a lack of sound theory, with much research being atheoretical in nature.

Clearly, despite the much cited importance of the socioeconomic aspects of tourist planning and development, there is some way to go before we can point to detailed critical knowledge and sound theoretical understandings of the relationships between tourism and the social environment.

Political

Politics affects travel patterns and processes. People's desire and ability to travel is affected by government ideology, policy and legislation. The destination appeal of countries may be reduced by political upheaval or general instability, stemming from such factors as civil war, protests and human rights. Governments may prohibit or limit the outbound travel of their people (e.g. Korea prior to 1989), or may prohibit or limit the inbound travel of people from 'certain other countries, because of fears of foreign ideologies and political values, or because of diplomatic disagreements' (Hall 1997: 3).

Some of the most dramatic events to impact on tourism have occurred in global political geography, a sphere in which great changes have occurred during the last three decades, including:

- global reductions in trade barriers;
- reduced barriers for individuals and corporations, namely multinationals, to enter countries;
- increased access to countries of Central and Eastern Europe following the collapse of the Soviet Union;
- the removal of the Berlin Wall and the reunification of Germany;
- the establishment of the European Community and a common currency, and relaxation of controls on visitors between member nations;
- the dramatic effects of the Gulf War in tandem with global economic recession sparked a decline in global tourist arrivals between 1990 and 1991;
- recent reforms in Vietnam have resulted in cuts to state subsidies, reduced centralised planning, rationalised exchange rates, growth in limited private enterprise, and liberalised foreign investment. Vietnam's economic growth potential places the country under a spotlight, as a serious contender in South-East Asia, both in respect to tourism and other economic pursuits;
- the opening up of China to the world;
- the encouragement of Japanese by their government to travel abroad;
- reductions to barriers for outbound Korean travellers;
- violence perpetrated on tourists for political and other gains (e.g. on 6 January 1998, the *Sydney Morning Herald* reported that 'Australians should not be deterred from travelling to Cambodia because of recent Khmer Rouge threats to kidnap and decapitate foreigners, the country's ambassador to Australia said yesterday... [Yet] The Khmer Rouge has offered about $13,000 to any of its rebels who take a foreigner hostage');
- the industrialisation of Asian and Pacific nations (Hall 1994; Butler 1995; Pigram and Wahab 1997).

Changes in global geopolitics can work negatively and positively for the development of tourism, and can present an opportunity to redirect the growth of tourism down the path of sustainability. Simultaneously, these changes contribute to the reconstruction of economies, societies and degraded environments (Wahab and Pigram 1997) in different ways. Government responses in industrialised nations generally focus on regional diversification, including the development of service-oriented industries, such as tourism.

Technological

Technological developments have had a very significant influence on travel decision-making and experiences. Transport developments in the air, rail and road sectors, allow for greater accessibility of destinations with greater speed and comfort, and generally at lesser cost, especially if travel time is costed in. Improvements in information technology (e.g. computer reservation systems) have substantially impacted on the travel distribution network, and therefore the supply and consumption of the tourist product. Indeed, Poon (1993: 13) argues that information technology:

> facilitates the production of new, flexible and high-quality travel and tourism services that are cost-competitive with mass, standardized and rigidly packaged options... It helps to engineer the transformation of travel and tourism from its mass, standardized and rigidly packaged nature, into a more flexible, individual-oriented, sustainable and diagonally integrated industry.

Tourists are now more discerning, and the simplistic notion of tightly controlled tourist packages, which allow little exploration or flexibility on the traveller's part, is being challenged as tourists seek more authentic nature and culture-based experiences. Planning for a diverse range of tourist opportunities, and recognition of the diversity of tourists themselves, requires a detailed understanding of the above issues and the resource base for tourism, as well as the development and application of appropriate planning approaches. The following sections examine the resource base for tourism, and introduce the concept of the Tourism Opportunity Spectrum as a framework for supplying diverse recreational and tourist opportunities.

The resource base for tourism

The complex pattern of tourism across the globe reflects the diversity of environments which constitute tourist resources, and the varied experiences which travellers seek. A common element is the contrast between the home region and the destination. If there were no perceived difference from place to place (natural or fabricated), tourism would not exist. Contrasts may be sought and discovered in the physical environment, the cultural and historic landscape, the people, artificially created attractions, and festival and events.

Notwithstanding the above discussion on socioeconomic, political and technological forces, perhaps of all the factors affecting the development of tourism, the most important are physical. Some of the strongest flows of tourists are from cool, cloudy regions, to places highly regarded for their warm, sunny climate. For many tourists, 'wanderlust' appears to take second place to 'sunlust' (see below). By contrast, the popularity of winter tourist resorts rests in great part on cool (though hopefully sunny) weather and the assurance of adequate and long-lasting snow cover.

Yet, one of the factors little considered in tourism literature, is the potential impact of climate change. For instance, studies on the impacts of climate change, due to an enhanced greenhouse effect, on the snow pack in Australia, suggest that climate change would increase the frequency of winters with little natural snow (Haylock *et al.* 1994 and Whetton *et al.* 1996, in König 1998).

Climate change due to an enhanced greenhouse effect is predicted to have the biggest impact on the Australian ski industry, and the highest resorts with the best natural snowfalls and the best conditions for snowmaking. This would create 'two classes' of resorts: (1) smaller resorts at lower altitude, which will lose their downhill ski operation first; and (2) larger resorts, at high altitude, where downhill skiing remains possible. However, in the long run (assuming a worst-case climate scenario for 2070), none of Australia's resorts will be snow-reliable (König 1998).

Another physical factor with obvious implications for tourist development is the appeal of the coast. Mercer (1972) explains the coastal location of many resorts in terms of the attraction of edges or junctions in the landscape – the coastline representing the interface between land and sea. The success of coastal resorts reflects the attraction of the beautiful setting, however, even away from the coast, the physical terrain holds great appeal for tourists.

One component of the physical environment which has more limited significance for tourism is the presence of mineral springs or spas. In historical times, conviction in the medicinal properties of mineral waters for drinking or bathing, stimulated the earliest visitors to places like Bath and Tunbridge Wells in Britain, and Spa, itself, in Belgium. Despite advances in modern medicine, 'taking the waters' at spas and similar health resorts continued to attract a considerable clientele. Increasingly, however, with the development of additional facilities close by for amusement and diversion, the function of spas became as much, if not more, social than therapeutic. One health resort in the United States, French Lick, in southern Indiana, even became the focus for thriving illegal gambling and liquor activities in the 'prohibition' era. However, tourism for health purposes remains important for many people, so that clinics and sanatoria continue to attract significant numbers of patrons.

A related phenomenon with implications for tourism, is the drawing power of religious shrines, like Lourdes in France and Knock in Ireland, based in part, on beliefs in the miraculous powers of water from local springs, which had their origin in visions last century. Spiritual reasons have always been a powerful stimulus to travel, and large numbers of pilgrims continue to visit Mecca and other Moslem holy places annually. Religious centres such as the Vatican, Jerusalem and Benares also attract pilgrims in large numbers.

Many tourists are genuinely interested in foreign places and people, so that aside from the physical environment, the opportunity to make contact with other people's culture and way of life is a strong influence on tourism. The appeal of traditional architecture, folklore, unusual customs, crafts and foods, is well-documented. Not all of these are authentic, and there is considerable potential for tourism to distort the cultural tradition of host communities.

Interest in past cultures is also the basis for historical tourism, whereby the primary focus is on inspecting the legacy of a bygone age. Features of historic interest have a proven fascination for tourists, whether these be the magnificent homes and castles of Britain and Europe, artefacts and ruins of the ancient world, sites of military battles, picturesque villages mirroring a past lifestyle, restored railways and steamships, or the collections of miscellaneous junk which pass for museums in some small, isolated settlements in outback Australia. Countries with a relatively short history (e.g Australia and New Zealand), often find it more practical and rewarding from a tourism perspective to re-create features and settlements of the past, and present these in something of an outdoor museum setting. Thus, Old Sydney Town portrays life in the First European Settlement in Australia, and Sovereign Hill promotes itself as a re-creation of one of the early goldmining towns in the State of Victoria. Historical theme parks also flourish in the United States, where attractions like Knott's Berry Farm and Disneyland in Los Angeles rely to a great extent on revivals of the past.

Clearly, tourism, nostalgia and culture can have a mutually beneficial relationship; interest in history stimulates tourism, which, in turn, makes historical (heritage) preservation possible. Handled correctly, preservation certainly pays in terms of tourism. As Newcomb (1979: 232) puts it: 'Our visible past is like a fire which... if we tend it carefully... will illuminate our pleasure and... touch our imagination and our hearts'.

Tourism systems

Tourism involves tourists themselves, the regions of tourist origins, the destination region and the linkages in-between. Some writers have suggested that a systems framework is the most suitable means of drawing these facets together for study.

Leiper (1979; 1981) proposed an open system of five interacting elements, encompassing a dynamic human element – the *tourists*; three geographical elements – the *generating region*, the *transit route* and the *destination region*; and an economic element – the *tourist industry* (see Figure 11.1). Leiper's model recognises that the central element of the system is people – the tourists themselves. They comprise the energising source, and their attributes and behaviour help define the role of other elements in the system. The generating region is the origin of potential tourist demand, linked by transit routes to the destination or focus of tourist activity. Subsumed within these three geographical elements are the industrial component and service infrastructure of tourism, comprising all the firms, organisations and facilities that are intended to serve the specific needs and wants of tourists, before departure, en-route and at the destination(s).

Leiper (1995: 26) suggests 'that any of the five elements can be used as a focal topic. Studying tourists involves considering tourists in relation to the other four elements... Studying places as tourist destinations involves considering that element in relation to the other four, and so on' (see Figure 11.2). In this way,

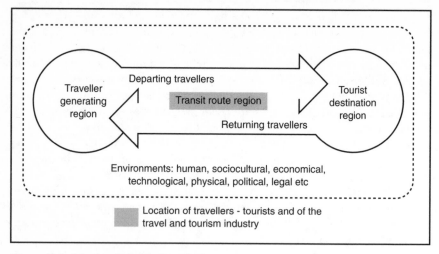

Figure 11.1 A basic whole tourism system

Source: Leiper (1995: 25)

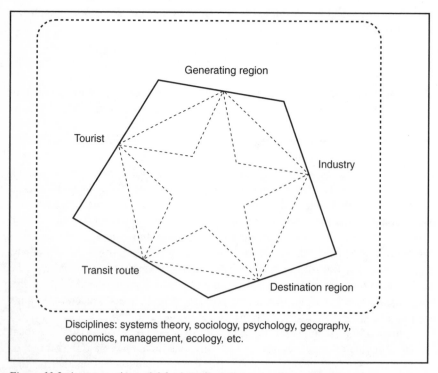

Figure 11.2 A systematic model for interdisciplinary tourism studies

Source: Leiper (1995: 26)

Leiper presents a means for the enhancement of tourism knowledge, and a basis for good scientific theory. He states:

> Evidence about tourism comes from looking at the elements (tourists, places, organisations), at their interaction with one another and with environments, and from making observations using appropriate techniques from a range of disciplines. This is balanced by the systemic structure explicit in the model for interdisciplinary tourism studies.
>
> (Leiper 1995: 27)

Leiper's model can be criticised, in that it might be argued that the destination region and its distinguishing characteristics should receive more prominence, and the generating region less. Obviously, the latter is the scene for a good deal of advertising and promotional activity, designed to stimulate tourism. Market research has also been directed to discovering what it is about the environment at the origin which helps generate an exodus of tourists. Unfortunately, however, tourists 'at home' are largely indistinguishable from the rest of the population, and even if they were distinguishable, their presence and often humdrum everyday existence, holds no special significance for the generating region. The destination, on the other hand, receives and reflects the full impact of the influx of visitors. This is where most tourism studies have been directed, and rightly so. Leiper (1979) concedes that it is the destination region where the most significant and dramatic aspects of tourism occur. Its attractions and facilities are essential to the tourism process, and it is the location of many of the important functional sectors of the tourist industry. To a significant degree, the environment and landscape of the destination region could be said to be an index of all the positive and negative features of modern tourism.

Tourist motivation

The subject of tourist motivation involves questions about why people travel. However, identifying clearly the relationships between an individual's motivations and selection of a destination is a difficult task. Krippendorf (1987), for instance, identified a number of tourist motivations, including:

- recuperation and regeneration;
- compensation and social integration;
- escape;
- communication;
- broadening the mind;
- freedom and self-determination;
- self-realisation;
- happiness.

Collectively, these motivations reflect that 'the traveller... is a mixture of many characteristics that cannot be simply assigned into this category or that one' (Krippendorf 1987: 28).

Tourist motivations have occupied an important place in tourism literature. One of the most widely cited publications on tourist motivation was that of Gray (1970), who presented two basic reasons for pleasure travel – 'wanderlust' (people's desire to leave familiar surroundings and experience things exciting and different) and 'sunlust' (seeking out places that have better attributes for specific purposes than are available locally, and which may literally mean a 'hunt for the sun') (Pearce 1987: 2). 'Wanderlust may be thought of essentially as a "push" factor whereas sunlust is largely a response to "pull" factors elsewhere' (Pearce 1987: 22) (also see Chapter 2 for a more detailed discussion of recreation motivations and choice).

The predominant approach to the study of tourist travel motivation has been to attempt to characterise 'push' factors as determinants of travel behaviour, such factors typically being conceptualised in terms of needs. For instance, the role of escapism was central to the work of Dann (1976), who argued that fantasy motivators form an important element of travel demand, and demonstrate its individualistic nature. As Leiper (1984) similarly argues:

> all leisure involves a temporary escape of some kind... tourism is unique in that it involves a real physical escape, reflected in travelling, to one or more destination regions where the leisure experiences transpire... A holiday trip allows changes that are multi-dimensional: place, pace, faces, lifestyles, behaviour and attitude. It allows a person temporary withdrawal from many of the environments affecting day to day existence.

The discussion of tourist motivations and behaviour, though brief here, clearly emphasises the critical role of market segmentation in tourism planning and development. Tourist areas cannot be all things to all people. They require careful planning and management, while providing appropriate resources and facilities for those travellers destination regions wish to attract. One means of developing appropriate resources, landscapes and facilities is the Tourism Opportunity Spectrum, derived from the Recreation Opportunity Spectrum (see below).

Tourism landscapes

Reference was made earlier to the landscape of tourism, not so much in the sense of attractive scenery, but in the association of distinctive physical and cultural features characteristic of tourist development. Used in this way, the term is analogous to agricultural or residential landscapes. The landscape of tourism reflects the imprint, both good and bad, of mass travel on the environment, and the relationship is inescapable. The landscape makes tourism and, in turn, tourism makes the landscape.

Given the diverse nature of resources and experiences which appeal to travellers, the range of recipient landscapes created for, and emanating from, tourism is wide. The natural beauty to be found in the west of Ireland, the glittering facade of Las Vegas, or the simulated atmosphere of the South Seas re-created in Hawaii, all represent particular landscape types orientated to tourism. Whereas it is easy to deplore the 'look-alike' landscapes spawned by mass tourism across the globe (Eckbo 1967), it is another matter to attempt to interpret and explain their evolution from a generic point of view (Price 1980; 1981). Some interesting work has been carried out on the townscapes of tourist destinations.

Lavery (1974) outlined the historical background to the development of holiday resorts in Western Europe (also see King 1997), and, in particular, alpine resorts, spas and seaside resorts. He proposed a typological classification of resorts, based on their function and the extent of their visitor hinterland. A hierarchy of eight categories was identified, encompassing: capital cities; select resorts; popular resorts; minor resorts; cultural/historic centres; winter resorts; spas/watering places, and day-trip resorts. Lavery concedes that the classification is subjective and that obvious omissions are, specifically, seaside resorts, religious/spiritual centres, and 'created' resorts such as Disneyland in Florida. Some resorts would also fit several categories, while others have progressed from one orientation to another.

Undoubtedly, tourist destinations, like resources in general, pass through cycles linked to fashion and tourist behaviour. The popular appeal of established destinations fluctuates as changed circumstances trigger new sets of interests and different clients. Innovative forms of tourism may emerge and lead to the eclipse of redundant tourist outlets and the discovery of fresh attractions and venues. Explanation of such cycles has been linked to the behavioural characteristics of travellers. Two major human polarities have been identified:

Allocentric persons – self-confident, successful, high earners and frequent travellers, who prefer uncrowded destinations and exploring strange cultures;

Psychocentric persons – unsure of themselves, low earners and infrequent travellers who seek the security of tours and familiar destinations.

(Plog 1972)

The great majority of people are mid-centric: they fall between these two extremes and favour budget tours, heavily-used destinations, familiar food and chain-type accommodation. According to this hypothesis, resorts tend to rise and fall in cycles which match their appeal to particular categories of tourists (see Figure 11.3).

They move through a continuum... appealing first to allocentrics and last to psychocentrics... As the destination becomes more popular, the mid-centric audience begins to pick up... (which) leads to further development of the resort, in terms of hotels, tourist shops, scheduled

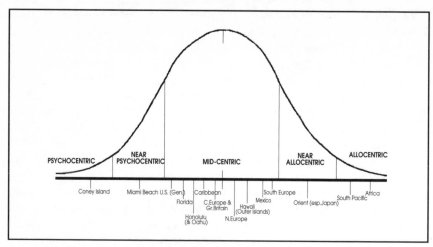

Figure 11.3 The cycle of tourist resort development

Source: Kaiser and Helber (1978: 8) (after Plog 1972)

activities for tourists and the usual services that are provided in a 'nature' resort area... continued development... carries with it the threat of the destruction of the area as a viable tourist resort... Destination areas carry with them potential seeds of their own destruction, as they allow themselves to become more commercialized and lose the qualities which originally attracted tourists.

(Plog 1972: 4)

It is important to note that decline of a resort is not inevitable. With appropriate planning and sound management, it is possible for success to be predicted, achieved and sustained. The possibility of rejuvenation is also stressed by Butler (1980), who cites the introduction of gambling casinos into Atlantic City, New Jersey, as an attempt to tap a new resort market. Other studies (eg. Christaller 1963; Hovinen 1981; 1982) have also examined patterns in the development of tourist destinations, while Butler's work is discussed in more detail later in this chapter.

From the point of view of landscape, it could be expected that each category of resort would develop its own recognisable blend of structures, activities and functions making up a tourist environment responsive to the requirements of the predominant type of visitors. The distinctiveness of tourist centres as special-purpose settlements is perhaps best seen in the morphology and townscape of seaside resorts, especially those of Britain and Western Europe.

In a study of English and Welsh seaside resorts, Barrett (1958) identified several common morphological features or characteristics. In particular, he noted the significance of the seafront in the structure and location of the commercial core, and a marked zonation of vacation accommodation and residential areas. Moreover, because growth along one axis was precluded or restricted, elongation of settlement occurred parallel to the coast. In Barrett's study, the core shopping and business

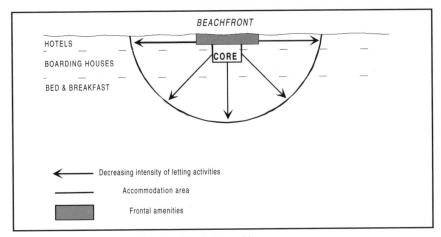

Figure 11.4 Schematic model of an English seaside resort

Source: Barrett (1958: 36)

district was offset symmetrically to a frontal retail and accommodation strip, which was the focus of resort activities, and which was functionally and socioeconomically distinct from the rest of the town (Figure 11.4). All these features were subject to modification, because of terrain and pre-resort transport and land use patterns.

Studies of New Jersey seashore resort towns also identified linearity in the various functional zones in response to location of principal routeways and proximity to the beach, and recognised a specialised frontal trading zone, termed the Recreational Business District (RBD) (Stanfield 1969; Stanfield and Rickert 1970). This zone was spatially and functionally distinct from the Central Business District (CBD), and comprised an aggregation of seasonal retail establishments catering exclusively for leisure-time shopping. Stanfield nominated the boardwalk as a uniquely American phenomenon, and grouped it with the British pier and promenade as a major contribution to the morphology of resort settlements.

Lavery (1974) put forward a schematic representation of a 'typical' seaside resort, with prime frontal locations occupied by the larger accommodation facilities, and a gradation in land values and tourist-oriented functions away from the seafront, the main focus of visitor attraction. Lavery also noted the spatial and functional separation of the CBD from the RBD, and associated the latter with the main route from the public transport terminal (e.g. railway station), in contrast to the emphasis given by Stanfield to vehicular access.

In Australia, an attempt was made to establish the extent to which the 'model' features found in British and North American seaside settlements were present in Australian beach resorts (Pigram 1977). The study was carried out at the Gold Coast on the Queensland/New South Wales border, which has become the focus of intensive tourist development catering to over two million visitors annually (Figure 11.5). Several interesting parallels can be drawn between the urban structure of Gold Coast settlements and that outlined above. The attraction of the coast and

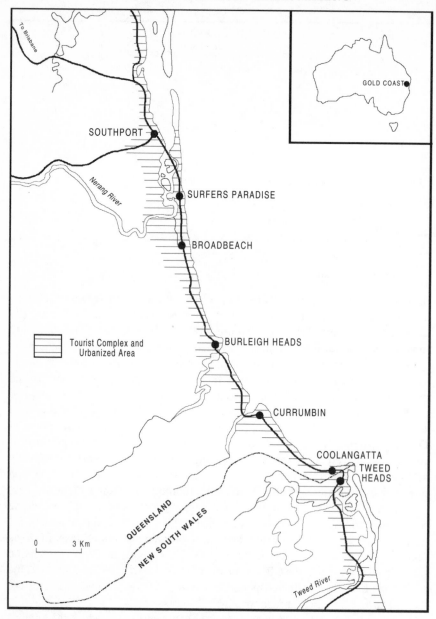

Figure 11.5 The Gold Coast Tourist Complex, East Coast of Australia

Source: Pigram (1983: 203)

beaches, the role of routeways and termini, the importance of topographical features, and the influence of pre-resort form and function are readily discernible.

An interesting aspect in the Australian study was the development of paired resort nodes at either end of the Gold Coast tourist complex. At the northern extremity, Surfers Paradise dominated the amusement and entertainment scene (RBD), whereas Southport is the regional and commercial centre (CBD). In the south, Coolangatta specialises in recreational business, while the CBD is across the State border in Tweed Heads. The end result is paired beach resorts which reflect, in part, the antecedents of European settlement in Australia, yet show clearly the effects of modern forces in shaping the tourist landscape.

More recently, in extending the work of those authors cited above, Meyer-Arendt (1990) developed a model of the morphology and evolution of seaside resorts in the Gulf of Mexico. In a departure from previous studies, Meyer-Arendt suggests that the development of seaside resorts actually resembles a T-shape pattern, with the initial beach access point becoming the main point (or locus) of tourist activities, eventually evolving into the recreational business district (RBD). This pattern also forms the basis of the beach resort model suggested by Smith (1992), who, in recognising the importance of second homes and low-budget accommodation, proposed eight stages through which a beachside resort develops.

Resorts are receiving increasing research attention (King 1997).

> Since resorts often aspire to being self-contained tourism destinations in their own right, the literature has studied the phenomenon from a diversity of angles, not confined to the study of resort facilities *per se*. These approaches include resort development (Dean and Judd 1985; Stiles and See-Tho 1991), planning (Smith 1992), the assessment of local attitudes (Witter 1985), marketing (King and Whitelaw 1992), the resort life cycle (Butler 1980), resorts as communities (Stettner 1993), architecture (England 1980), landscaping (Ayala 1991a, b) and key success factors (Wober and Zins 1995). A significant dimension of this literature is the role of resorts as 'enclave' developments, separated from the reality of daily life in adjacent areas or regions.
>
> (Freitag 1994, in King 1997: 11)

Integrated resorts are properties which incorporate a wide range of recreational facilities and accommodation types. 'Historically the evolution of tourism has been closely identified with the beginnings and subsequent development of resorts' (Medlik 1993: 126). 'The resort concept is based on providing leisure and recreation opportunities. Many resorts are self-contained destinations providing accommodation, food service, shopping and developed recreation opportunities. Some resorts rely on the natural resource base of the area for access to recreational opportunities' (Gartner 1997: 135). Resorts can be classified in many ways according to their specific location, their season of use, and/or the recreational opportunities they offer – island resorts (see King 1997), seaside or beach resorts, mountain or ski resorts, or health resorts. The term 'resort' has been utilised to describe tourist destinations at different scales, each locale combining specific locational, seasonal and recreational characteristics.

Butler (1980) has modelled the evolution of tourist resorts/destinations, using product life-cycle analysis, and identifying the links between the development or otherwise of a tourist resort/destination area and the nature of the travel market (see Figure 11.6). In the early stages of the life cycle, few people visit the area, and most services are locally provided. As the area increases in popularity, the extent of tourist development increases, and the nature of that development changes as there is a shift from natural physical and cultural attractions to ones which are more contrived and less authentic. In the development phase:

> ... local involvement and control of development will decline rapidly. Some locally provided facilities will have disappeared, being superseded by larger, more elaborate, and more up-to-date facilities provided by external organisations, particularly for visitor accommodation. Natural and cultural attractions will be developed and marketed specifically, and these original attractions will be supplemented by manmade imported facilities. Changes in the physical environment of the area will be noticeable, and it can be expected that not all of them will be welcomed or approved by all of the local population.
>
> (Butler 1980: 8)

Eventually, resort areas reach maturity, where options for planning and development arise and become more pressing. The resort may enter a stage of decline or stagnation, and may then be rejuvenated. As development increases, infrastructure keeps pace with the rising level of visitors. More activity options

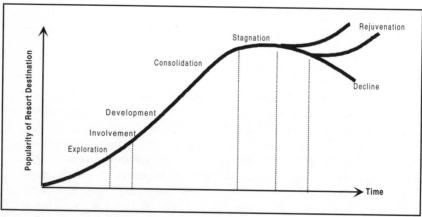

Figure 11.6 Hypothetical model of resort life cycle

Source: Adapted from Butler (1980)

are also added. Environmental and social impacts result from facility, attraction and infrastructure development. Roads may be built into scenic areas to offer more sightseeing opportunities; seaside resorts may experience a spread of development along the coastline; ski resorts may expand the number of their trails, and so on.

Mathieson and Wall (1982: 121) identified four types of transformation that occur during the development stage: architectural pollution; ribbon development and sprawl; infrastructure overload and traffic congestion. However, this analysis tends to focus on the negative impacts of tourism on the environmental quality of the area, and does not give due regard to the complex interrelationship between the environment and tourism, which are inextricably linked, and which can produce benefits for each other (see Chapter 12). Nonetheless, mass tourism, in particular, may have negative impacts: urban sprawl, diminished aesthetic values of the natural landscape; and decline in the levels of local ownership (Kariel 1989). Unrestricted tourism development presents problems. Therefore, tourist development must be well planned, professionally managed and set in a broader context of development (see Pearce 1989).

Tourism opportunity spectrum

With the rapid increase in the popularity of outdoor recreation, and particularly such activities as adventure, high-risk, and nature-based tourism, 'the impacts of the increased commercialisation of these travel opportunities have placed great pressure on unique and significant natural resources' (Butler and Waldbrook 1991: 2). A planning strategy, the tourism opportunity spectrum (TOS), useful in assessing the overall development of an adventure-oriented tourism destination, was presented by Butler and Waldbrook (1991). The tourism opportunity spectrum was presented as an extension and development of the recreation opportunity spectrum (RSO) (see Chapters 5 and 6).

In their strategy, Butler and Waldbrook (1991: 5) link:

> Butler's cycle of destinations to types of tourists, increasing levels of packaging and sales, and decreasing levels of adventure available to the explorer.
>
> A destination based on natural resources may be able to service several markets concurrently. Haywood (1986) describes the possibility of sequential entry of distinct segments, each with its own degree of institutionalisation (commercialisation). Cohen argues further for a 'multiplicity of types' and a 'multilinear approach' to describing tourists (Cohen 1979: 20). It is reasonable to assume that not all destinations will proceed simply through such a cycle but may instead experience a transition along a continuum of product development.

The ROS was, of course, developed as a tool for wilderness or remote area managers, but it has been more widely applied. In the same way that the key underpinning principle of the ROS is the supply of diverse recreation opportunities, a wider market penetration, with greater compatibility among elements, can be obtained by implementing the TOS (Butler and Waldbrook 1991: 6).

The six factors of the ROS are utilised in the TOS, namely: access; other non-recreational (adventure) uses; onsite management (tourism plant); social inter-action; acceptibility of visitor impacts; and acceptable level of regimentation. For each of these factors, various agency and other responsibilities for tourism development can be identified and allocated (see Table 11.1). According to Butler and Waldbrook (1991: 11), the TOS 'represents a new and potentially useful tool for the planning and managing of tourism resources and could ensure a more coordinated approach to presenting a destination to a broadened adventure travel market'. An important feature of the TOS, like its predecessor, the ROS, is that it accounts for demand and supply elements, while resident attitudes towards development, business opportunities and constraints should be examined and placed within the context of the spectrum. Economic impacts are also recognised. Perhaps, also, the concept of the TOS, like the ROS, has potentially wider application than remote areas, because it:

> offers a context in which the present functioning of the system can be understood and above all in which the likely implications of development can be anticipated... Spatial proximity facilitates combined use and there-fore increases the opportunity spectrum in a particular site. Obviously this concept, mainly based on the distribution characteristics of the supply side, also refers to the actual pattern of use, the demand side. Not all combinations of facilities equally increase the quality of the tourist experience. A full understanding of tourist behaviour (in time and space) and motivation patterns is needed.
>
> (Jansen-Verbeke and Van de Wiel 1995:140)

The TOS requires reliable, current and comprehensive data. It encompasses demand and supply (including infrastructure) elements, resident attitudes, business opportunities, and economic aspects of various adventure tourist types in terms of appropriate marketing strategies. It presents a means of maximising returns from visitors while minimising negative impacts. Linking the concept with the destination life cycle provides explicit recognition of the need to control market and product development in line with sustainable development parameters. This is a major planning challenge.

Summary

While many forces influence tourism, tourism itself is a powerful agent of social, economic and physical change, even disruption. Many of the problems and

Table 11.1 Responsibility for tourism development under the tourism opportunity spectrum (Adapted from Clark and Stankey 1979)

1 Access	
Access System	
Transportation	• N.W.T. Department of Public Works (Highways and Marine Divisions)
	• N.W.T. Department of Economic Development and Tourism
	• elected members of the Territorial and Federal Legislative Assemblies
	• Canadian Transport Commission
	• Air Canada and Canadian Airlines officials (also includes local feeder airlines)
	• local lobby groups including aboriginal groups, the Delta/Beaufort Development Impact Zone group (monitor oil impacts)
Marketplace	• local tour operators
	• non-local travel trade
	• market interest and desire to seek out information
Means of Conveyance	
Transportation	• individual ownership of mode of transportation
	• commercial sales and rentals of transportation in the region
Information	• degree of customer interactions
Channels	• availability of books on the region at libraries, bookstores, and actual numbers published
	• N.W.T. Department of Economic Development and Tourism (Travelarctic) production and distribution (mail, telephone, fax) of materials
2 Other non-adventure uses	• private sector oil companies (Esso, Amoco, Gulf)
	• Mackenzie/Beaufort Regional Land Use Planning Commission
	• Western Arctic Visitor's Association
	• Department of Economic Development and Tourism
	• Aboriginal Land Use Corporation
3 Tourism plant	• private sector operators
	• Department of Economic Development and Tourism (Inuvik Region)
	• Western Arctic Visitor's Association
	• Municipal Government
	• Aboriginal Land Use Corporations
	• Hunters and Trappers Associations
4 Social interaction	• consumers will regulate themselves as they react to crowding levels and host behaviour
	• Department of Economic Development and Tourism (Inuvik Region)
	• Western Arctic Visitor's Association
	• Aboriginal Cultural Agencies
5 Acceptability of visitor impacts	• internal market regulation based upon customer perception of region, depending upon the availability of market information, local tourism associations will become involved
6 Acceptability of regimentation	• local and non-local tour operators
	• market reactions
	• unpredictable weather patterns

Source: Butler and Waldbrook (1991: 11)

undesirable features associated with tourism flow from inadequate attention to the planning and design of tourist developments. At first sight, planning for tourism might seem a contradiction in terms, and likely to inhibit the spontaneity identified with pleasure/leisure travel. However, planning for tourism is as essential as planning within other sectors of an economy. It is the absence or weakness of planning which allows the development of types of tourism incompatible with natural and other (e.g. economic) systems, and which permits the expansion of tourism at a rate inconsistent with the capacity of the infrastructure and society to cope with the pressure.

The next chapter examines the relationship between tourism and the environment. It is argued there that tourism can contribute to substantial upgrading of the environment, and to economic and social development; this can add to visitor enjoyment, if development is of an enlightened, sustainable form. Environmental changes stemming from tourism can be positive.

Guide to further reading

- There are many comprehensive *introductory texts to tourism*. Readers should consult: Mathieson and Wall (1982); Mill and Morrison (1985); Pearce (1989); Hall (1995); Pearce *et al.* (1998).
- *Tourism in the Pacific and Asia*: Hall and Page (1996); Hall (1997).

Review questions

1 Compare and contrast definitions of 'tourist', 'tourism' and 'tourist industry'. Can you readily identify the 'tourist industry'? Examine contesting arguments about the existence of such an industry.
2 Critically examine the main forces affecting global tourism patterns and processes.
3 Conduct an inventory of the resource base for tourism and recreation in a place of your choice (perhaps your local area). In that inventory, identify the agencies responsible for the management of those resources. Have there been any recent, notable conflicts among those agencies, with respect to recreational and tourist use of resources? Based on the resources you have identified, can you identify any potential recreational and tourist opportunities yet to be identified or explored by management agencies?
4 Critically assess the Tourism Opportunity Spectrum as presented by Butler and Waldbrook (1991). Are there any deficiencies in this framework? If so, explain why they exist, and how they might be addressed by refinement of the framework.

12

TOURISM AND THE ENVIRONMENT

> Rather than opposing change, or merely accepting and
> accommodating change, the tourism industry must manage change
> to its advantage and that of the environment which nurtures it.
> Endorsement and application of the concept of sustainability and
> of best-practice environmental management offer compelling
> evidence of how change can be harnessed to contribute towards
> the achievement of environmental excellence. Although tourism
> flourishes best in conditions of peace, prosperity, freedom and
> security, disturbance to these conditions is to be expected. The
> industry response must be sufficiently resilient to generate oppor-
> tunities for the growth of tourism in keeping with the dynamics of
> a changing world and increasing concern for ecologically
> sustainable development.
>
> Pigram and Wahab (1997: 17)

The environment is the aggregate of resources available to human beings. It
comprises the natural environment of earth – water, air, flora, fauna and ecosystems and their associated processes – and the social, cultural and economic
environments of human beings. Clearly, the environmental impacts of tourism
must be viewed in the broadest sense, taking into account the full range of economic,
ecological, social and cultural factors.

Recreation, tourism and the environment are interdependent, but their relationship is not constant, varying over space and time. The forces underpinning the
growth and changing patterns and processes with respect to leisure, recreation
and tourism (see Chapters 1 and 11), coupled with such factors as greater environmental awareness, industrial growth, pollution, limited knowledge about ecological/
environmental processes, and other factors, mean that this relationship is very
complex and dynamic. Recreation and tourism can cause negative and positive
environmental change, with debate about acceptable limits of change likely to
emerge. Critical concerns, then, are understanding the relationship between
recreation, tourism and the environment, and identifying ways of managing resources in harmony with the attractiveness of many recreational and tourist activities.

Tourism – environment interaction

Tourism can have beneficial and negative consequences for the environment; tourist development can contribute to substantial upgrading of the recreational resource base, and thus add to visitor and local resident enjoyment. It can also lead, for example, to improved transportation systems (an important component of the tourist experience) through advances in vehicle and routeway design (Gunn 1994). This allows greater opportunity for pleasurable and meaningful participation in travel, and, simultaneously, creates external economies. 'Improvements in transportation networks, water quality and sanitation facilities may have been prompted by the tourist industry, but benefit other sectors of the economy. An international airport... provides improved access to other regions for locally produced goods' (Vanhove 1997: 67) (also see Gunn 1994; Page 1994).

Enhanced understanding of the resource base is another positive outcome of pleasure travel, brought about by the application of various management techniques to interpret and articulate the environment to visitors (e.g. see Hall and McArthur 1993; 1996). Beneficial modifications or adaptations to climate in the form of recreational structures, clothing and equipment, have also been developed in response to the stimulus from tourism.

Improved habitats for fish and wildlife, and control of pests and undesirable species have become possible through the economic support and motivation of increased use. According to McNeely (1988, in Lindberg 1991), African fauna (e.g. african elephants, lions, mountain gorillas and rhinos) are protected and managed as tourist resources. Lions and elephant herds were individually estimated to be worth about $27,000 and $610,000 per annum, respectively. Further positive response can be seen in the broadening of opportunities to view and experience both the physical and cultural world. Ready examples are the opening of national parks, wilderness areas and forests for recreational use, and continued agitation for better access to water-based recreational resources along streams and coastlines.

The many ramifications of tourism give much scope for interaction with the environment. Some observers (e.g. Eckbo 1967; Relph 1976), while conceding beneficial spin-offs in the economic, political and cultural spheres, remain convinced that, 'in the long run, tourism, like any other industry, contributes to environmental destruction' (Cohen, 1978: 220), or conserves only the things that are of potential and actual tourist interest (i.e. flora, fauna, cultures and landscapes that tourists want to see). Gartner (1997) points to critics who oppose tourist development on the grounds of declining water quality, stemming from visitor overuse in the Mediterranean, the Adriatic and other popular destinations (for a detailed discussion on seacoasts and tourism, see German Federal Agency for Nature Conservation 1997). Others, like the authors of this book, believe that with proper planning and management, incorporating sustainable environmental goals, tourism can help maintain, or even enhance the environment, and be a positive influence in the process of cultural dynamics, while simultaneously contributing to a region's economic development. Despite the protracted debates about tourism and its economic significance in developed and developing countries, there is considerable disagreement as to whether the incidence and magnitude of the effects of tourism can be accurately measured (e.g. the problems in determining carrying capacities that are noted in earlier chapters).

Unfortunately, the tourism–environment relationship is generally expressed in terms of opposing alternatives – either protecting the environment for tourism, or protecting the environment from tourism. However, these objectives need not be mutually exclusive. There appear to be several modes of expressing the impact of tourism: the net effect may well be tourism-related *and* environmental enhancement. A balanced appraisal of the growing links between heritage and tourism in European cities (Ashworth and Tunbridge 1990) identified both the benefits and problems in such links, and the importance of delicate planning and management if the two were to exist in a mutually beneficial manner (also see Hall and McArthur 1993; 1996). Increasing cultural awareness and consciousness has stimulated restoration of historic sites and antiquities. Referring particularly to Europe, Haulot (1978) raises questions about how this heritage – the landmarks, castles and artifacts of past eras – could be kept intact, if its existence and preservation had not become the ongoing concern of a great audience of tourists, resulting both in substantial financial contributions from the visitors themselves and generous state support.

Design of contemporary tourist complexes appears to benefit from the demands of a more discerning tourist population. While there remain many examples of unfortunate additions to the tourist landscape, modification of the built environment for today's tourist is marked increasingly by quality architecture, design and engineering. Higher standards of safety, sanitation and maintenance also help to reduce the potential for pollution. These advances demonstrate that tourism need not destroy natural and cultural values, and, in fact, can contribute to an aesthetically pleasing landscape (e.g. Harrison and Husbands 1996; Murphy 1997).

The potential for enlightened development is exemplified in the many sophisticated tourist developments in various countries of the world (e.g. see Harrison and Husbands 1996). However, there are environmental costs in tourism development, the distribution of the economic benefits of tourism is uneven, and ecological damage and destruction are widespread.

The environmental benefits and costs of tourism raise the question of externalities – 'those costs or benefits arising from production or consumption of goods and services which are not reflected in market prices. Because of this there is little incentive for firms to curb external costs, since they do not have to pay for them' (Tribe 1995: 244). Examples include aircraft noise disturbing residents around airports serving tourist routes; the loss of a mangrove swamp when a tropical island resort is built; and inadequate fauna preservation through establishment of African safari parks (e.g. see Bull 1995: 163). Tribe (1995: 245) illustrates the point with reference to sewage discharges into the sea:

> Whilst there is little marginal private cost to the water companies for pumping sewage into the sea, it represents a loss of well-being to people who want to use the sea. There is a considerable marginal external cost,

which takes the form of cleaning costs to surf equipment, medical costs to treat infections and loss of earning caused by sickness. These are readily identifiable costs to which must be added the general unpleasantness of contact with sewage.

'By analysing externalities, the public or social benefits and costs of tourism may be added to, and subtracted from, its commercial market value to an economy' (Bull 1995: 163). Clearly, then, 'Recognition of the importance of unpriced values and externalities in tourism at least warns us to treat carefully any statistics on the commercial importance of this sector to an economy' (Bull 1995: 176). Rarely, however, are the external benefits and costs of tourism accounted for in financial terms.

Although it can be conceded that tourism has much (perhaps unrealised) potential for environment enhancement, negative impacts do occur in a number of areas, particularly from the predatory effects of seasonal migrations of visitors, and resulting disturbance to, or destruction of, flora and fauna. The most obvious repercussions are likely to be in natural areas, but the built environment and urban areas may also be impaired, and the social fabric of communities can be widely disrupted.

Pollution, both direct and indirect, and in all its forms (from aircraft emissions, to architectural insensitivity, to destruction of ecosystems), is a conspicuous manifestation of the detrimental effect of tourism (Young 1973; Gunn 1994). Erosion of the resource base is a particularly serious environmental aspect (Wall and Wright 1977). This can range from incidental wear and tear of flora and structures, through soil erosion, to vandalism and deliberate destruction or removal of features which constitute the appeal of a setting. This erosive process is accelerated at times by use of incongruous technological innovations and by inferior design and inappropriate style in the construction of tourist facilities. Tangi (1977), for example, described some of the tourist resorts of the Mediterranean as architectural insults to the natural or historic sites where they are located. It is as well to remember, of course, that the strange architecture of today, which may be challenged by so many, may become the heritage of tomorrow and challenged by few (as is the case with the Sydney Opera House, Australia).

In many destination areas, the environment must serve not only conflicting tourist uses, but also the resident community, many of whom take a proprietorial attitude towards their surroundings. Congestion and overtaxing of infrastructure and basic services, which are particularly prevalent in high seasons, can generate dissension between visitors and the local population, the latter coming to resent the intrusion of tourism (e.g. see Doxey 1974; Pearce 1978; 1979). In a related study, Rothman (1978) listed municipal services and facilities, access to recreational sites, and personal and social life, as features of the sociocultural environment, seasonally curtailed by vacationers. Social interaction between residents and tourists was reported as minimal, with large numbers of visitors opting for an exclusive environment requiring the least cultural adjustment on their part. In developing countries, too, aspects of tourism may have long-term disruptive effects on the life-styles and employment patterns of host communities, especially where the actual dispersal of visitors is minimal and they stay largely within the confines of 'the resort'.

The potential for tourist activity to disrupt host communities often varies seasonally. The subject of seasonality with respect to tourism is complex, but its causes and effects have received insufficient critical attention. As Butler and Mao (1997) point out, 'The nature of the relationship between seasonality and the motivation of visitors is not known, and issues such as whether dissatisfaction with conditions in the origin region, or desire for the attractions of the destination, play a greater role in shaping the seasonal patterns of tourism is also a mystery'. It is also not known why tourists travel in peak seasons, because a number of forces are likely to be acting on tourists' motivations and choice at any one time.

With so many variations on the theme, it is difficult to generalise on the relationship between tourism and the environment. The relative importance of each influential factor varies with the location and situation, and negative effects need to be balanced against positive impacts. Certainly, the ugly face of tourism receives wide exposure, and the relationship depicted in Figure 12.1a could well apply, with an increase in tourism bringing about a decrease in environmental quality. However, change does not necessarily equate with degradation, and tourism and environmental quality are not mutually exclusive goals. The net effect may be marginally negative (Figure 12.1b), or the two may be organised in such a way that both benefit and give each other support (Figure 12.1c).

Clearly, tourism and protection of the resource base are more alike than contradictory; the demands of tourism, instead of conflicting with conservation, actually require it (Gunn 1972). If this is not the case, the very appeal which lures the visitor to a site will be eroded, and with reduced satisfaction, any chance of sustained viability for the destination will disappear.

Environmental influences

The nature and extent of tourism's impact on the physical environment are determined by many factors, including:

- *the length of time since tourist development was initiated and the aspirations of developers* – short-term goals characterise much tourist development, which is largely speculative in nature, and which is facilitated by entrepreneurs who are often either ignorant of, or blatantly ignore, the consequences of their actions and their cumulative effects;
- *the number of tourists and the intensity of on-site use* – all things being equal, as visitor numbers increase, it is likely that there will be greater transformation of the environment. Of course, the resiliency of the ecosystem (see below) and management regimes will affect this relationship so that it is non-linear;
- *the nature and resilience of the destination's ecosystem* – certain types of vegetation and soils can withstand greater visitor numbers, while climate and

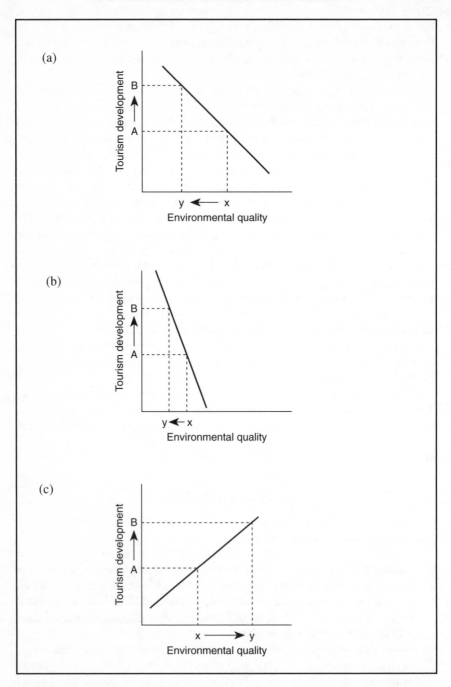

Figure 12.1 Tourism and environmental quality

Source: Pigram (1983: 210)

seasonal variations in temperature, rainfall and humidity, influence plant growth rates. Similarly, Western cultures are perhaps less likely to be influenced by Western visitors' attitudes and behaviour than less developed or Asian countries and *vice versa*. All in all, there is great variation among ecosystems and cultural systems even at the local level;

* *the dynamic nature of tourist demand and the dynamism of tourist development* – tourist motivations and choices change over time, so that destinations rise and fall in popularity, while the nature of attractions, services and facilities at destinations change.

Aside from the above factors, a broad range of environmental factors and their relationship to tourism warrant discussion.

Weather, climate and environmental uncertainty

The relationships between tourist and recreation activity and the environment (e.g. aesthetics, temperatures; topography and wildlife) have received insufficient attention. Weather and climate may be regarded as inputs to the amenity index of a place or region (Pigram and Hobbs 1975). In this context, they are important influences on tourist choices and behaviour. 'Environmental changes, including... shifts in climatic patterns, have the ability to affect destinations, by directly changing some attributes of the tourism product itself and by altering locational advantages of different types of destination' (Craig-Smith and Bull 1990, in Bull 1995: 238) (also see Chapter 11). Acid rain, global warming and ozone depletion are indications that the atmosphere may be changing at an unprecedented rate, with implications for flora, fauna, and human activities, including tourist development. Extreme events such as floods, droughts and hurricanes may increase or decrease, and the insurance industry is playing a growing role in influencing what can be built where. Many recreation activities take place on coasts and in mountains, both particularly vulnerable locations in an era of change.

> Global climate change, including ozone depletion, may modify tourism and recreation resources and how the potential clientele perceives them and thus require businesses and destination areas to adjust to changing circumstances... with respect to the climate, the past may no longer be an adequate guide to the future... climate change may be one factor among many worthy of inclusion in tourism planning and investment decisions.
> (Carmichael *et al.* 1995: 513)

Despite recent indications that many tourist areas may be affected by the warmer temperatures stemming from global warming, research concerning (1) the relationships between tourist and outdoor recreation activity and environmental (and, more specifically, climate) change, and (2) industry response strategies, is lacking. Outdoor recreation is an activity that will probably adapt to different conditions as tourists substitute one activity for another, or one resort for another, as conditions and seasons change. Downhill and cross-country skiing are two activities likely to be most affected by global warming (see Chapter 11), not to mention coastal areas and associated activities and resources, if sea levels rise. Other activities, too, could be significantly affected. As the relationship between tourism and the environment becomes more widely recognised, so, too, does the relationship between tourism and nature conservation, and the need for environmentally sustainable forms of tourist development.

Tourism and nature conservation

Conservation is a philosophy which is directed at the manner and timing of resource use (O'Riordan 1971: 8), and may be defined as managing the resources of the environment – air water, soil, mineral resources and living species, including man – so as to achieve the highest sustainable quality of life.

Nature conservation is a dynamic concept which is subject to diverse understandings and interpretations, spatially and temporally, and which is supported for many different reasons (e.g. ethical reasons; encouraging environmental sustainability; maintaining genetic diversity; recreation; scientific research; future choices and utility; education; and political reasons).

Recognition of the importance of nature conservation can be seen in the relatively recent rise of the environmental movement, and, simultaneously, the development of a conservation ethic in modern society. That recognition is tangibly evident in (1) the creation and resourcing of public and private sector agencies and interest groups, (2) related legislation and public policy, and (3) the establishment of resource management units such as national parks and wilderness areas, which often serve as important tourist attractions.

Tourism and nature conservation are interdependent, and their relationship has been a lengthy one. Tourism often stimulates measures to protect or conserve nature, but, at the same time (and somewhat paradoxically), presents a significant environmental risk, especially because of its demands on the natural environment, and therefore on responsible agencies. These risks are intensifying as domestic and international tourist demand for natural areas grows in many developed and less developed countries. Furthermore, the nature of that tourist demand is such, that tourists are seeking more spontaneity, independence and participation in their travel experiences (e.g. the growth of nature-based tourism – see below).

Budowski (1976) noted three different relationships with respect to conservation and tourism – conflict, coexistence or symbiosis – which can exist between those promoting tourism and those advocating conservation of nature. Conflict occurs when conservationists see that tourism can have only detrimental effects on the environment. Coexistence is noted when some, though possibly little, positive contact occurs between the two groups (conservation and tourism). Symbiosis is reached when the relationship between tourism and conservation is organised in such a way that both derive benefit from the relationship. Conflict and coexistence are common. Symbiosis is perhaps the least represented relationship in the national and international perspective.

Tourism can cause environmental degradation, but it can also contribute to substantial enhancement of the environment. Sustainable tourism requires the conservation of nature, and thereby leads to the maintenance or substantial enhancement of natural areas, and subsequently to increases in visitor satisfaction.

As noted above, tourism provides an economic impetus for the conservation of the environment, due to the fact that protected and/or scenic areas are major attractions for domestic and international tourists. Tourism can also contribute to a wider appreciation of nature conservation by promoting and making more accessible specific sites and aspects of nature.

The role of tourism as a consistent contributor to nature conservation is often debated, because, among other things, tourists trample vegetation, disturb wildlife, carry pathogens and weeds, and do not always behave in ways which promote the symbiotic relationship desired between the industry and conservation (e.g. when engaged in vandalism or littering). Tourism, too, has fostered the intensive viewing (with resulting disturbance or damage) and export of protected and/or endangered species.

The interrelationships between tourism and nature conservation are thus extremely complex and dynamic, with conflict being most acute where tourist development occurs rapidly and without strategic planning. Unfortunately, tourism and recreation research has developed few strong concepts or theories to guide the role and management of tourism in nature conservation. Many studies focus narrowly on the physical impacts of developments at a particular site and neglect the human element, few have a longitudinal basis, and most are reactionary. Even when impacts are identified or speculated, the research focus is largely limited to the effects of tourism on vegetation and, to a lesser extent, on wildlife, with impacts on air and water quality, soils and ecosystems relatively neglected. Therefore, a number of methodological problems concerning research on tourism and the natural environment (and thus nature conservation) can be identified:

- the difficulty of distinguishing between changes induced by tourism and those induced by other activities;
- the lack of information concerning conditions prior to the advent of tourism and, hence, the lack of a baseline against which change can be measured;
- the paucity of information on the numbers, types and tolerance levels of different species of flora and fauna;
- the concentration of researchers upon particular primary resources, such as beaches and mountains, which are ecologically sensitive (Mathieson and Wall 1982: 94).

Tourism development must be environmentally sensitive and consistent with long-term nature conservation, otherwise it presents risks to the sustainability of the industry itself, and more generally the natural environment. Tourist pressures on nature conservation will continue to grow, and a clearly established and widespread balance between tourism and nature conservation will never be universally accepted, this being perhaps most problematic in wilderness and very sensitive areas. The relationship between tourism and nature conservation is thus a highly political issue, which is in need of much greater research attention if the natural resources upon which tourism so heavily relies, are not to be degraded or destroyed. Successful integration of tourism and nature conservation objectives is of increasing importance, because it can enhance the choices of people and help maintain or even enhance the quality of the environment.

Economic and social perspectives

The role of tourism in the economy is seldom perceived clearly. Rarely are specific economic sectors focused on tourist activities, and the attributes of the tourist 'industry', which would facilitate an assessment of its overall economic significance, are difficult to establish. Yet, without such assessment, public and private sector decisions regarding tourism at various scales must be made in a vacuum. Allocation of resources to tourist enterprises by governments, investors, consultants or planners, is too important to be based merely on intuition or value judgments that tourism is a good or bad thing. In the harsh environment of commercial risk-taking and in the disbursement of public funds, economic worth must be proven, not merely inferred.

The term 'tourist industry' is a generic expression, yet the industry is anything but homogenous. It includes activities which are clearly of primary importance to tourism (e.g. travel agents, accommodation and transport), as well as other enterprises, where the degree of involvement with tourists is indirect or secondary, if not dubious or unknown (e.g. some festivals and events, restaurants, and recreational facilities). On this basis, the tourist industry comprises accommodation outlets; food and catering establishments; travel agents and wholesale tour operators; transport services; recreation and entertainment facilities primarily for travelling, and public agencies concerned with tourism. The impact of tourism at an international level was discussed in Chapter 11. However, it is on the domestic scene, where the economic effects may be demonstrated most effectively, both at the national scale, as well as regionally and sectorally.

The most obvious economic impact of tourism is associated with tourist expenditures, which become a source of income and employment to local communities. Tourism can also play an important role as a means of decentralisation, and of boosting employment in otherwise depressed regions. In less developed regions, resources may become unproductive for agriculture or industry, but can become a source of wealth through tourism. The generation of resultant economic activity may prevent further erosion of population, and may also lead to a fuller utilisation of existing resources, facilities and services. As a consequence, enterprises previously operating at less than capacity are able to expand production to the point where economies of scale can be realised, and, in so doing, can function more efficiently, as well as develop a better quality product or service. The injection of new spending, in turn, can lead to renewed growth from an expansion of economic activity generally.

From the point of view of employment, tourism makes up a large segment of the service industries and requires a large, diversified and dispersed labour force. Therefore, employment generation is an important benefit accruing from tourism. The two broad types of employment created are:

- *direct employment* – including occupations created in the tourism industry such as in accommodation, travel agencies and transport operators;
- *indirect employment* – the additional jobs generated by the need to increase the service and physical infrastructure of an area to support tourism and the tourist industry (e.g. retail sales, infrastructure such as road construction, water and sewage).

Moreover, the travel market provides jobs for supplemental workers – part-time, casual and seasonal labour, and, in particular, the less-skilled. In general, tourism is labour-intensive rather than capital-intensive, and is a growth industry in terms of employment. Hence it receives considerable attention from governments in the formulation and implementation of fiscal and monetary policies.

Direct or primary expenditure is not a true and complete measure of the prosperity created by tourism, because the initial recipients, in turn, recirculate a certain proportion, thus generating additional income and employment. At each round of transactions, some money leaks out of the system, so that the income-creation effects are successively reduced and ultimately exhausted. This subsequent indirect impact of the initial tourist spending results from an important concept known as 'the multiplier effect'. Furthermore, as incomes rise within a region, local consumption expenditure increases, and this may induce an even greater impetus to the regional economy. 'Together the indirect and induced changes are called secondary effects and the multiplier is the ratio of the primary plus secondary effects to the primary direct effect alone' (Archer 1973: 1).

The magnitude of the multiplier effect depends on the degree to which a regional economy is able to retain as income, the money spent by visitors. This, in turn, is a function of the ability of the local economy to produce the various items and services consumed by tourists. The smaller the size of the region's economic base, and the fewer the intraregional linkages, the greater the number of goods and services which have to be brought into the region. In these circumstances, the greater will be the leakage, and hence, the lower the value of the multiplier.

The multiplier concept has incurred much criticism, and its theoretical and practical limitations are well-documented. The shortcomings rest not so much with the model itself, but with the ways in which it has been misapplied. It is especially difficult to generalise from one situation to another; the multiplier effect varies from project to project, from region to region, and from one form of tourist activity to another. Despite the criticisms and deficiencies inherent in the concept, some attempt needs to be made to derive an estimate of the overall significance of tourism.

The economic significance of tourism varies sectorally with the type of activity and the characteristics of travellers and host communities. It also varies spatially with a region's degree of self-sufficiency. In these circumstances, economic analysis becomes an important tool for establishing and justifying the type and direction of public investment and assistance to the industry, as well as being a guide for private promotional initiatives and marketing strategies. For Asian and Pacific tourist destinations, such economic guidance and assistance with decision-making are vital, especially during a period when new pricing structures and advances in transportation are helping to reduce isolation and expose countries to the international travel market. For developing countries, the need to clarify the role of tourism as a strategy for development is critical.

Tourism and developing countries

Tourism in less developed nations grew rapidly after World War II 'because land and tax incentives encouraged Western capital to invest in the essential infrastructure to attract foreign visitors. Westerners held managerial positions, hiring and training local labour for routine services' (Smith 1994: 163). According to Turner (1976), tourism seems tailor-made for the Third World, and a growing number of developing countries are placing emphasis on tourism in their development plans. Reasons are not hard to find. A ready market is available for the attractions these destinations can offer: appealing climates, exotic scenery and rich heritage. Land and labour costs are comparatively low, and in the absence of significant mineral production or an export-oriented agricultural sector, tourism is a potential source of foreign exchange and can generate new opportunities for employment, as well as stimulating demand for local products and industries. Tourism is also said to make possible improvements in local infrastructure, by way of the provision or upgrading of roads, airports, harbour facilities, accommodation, shopping, entertainment, communications, health services, power, and water supplies and sanitation.

Research concerning the role of tourism in Third World development has increased dramatically over the past two decades (e.g. see de Kadt 1979; Britton 1982; Lea 1988; Opperman and Chon 1997). Yet, 'For more than forty years, tourism has been considered as an economic panacea for developing countries... thought to be a vital development and an ideal economic alternative to more traditional primary and secondary sectors. International tourism, particularly from the developed to developing countries, is seen as generating crucially needed foreign exchange earnings, and infusing badly needed capital into the economy of developing countries' (Opperman and Chon 1997: 1). For some countries, tourism appears to have fulfilled its promise, and parts of the less-developed world have received significant financial benefits.

Nevertheless, misgivings have been expressed as to whether tourism is the most appropriate form of investment for developing economies. Typically, the questions raised include:

- Is tourism the best way towards economic independence and a better quality of life?
- To what extent is the nature of tourism so seasonal that it leads to saturation levels of visitors?
- Can or should economic control be in local hands?
- To what extent does the multiplier apply?
- Does tourism help revive and sustain cultures or does it destroy traditions?
- Does tourism contribute to anti-social activities, including gambling and prostitution?
- Does tourism promote understanding between hosts and guests, or misunderstanding and prejudice?

These and other related questions need to be addressed, before even qualified endorsement can be given to tourism as a strategy for regional development (de Kadt 1979).

In the first place, foreign exchange earnings are offset by considerable expenditure, caused by tourism, on increased imports, including food and drink, and on development of the necessary infrastructure for international travel. In Fiji, for example, it was noted by Britton (1980) that 53 per cent of hotel food purchases, 68 per cent of standard hotel construction and outfitting requirements, and more than 95 per cent of tourist shop wares were supplied from imports. A further drain on foreign exchange is attributed to repatriation of the profits of foreign-owned corporations, and payment of salaries to higher-skilled personnel imported to the better-paid positions in the tourist industry. A lack of capital and substantive investment leakages may then lead to a dual economy, intersectoral competition and inflation, and regional imbalances (Pearce 1989; Mercer 1995). 'These issues need to be weighted carefully, based on accurate assessments of the actual economic effects. Too often, multiplier effects are overestimated, leakages are misjudged, and costs for infrastructural developments and induced leakages through demonstration effects are not considered' (Opperman and Chon 1997: 109).

Too great a commitment to tourism, brings with it the twin dangers of loss of control over resource allocation and decision-making, and dependence on a single, fluctuating economic base. Tourism is notoriously open to pressures from inflation, fuel crises, political conflict and industrial troubles, security fears, and the fickle nature of resort popularity. If local populations have deserted traditional occupations for work in tourist undertakings, any downturn in international travel can be doubly unfortunate. All too often, the rate and direction of tourist development are in the hands of multinational corporations, especially in the areas of transport, hotel chains, tour wholesaling and marketing. In the South Pacific, for example, no island nation owns any of the companies which operate cruise ships, and few governments can afford a national airline. The extent of external control is likely to grow, with the tendency towards vertical and horizontal integration in the tourist industry (Bull 1995; Hall 1995).

Even where economic benefits can be demonstrated, these must be balanced against adverse sociocultural consequences. Costs and benefits are not evenly distributed between local residents and tourists, nor within host communities (UNESCO 1976; Richter 1989). The confinement of tourists and their expenditures to air-conditioned enclaves offering carefully programmed travel experiences, can become a source of frustration and resentment, and merely accentuates the gulf between affluence and poverty. Imported goods and services compete with local enterprises, and conflicts can occur as commercialisation of land and resources intrudes on traditional values.

Tourism can also serve as a powerful agent of social change and disruption. The excesses of tourism, and the unregulated behaviour often associated with it, can act as an affront to host cultures. Biddlecomb (1981) nominates conspicuous consumption, eccentric clothing, unacceptable behaviour (e.g. nude bathing and illegal activity, including the use of drugs) as sources of inter-personal and inter-cultural tensions. An increase in crime and anti-social behaviour is only one of many possible undesirable consequences of tourism, even in developing countries (Walmsley et al. 1981; Opperman and Chon 1997). According to Butler (1975), the extent to which tourism is capable of inducing change in the host society, is a function of the characteristics of the visitors and of the destination. He contrasts the localised impact, of, say, a cruise ship, with that of large numbers of visitors on an extended stay, especially where there are sharp economic and cultural differences between tourists and indigenous people. On the other hand, a strong local culture and pronounced nationalistic outlook can act as a buffer to distortion of the social fabric.

A similar anthropological theme is to be found in the study of 'Hosts and Guests', edited by Smith (1978), where the potential of tourism to foster cultural disruption and transformation of lifestyles is illustrated in individual case studies drawn from around the world. Further examples of societal strain as the result of incursions of tourists were presented to a UNESCO Workshop on Tourism in the South Pacific (Pearce 1980). The workshop conceded that tourism was only one factor in social change, but recommended closer examination of alternative forms of tourist development that would minimise adverse effects on the islanders' way of life.

This approach was summed up some 20 years ago in the Cook Islands Government policy emphasis on indigenous tourism:

> Tourism should not be the means for us to change our way of life but an incentive to make us more aware of what we are in terms of our culture, customs and traditions. This should not be interpreted negatively to mean that all changes which affect our way of life must be avoided. Change is

inevitable. Instead, a positive rate and direction of change and how we manage that change and its conflicts are more important.

The guiding principle should be: preserve that which is good, modify or destroy the bad and adopt the new to strike a balance.

(Okotai 1980: 173)

Despite these laudable statements and increasing attention to the nature and extent of tourist development in developing countries, the problems and issues cited previously continue to plague tourist development in these same countries. In short, little affirmative and effective action has been forthcoming (see Lea 1988; Opperman and Chon 1997). Indeed, a new dimension has arisen with respect to Third World development, where

perhaps lies the great irony of the socio-cultural effect of mass travel; the cultural gap between different worlds is gradually closing, tourist and host native are exchanging ambitions and ways of life, while the tourist from the industrial world seeks escape, the native travels to learn how he can satisfy the demands for the good life that he imagines the affluent tourist has achieved.

(Swinglehurst 1994: 102)

The relationship between tourism and developing countries, perhaps highlights the powerful agents of change associated with tourism more than is the case in developed or industrialised nations. However, wherever we look, the case for environmentally compatible tourism, in its widest sense, is a laudable goal, and one very close to the heart of sustainable tourism.

Tourism and environmental compatibility

Environmentally compatible tourism describes a situation whereby tourism and the environment are able to exist in harmony, so that tourism does not detract from, or harm the environment nor *vice versa*. Increasing environmental awareness and conservation activities around the globe, have contributed to efforts to establish environmentally compatible tourism. In some instances, the tourism industry has entered into partnerships with environmental and other groups, has consulted effectively with host communities, resource management agencies and governments, and has directly supported conservation or preservation objectives. However, these situations are all too rare, and with the recent rise of nature-based travel, industry support for environmental conservation or preservation is often merely a means of marketing and promoting business operations. The recent development and promotion of ecotours by many tourist operators, is one example where industry regulation with respect to restricting undesirable business and tourist behaviour, and environmental impacts, is lacking, and where suspicions have been cast on the environmental compatibility of tourist operations with the host environment.

In order to reduce the conflicts, and enhance the relationship, between tourism and the environment, environmental impact statements are now required in many countries as part of the approval and monitoring and evaluation processes for tourism projects. This is particularly the case if projects are large, or located in, or adjacent to, environmentally sensitive areas such as protected areas, rainforests, coastlines or estuaries. Further developments in environmentally compatible tourism approaches can be seen in the implementation of environmental auditing processes in the public and private sectors.

The key to achieving environmentally compatible tourism and, ultimately, a sustainable tourist industry, is recognition of the need for environmentally sensitive policy-making, planning and development. The integration of tourism and the environment is being carried out at different levels in a number of places and for a variety of reasons, with various mechanisms being utilised. Strategies and related activities range in size from small-scale to large-scale projects, and include various economic, nature conservation, cultural, social, heritage, spatial/regional and political benefits and costs. On a broader national and global scale, approaches to integrating tourism and environmental objectives are being developed and promoted by international and national tourism agencies, and to a limited extent by multinational corporations. The development and promotion of nature-based tourism and ecotourism are notable responses.

Nature-based tourism

Nature-based tourism may be defined as 'domestic or foreign travel activities that are associated with viewing or enjoying natural ecosystems and wildlife, for educational or recreational purposes' (e.g. see HaySmith and Hunt 1995: 203). Tourism industry leaders and natural resource managers face significant challenges in promoting sustainable development of tourism in protected areas, and in managing impacts on flora and fauna (HaySmith and Hunt 1995). Nature-based tourism, encompassing ecotourism, adventure tourism, outdoor-oriented educational tourism, as well as a whole host of other outdoor-oriented, non-mass tourism experiences, is arguably the fastest-growing segment of the tourist industry in many countries (McKercher 1998: ix), and one which holds much promise to environmentally compatible tourism objectives.

A review of the principles of ESD [Ecologically Sustainable Development] offers valuable insights into how the tourism industry must act in relatively undisturbed areas. Underlying the entire ESD philosophy is a commitment to operate within the social and biophysical limits of the natural environment. To abide by this tenet, tour operators may have to trade off economic gain for ecological sustainability and, indeed, will have to accept that there are some places where tourism should be excluded.

(McKercher 1998: 191)

Nature-based tourism can only survive when the resources on which it depends are protected. Ecotourism was first described by Hector Ceballos-Lascurain (1987, in Boo 1990: xiv) as 'Travelling to relatively undisturbed or uncontaminated natural areas with the specific objectives of studying, admiring, and enjoying the scenery and its wild plants and animals, as well as any manifestations (both past and present) found in these areas'. According to Whelan (1991: 4), 'ecotourism, done well, can be sustainable and a relatively simple alternative. It promises employment and income to local communities and needed foreign exchange to national governments, while allowing the continued existence of the natural resource base'. This last point gives implicit recognition to the need for adequate and appropriate management regimes (also see Valentine 1991: 5), which foster environmental and cultural understanding, appreciation and conservation (e.g. see Richins *et al.* 1996).

An important element in the development of any management regime or programme, is appropriate research. Yet, much tourist activity in natural areas is permitted without a great deal of understanding of tourism's impacts on the ecosystem. With respect to both flora and fauna, and the landscape itself, this is a critical point. For instance, the impacts of tourism on wildlife are well-documented but largely site-specific, and related findings and management strategies are difficult to apply universally. As HaySmith and Hunt (1995: 206) point out:

Impacts on wildlife from nature tourism are varied, and are often difficult to observe and interpret. Reactions of animals to visitors are complex. Initially, some species or individuals of a species retreat from visual or auditory stimuli caused by humans but become habituated over time. Other species or individuals that are more sensitive may alter their behaviour and activities to completely avoid contact with visitors, with potentially long term effects. Other animals cannot escape the disturbance and may be negatively affected, directly injured... or killed.

Nature tourism can be blatantly invasive toward wildlife when hundreds of observers congregate to view one rare animal or group of animals, when artificial feeding is used to draw animals for tourist viewing and entertainment, and when relationships between species are disturbed (for a more detailed discussion, including case studies, of the relationship between recreation, tourism and wildlife, see Knight and Gutzwiller 1995) .

Nature-based and related forms of tourism will only be successful if comprehensive planning strategies include appropriate and extensive research programmes. Any arguments that nature-based, or any other form of tourist activity, has a particular beneficial or negative relationship with the environment, cannot be sustained without related research. Those who choose to argue one way or another could be easily challenged by questions about the precise nature of the tourism–environment relationship.

Tourism and sustainability

In a more environmentally conscious world, the tourism industry faces increasingly stringent conditions on development; this reflects the concern for sustainability, and the long-term viability of the resources on which tourism depends. The challenge for the industry is to justify its claims on those resources with a commitment to their sustainable management.

Environmental auditing, whether by regulation or legislation, or when undertaken as a self-regulatory initiative, can be a useful management tool to help achieve sustainable development. As global demands on space and resources grow, with increased population, technological change, and greater mobility and awareness, pressure will increase on the tourism industry to implement appropriate steps for monitoring and evaluating its environmental performance. The task ahead is to formulate and implement effective, self-monitoring procedures in order to promote greener, more environmentally compatible forms of tourism, and to avoid the imposition of mandatory compliance measures.

Originally, facilitating travel was the primary focus in tourism planning, with the focus largely on tourism promotion. 'Subsequently, policies broadened to include spatial planning, but the emphasis remained on maximising economic development' (Getz 1986, in Godfrey 1996: 59). Since the publication of the World Conservation Strategy by the International Union for the Conservation of Nature, many countries and regions have begun working towards the goal of sustainable resource development (World Commission on Environment and Development 1987). Sustainable resource management is now widely accepted as the logical way to match the needs of conservation and development.

The era of environmental concern, ushered in by the World Conservation Strategy, is of immediate relevance to tourism. The environment represents not merely a constraint for tourism development, but a resource and an opportunity. Ideally, satisfying tourism settings grow out of complementary natural features and compatible social processes. At the same time, modern tourism amply demonstrates the capacity of human beings to manipulate the environment for better or for worse. Yet, the consequences are not easily predictable. Tourism can certainly contribute to environmental degradation and be self-destructive; it also has the potential to bring about significant enhancement of the environment. With tourism-induced change, an important issue is irreversibility, which, in turn, is a function of factors outlined above.

Much attention is given by government and industry to the development of sustainable tourism (e.g. Butler 1990, 1991; Pigram 1990; Inskeep 1991; Bramwell and Lane 1993; Gunn 1994; Godfrey 1996).

Sustainable development is positive socioeconomic change that does not undermine the ecological and social systems upon which communities and society are dependent. Its successful implementation requires integrated policy, planning and social learning processes; its political viability depends on the full support of the people it affects through their governments, their social institutions, and their private activities.

Rees (1989, in Gunn 1994: 85)

Sustainable development stresses that economic development is dependent upon the continued well-being of the physical and social environment (Dasmann 1985; Barbier 1987; Butler 1991). Dutton and Hall (1989) identified key mechanisms by which sustainable development could be achieved:

- developing cooperative and integrated control systems;
- developing mechanisms to coordinate the industry;
- raising consumer awareness;
- raising producer awareness;
- planning strategically to supersede conventional approaches.

While sustainability is an extremely influential concept in tourism planning, in practice it is fraught with problems (see Hall *et al.* 1997). Perhaps there is some particular merit in Ashworth's (1992: 327) rather cynical observation that the:

tourism industry is tackling the criticisms being made of it, not the problems that cause the criticisms. If there is no resource or environmental problem, then it does not need to be defined nor do solutions need to be found. The problem is seen as one of promotion, and promotion is what the tourism industry is particularly good at. Buying off the grumblers with a few 'commitments' and 'mission statements'... is easier than the alternative [of sustainable tourism planning].

These comments aside, and despite difficulties in achieving sustainable tourism, the integration of economic, sociocultural and environmental planning goals, is increasingly being recognised as a vital component of longer-term tourism development that maintains cultural identity and biodiversity.

The realisation that more than one form or manifestation of tourism is possible, has prompted the development of alternative typologies seen as achievable and desirable, depending upon the circumstances. The term 'sustainable tourism' can be used to refer to tourist typologies, options or strategies preferable to mass tourism. This has led to some confusion as governments and industry attempt to avoid the mass tourism label. As Godfrey (1996: 60–1) pointed out:

In trying to be different, common phrasing and synonyms such as soft (Kariel, 1989; Krippendorf, 1982), postindustrial (SEEDS, 1989), alternative (Gonsalves and Holden, 1985), responsible (WTO, 1990), appropriate (Singh *et al.* 1989), green (Bramwell, 1991), rural (Lane, 1989, 1990), low impact (Lillywhite and Lillywhite, 1991), eco- (Boo, 1990), and nature-based (Fennel and Eagles, 1990) have all been applied.

Differences in conceptualisations and applications of sustainable tourism plans and policies are to be expected. Depending on a person's viewpoint, sustainable tourism may represent particular markets, may be about planning and policy processes, or it may be a governing principle (see Godfrey 1996). Whatever the case, positive elements in a strategy for sustainable tourism typically include:

- development within each locality of a special sense of place, reflected in architectural character and development style, sensitive to its unique heritage and environment;
- preservation, protection and enhancement of the quality of resources which are the basis of tourism;
- fostering development of additional visitor attractions with roots in their own locale and developing in ways which complement local attributes;
- development of visitor services which enhance awareness, understanding and development of local heritage and environment; and
- endorsement of growth when and where it improves things, not where it is destructive or exceeds natural and social carrying capacities, beyond which the quality of human life is adversely affected (Cox 1985: 6–7).

Summary

The concept of 'environment' is very broad, as is the concept of a 'tourist industry'. The relationship between tourism and the environment is extremely complex and dynamic, and not well understood. Research concerning the impacts of tourism on the economic, physical and social environments is lacking. So, too, is research on the impacts of tourism on the physical environment.

Integration of tourism and the environment is occurring at the global to site-specific levels, with benefits accruing to conservation of natural and cultural environments. Increasing attention is also being given to tourism's potential to contribute to regional economic development, even in national parks (see Chapter 10). Unfortunately, however, there are still numerous cases where the effects of tourism on the environment are negative, and unnecessarily so. The emergence of sustainable approaches to tourism development, encompassing notions of environmental compatibility, are laudable, but even then, the concept of a sustainable industry is open to challenge.

Travellers are becoming increasingly sophisticated and discerning. Many such travellers are looking to high-quality, authentic, natural and cultural environments, where the likelihood of recreational satisfaction is high. A great responsibility rests with the industry and governments to develop sustainable industry practices which conserve the natural and cultural environments, and which, subsequently, will hold tourist appeal. Somewhat ironically, economic arguments relating to the generation of tourist revenue, often hold the key to the conservation of resources, whose 'real' values are intangible now and in the long term.

Guide to further reading

- *Seasonality and tourism*: Baron (1975); Bonn *et al.* (1992); Calantone and Jotindar (1984); Chon (1989); Snepenger (1987); Uysal and Hagan (1994); Uysal *et al.* (1994); Butler and Mao (1997).
- *Sustainable tourism*: Barbier (1987); Pigram (1990); Bramwell and Lane (1993); Butler (1991); Hall (1995); Butler *et al.* (1998).
- *Nature-based tourism*: Valentine (1992); Harper and Weiler (1992); McKercher (1998).
- *Ecotourism*: Boo (1990); Whelan (1991); Cater (1993); Wight (1993); Lindberg and Hawkins (1993); Richins *et al.* (1996).
- *Tourism in Developing Countries*: Lea (1988); Richter (1989); Opperman and Chon (1997); Mowforth and Munt (1998).

Review questions

1 Define sustainable tourism. Discuss the relevance of sustainable development principles to tourism planning and management.
2 Discuss the relationship between tourism and the environment. Present an overview of case studies where tourism has contributed to conservation of the natural and built environments.
3 What approaches have been utilised to study and manage seasonality in tourism?
4 Why have some less developed nations utilised tourism as a means of economic and social development? Overall, would you consider tourism has brought many benefits to such countries? Explain your answer with reference to case studies.
5 What are the main factors affecting tourism's potential to impact on the physical environment? What planning measures have been utilised to manage tourism's physical impacts in one or more natural areas in your country or local area? What have been the outcomes of these measures with respect to visitor management (e.g. satisfaction) and visitor impacts on the physical environment?

13

PLANNING FOR OUTDOOR RECREATION IN A CHANGING WORLD

At first sight, the notion of planning for outdoor recreation might seem a contradiction in terms, and likely to inhibit the spontaneity and freedom of choice associated with leisure activities. However, planning for outdoor recreation should be seen as essential as planning for other human needs such as health and welfare, transport and education.

By definition, planning should be proactive and forward looking, not relying merely on prohibitions and the *ad hoc* imposition of restrictions in reaction to problems as they arise. Emphasis in the planning process for outdoor recreation should be on the creation of physical and social settings in which people can exercise choice and satisfy their demands, within prevailing laws, economic limitations and resource constraints. It is in the expansion of choice through the provision of a diversity of opportunities for recreative use of leisure, and in the satisfaction of recreational participants, where the planning and management of recreation resources make an essential contribution.

In one sense, planning can be thought of as the ordering of space through time. In the planning of recreation space, the aim should be to provide a range of functional and aesthetically pleasing environments for outdoor recreation, which avoid the friction of unplanned development, without lapsing into uniformity and predictability. New spatial forms and settings need to be kept as open and flexible as possible, in keeping with a diverse array of interests and dynamic physical, political, economic, social and technological circumstances. Recreation is generally marked by voluntary, discretionary behaviour. People choose to take part or not, and decide the location, timing, activities and costs to be incurred. Any one of these attributes can be modified or dispensed with by unforeseen or uncontrollable factors. Moreover, the process of choice is imperceptibly influenced by such factors as family relationships and personal characteristics, and pervasive adjustments to changes in income, education, lifestyle, social mores, traditions and culture.

Against such a background of change, planners seeking to cater for outdoor recreation demands into the next century, must somehow anticipate a future influenced by a bewildering set of forces, many of which are, and will be, difficult to predict. Given this uncertainty, planning initiatives become even more important, to help underpin forms and patterns of outdoor recreation resilient enough to respond readily to environmental changes. This concluding chapter explores some of the implications of change for the planning and management of recreation resources, introduces a strategic planning approach as a means of adapting and coping with change, and looks towards what kind of future recreation planning should be directed.

Trends in leisure and outdoor recreation

The only things that are certain about the future are uncertainty and the inevitability of change and need for adjustment. Forecasts about possible leisure scenarios range from the fanciful prophecies of science fiction to more considered statistical predictions based on short-term projection or extrapolation of current trends. Such forecasts can only be expressed in terms of probability, and without the benefit of insight into innovations, changes in social circumstances and public policy, or technological breakthroughs. The demand dimension (e.g. population charact-eristics and recreation propensities) and the supply side of the equation (e.g. futuristic possibilities regarding the availability and use of recreation space) both lack clear definition. The ways in which demand and supply factors interact in terms of environmental impacts and recreation decision-making are not well understood, except in specific case studies. Moreover, any planning initiatives must be undertaken against a background of increasing environmental awareness and constraints on freedom of choice, because of concern for repercussions on nature and society. The travel industry, for example, is grappling with specific issues such as air and noise pollution, which have to be solved regardless of cost in money or efficiency terms. 'Consumerism', too, is imposing greater demands on the recreation planner to provide quality products and experiences that do not always coincide with the earlier trends to mass participation and packaged tours. Currently overshadowing all these factors are economic considerations in public sector planning, particularly with respect to global financial markets.

Economic fluctuations

Globalisation has brought with it the realisation that no part of the world can be quarantined from the shock of economic reversals and the often painful restructuring which ensues for national, regional and local economies. Reductions in government intervention and public sector expenditure, already hallmarks of 'economic rationalism', are accelerated by such things as a downturn in economic prospects, or by national and regional budget deficits. As one or a combination of factors such as inflation, unemployment, recession and outright poverty bite into scope for individual choice and relatively unfettered decision-making, many people's recreation opportunities inevitably contract. Curtailment of living standards, frustration of aspirations towards self-betterment, loss of self-esteem, destruction of long-held values, erosion of faith in 'the system' or government, and personal stress leading to emotional and behavioural trauma, can all be the outcomes of economic instability. These, in turn, have serious implications for recreational patterns, opportunities and planning processes.

Predictably, any rearrangement of priorities in a time of financial stringency, is likely to see recreation decline in importance. This holds equally for individuals, households and governments. Thus, pleasure travel, generally, is curtailed, purchases of recreation equipment are postponed, and participation in recreation, in so far as it involves spending, or even the use of resources (including time), which could be income-producing, is minimised. Moreover, governments and providers of recreation opportunities in the private sector, also experience difficulties in meeting their commitments during periods of inflation, economic recession, high interest rates, or any combination of these.

In times of adversity, the availability of recreation outlets takes on renewed urgency in helping to mitigate the effects of economic hardship. Recreation, in the sense of revitalisation, can act as a compensating mechanism in allowing people to forget their worries, or at least to cope better. Fresh interests can be developed, and hitherto neglected, or simpler pursuits rediscovered, in order to occupy an excess of leisure time in a less cost-intensive manner. New skills and attitudes can be acquired which will enable disadvantaged sectors of the population to maintain their self-confidence, pride and hope.

At the same time, there is a brighter side to economic and social hardship (e.g. during periods of excessive inflation and during recession). When governments are forced to withdraw from, or reduce their involvement in, the field of recreation, communities have an opportunity to query the need for continued dependence on public funding, and the stimulus to examine the potential of self-help, co-operation and other means of economising. Thus, hard times can become a vehicle for bringing communities together. People can share frustrations and problems, substitute voluntary effort and talent for that previously provided and, in so doing, achieve a satisfying, cost-effective recreation programme at the neighbourhood or community level.

Of course, some governments and public agencies do not need the excuse of budgetary constraints to opt out of any responsibility for recreation. Even in 'normal' times, there are wide disparities at the national, state and local level in the commitment of funds and resources in this area. Some authorities maintain that recreation is not a legitimate field of interest, or priority, for publicly-elected bodies, and that private enterprise can best fill the gap. Others justify reductions in funding on the grounds of past excesses and waste. The notion that public provision for leisure and recreation is somehow dispensable, or at least low in priority, can only be overcome by a well-directed campaign from those affected (i.e. the community) to convince legislators that recreation is no longer a luxury or a privilege, but a right. In the meantime, competition for scarce public funds, overuse of available recreation resources, and intensified conflicts over shrinking recreation space, can only make the relevance of planning and management of outdoor recreation opportunities an even more urgent social issue.

Societal change

A second group of factors is that of changing lifestyles related to alterations in demographic patterns and social values. The changing age distribution of populations, the postponement of marriage and children (perhaps indefinitely), the liberation of women, the more frequent breakdown of family relationships, the proliferation of communication and other technological advances, and the change in societal attitudes to particular recreation forms and activities, are seen as some of the important influences on leisure and outdoor recreation in a changing world.

Societal changes normally do not exhibit the same cyclical characteristics as those of an economic nature. Rather, they are evolutionary and cumulative, and, sometimes, almost imperceptible. Ageing of Western populations has been accompanied by a measure of affluence, increased unobligated time, and a desire to remain active among older age groups. For example, it is no longer unusual to find 'elderly' people undertaking strenuous forms of outdoor recreation, uninhibited by misperceptions, restrictions and taboos of a time past.

Coupled with sharp reductions in the hours and periods of work, and a change in attitudes to work and play, these trends are helping to shape a different set of recreation patterns and demands in North America and similar developed economies. The US shares with Canada, Britain and, increasingly, Australia, strong overtones of cultural pluralism in its population characteristics. Once again, a reaction can be discerned in leisure behaviour, as new groups of people are assimilated to a greater or lesser degree into the population as a whole. A heterogeneous society offers a richer spectrum of recreation opportunities, but at the same time generates difficulties for governments in providing a sufficiently diverse array of recreation experiences for a multicultural population.

Technological innovations

Outdoor recreation, as is the case for many other human activities, is feeling the shock of technological change. Improvements in transport and communications have decreased the friction of distance and made the greater part of the globe accessible. In tourism, for example, the advent of long-distance, large-capacity aircraft has made mass participation an international reality, while high-speed, computer-based communication facilities are now an integral part of the global tourism network. Not only do these facilities enable instantaneous links across the world, they have also added immeasurably to levels of awareness, both of tourists and those servicing the travelling public. With awareness comes stimulated demand for hitherto little-known sites and destinations. This, coupled with the ability to move vast numbers of people great distances in relatively short periods of time, means that few parts of the planet can be regarded any longer as out of reach for tourism and outdoor recreation.

Technological advances in motor vehicles, along with improvements to routeways and servicing, have increased the range and accessibility of places for recreation. The development of all-terrain vehicles, including the four-wheel drive, allows the recreating public to penetrate remote and possibly fragile environments. This brings with it potential problems of ecological disturbances, resource degradation, litter and overcrowding. At a larger scale, the prospect of wider introduction of high-speed rail transport is likely to impact on the recreation travel market and take business away from competitors. The Channel Tunnel is one example, and there are plans for fast trains to link airports to cities, and to cut travel time drastically between destinations. This could bring a whole new dimension to the limits on day and weekend recreation trips, and could reduce the obligated travel times associated with the journey to work. In the process, greater pressure is likely to be felt on recreation resources, both in near-urban areas and at more remote and sensitive sites, making the need for planning and resource management even more apparent.

New technology is also enabling people with disabilities to participate in outdoor recreation activities such as camping, sailing, fishing, rock climbing and snow skiing. At the other end of the spectrum, technology has opened up new opportunities in risk recreation. Examples include high-performance mountain bikes; lighter and more hydro-dynamic surf boards; stronger, more flexible climbing equipment; lighter, inflatable and more durable kayaks; and helicopter access to remote sites. Whereas advances in technological equipment and materials enhance recreation options, negative impacts such as loss of self-sufficiency, and diminution of uncertainty and risk, must be addressed by recreation managers and planners (Hollenhorst 1995).

Perhaps the most exciting and powerful tool of technology to impact on outdoor recreation is the Internet's World Wide Web (Kirkby 1996). The potential of the rapidly expanding Internet to create and disseminate knowledge relevant to leisure activities is receiving increasing attention (Hawkins 1997). Interactive media, accessed through the Internet, creates new opportunities for recreation and tourism enterprises and services. It is estimated that more than 100 million people will be connected to the Internet by the year 2000, with instant access to relevant, in-depth, up-to-date information about any country or recreation destination. Used interactively, the Internet offers the facility to select sites and activities based on product information, images and the promise of a fulfilling leisure experience.

In the context of recreation planning, Geographic Information Systems (GIS) are a most exciting technological development. GIS use digital data collected from various sources that are processed and analysed by high-speed computers, and presented to planners and decision-makers for action. Some potential applications in outdoor recreation include: locating new trails with less potential for environmental damage; specifying fire hazard zones within a park; locating public facilities given constraints of access and proximity to park attractions; locating waste management facilities; and monitoring the environmental impacts of recreation use over time.

Given their multi-variable, multi-dimensional modelling capabilities, GIS appear to be valuable in site locational analysis for large recreation complexes and tourist developments. Moreover, GIS can integrate data on spatial attributes and projected demand structures for a number of competing sites, for the purpose of analysis, evaluation and choice of optimum location. Coupled with a further advance in computer technology known as Digital Visualisation, the recreation planner is able to visualise a 'virtual landscape' undergoing a simulated change of use (e.g. a change from agricultural land to theme park) or degradation (e.g. the effect of traffic in a park). The planner can more easily assess the possible outcomes, according to estimated changes to different variables (e.g. number of users and patterns of use), before the event has happened, and thus formulate and implement appropriate management responses in a proactive manner.

Undoubtedly, the availability of technological advances is expanding recreation options and opening up access to more recreation opportunities. Whereas these developments may bring problems of conflict and resource degradation, the new tools, in the hands of planners and managers, will equip them to deal better with the uncertain dimensions of the outdoor recreation scene in the twenty-first century. The prospective changes and change agents, reviewed in this and previous chapters, should not be considered in negative terms when set against the promise of technology. Change is a powerful force when harnessed constructively. In the context of outdoor recreation, change will challenge both recreationists and planners to respond positively to the new opportunities it offers.

Planning for the future

> ... planning is a process, a process of human thought and action
> based upon that thought – in point of fact, thought for the future –
> nothing more or less than this is planning, which is a general human
> activity.
>
> (Chadwick 1971: 24)

The inevitability of change, and the need for flexibility mean that it is unwise and impractical to base planning for future patterns of recreation demand and resources on past experience. Many forms of outdoor recreation are responsive to variations in such factors as the cost of participation, the availability of transport, or even seasonal variations in opening times. As noted above, they are also likely to react to technological change, to rates of growth in population, to economic prospects, and to policies and priorities set by government and the community. Moreover, constraints on recreation planning can be imposed by unforeseen events, such as changes in international geopolitical or economic circumstances, or natural hazards (e.g. cyclones, earthquakes, tidal waves, floods or drought). None of these variables can be forecast with any real assurance (though they can be recognised as potential threats in contingency plans, in areas where they are likely to occur), nor can it be predicted how they might combine to influence opportunities for outdoor recreation.

Therefore, the preferred approach to meeting future demands for recreation is to plan strategically in terms of recognising a range of possible outcomes, in light of clear aims and objectives. This would encompass a limited number of flexible policy choices (options) and trade-offs, linked to a degree of acceptable risk in the assessment of those outcomes. The time frame can also be important, given the cost and long lead time frequently involved in the acquisition of recreation space and the development of recreation facilities. In some circumstances, one year into the future might be too far away for planning purposes; for others, 2020 might be too close.

By adopting the alternative futures approach, the planner can develop combinations of strategies for meeting anticipated scenarios of future demand, relative to possible fluctuations in variables affecting recreation participation. Seeking technical solutions in 'knee-jerk' fashion to single-issue problems is not planning, particularly in the context of Chadwick's above definition. Increasingly, the recreation planning process will be expected to respond to a changing societal context; one that is marked by dynamic interaction between emerging technology and organisational change, political priorities and economic realities, as well as environmental constraints, to meet a complex array of recreation demands.

A strategic approach to outdoor recreation planning

Earlier in this chapter, it was suggested that planning could be thought of as the ordering of space. With recreation planning, that rather bland description needs further elaboration. Paraphrasing Getz (1987), recreation planning can be seen as a process, based on research and evaluation, which seeks to optimise the potential contribution of recreation to human welfare and environmental quality. Getz was actually focusing on tourism planning, but the message is the same. Rather than developing and promoting recreation for its own sake as a perceived desirable aspect of growth and change, the emphasis should be on socioeconomic and environmental enhancement, and the use of recreation to achieve those broad goals.

Getz (1987) advocates an integrative approach to tourism planning. Translating that approach to recreation, integrative recreation planning should be characterised as:

- *Goal-oriented*, emphasising the role to be played by recreation in achieving specified societal goals.
- *Democratic*, with meaningful input from the community level.
- *Integrative*, placing recreation planning issues within mainstream planning for other purposes.
- *Systematic*, based on research, prediction, evaluation and monitoring of outcomes.

Much earlier, Driver (1970a) pointed to key activities in a management process, in which 'planning' is but one activity:

- *The democratic process*: by which representation of interests and values are built into the political process of democracy.

276

- *The decision process*: of choosing among alternatives.
- *The administrative process*: whereby agencies created by decision-makers carry out the functions assigned to them.
- *The planning process*: accomplishing goals and providing information for decision-making, and the formulation, implementation and control of plans.

The conventional strategic management process (the terms 'strategic management' and 'strategic planning' will be used interchangeably) can be adopted by planners and resource managers. Such an approach accommodates Getz's integrative approach, Driver's concept of planning, and the concept of planning for alternative futures. By its very nature, strategic planning requires planners and their respective organisations to consider their existing operating environments (inside and outside of their organisations), and potential change in those environments.

Strategic outdoor recreation management is:

> The process of identifying, choosing and developing outdoor recreation activities that will enhance the long-term provision of satisfying recreational experiences by setting clear directions, and by creating ongoing compatibility between available skills and management resources, and the changing internal and external planning environments within which planners and managers operate.
>
> (adapted from Viljoen 1994)

As noted above, the environment of recreation planning is inherently less predictable than in the past. Strategic management offers a means of dealing with change. The following discussion provides a brief introduction to the strategic management process. Readers are referred to the Guide to Further Reading at the end of this chapter for more references to comprehensive discussions of strategic management.

Strategic management is a process encompassing a range of interrelated activities, through which planners and managers move back and forth over time. At its broadest level, strategic management requires a clearly articulated mission statement. The 'mission' guides the organisation through several interrelated processes: strategy analysis, direction setting, strategy choice and strategy implementation, evaluation and control (see Figure 13.1) (Viljoen 1994: 40–3). The nature of the mission statement and the basic parameters of these four processes are discussed below.

Establishing a mission

The establishment of an organisation's mission is critical to its operations because:

> The mission statement, *inter alia*, articulates the overall purpose of the organisation and its distinctive characteristics... This is important to guide

277

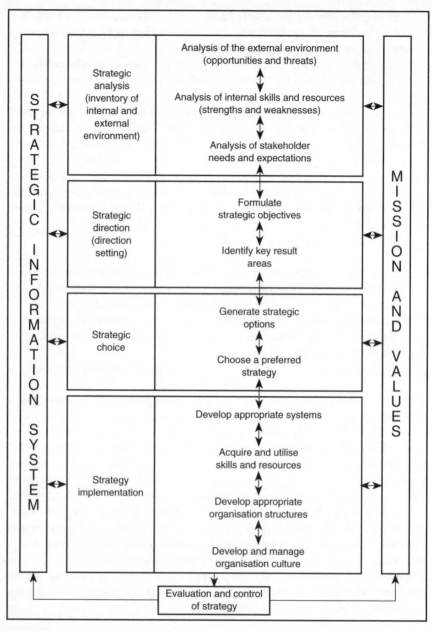

Figure 13.1 The strategic management process

Source: Adapted from Viljoen (1994: 43)

the activities of strategic analysis (we should only analyse the environment and our resources in relation to our stated overall purpose), strategic choice (we should only choose strategies that are consistent with our overall purpose), and strategy implementation (we should implement strategies in a way that will help us better achieve our overall purpose). All aspects of strategic management, therefore, should be referenced to the mission of the organisation.

(Viljoen 1994: 42)

Mission statements depict the vision for an organisation, defining its purpose and outlining what it intends to accomplish in the larger environment (Rossman 1995). According to Drucker (1974: 94, in Hall and McArthur 1996), 'Defining the purpose and mission of the business is difficult, painful and risky. But it alone enables a business to set its objectives, to develop strategies, to concentrate on resources and go to work. It alone enables a business to be managed for performance'.

Strategy analysis

Strategy analysis involves the gathering and use of information to ascertain the strategic position of the organisation and the situations likely to be faced in the conduct of its activities. Strategy analysis requires managers to inventory all the major forces affecting their recreational product and to determine whether these represent opportunities or threats. These forces may include environmental forces (political, economic, social, technological and physical) influencing the availability of recreational opportunities, as well as the skills and resources (financial, human, physical and intangible) available to manage those opportunities. More specifically, strategic analysis involves several types of analyses such as those presented in Table 13.1.

Establishing strategic direction

Based on the strategy analysis, it becomes necessary to establish goals, objectives and strategies that best suit the organisation. 'Strategic objectives should be derived directly from an analysis of the internal and external environment and the requirements of key stakeholders' (Viljoen 1994: 304). These objectives are tied to Key Result Areas (KRAs) where actions are required. 'Each KRA becomes the focus of the subsequent strategic management processes of strategy choice and implementation' (Viljoen 1994: 41). If the organisation is a local government agency, it may well be influenced more by social equity considerations – the provision of a range of recreational activities accessible to a broad sector of the community – than financial returns in recreation plans and programmes. Thus, an equitable distribution of accessible recreation space may be the overriding strategic objective.

Strategic choice

Strategic choice 'involves the generation, evaluation and choice of a strategy that best suits the needs of the organisation. This process must be built on the previous phases of strategy analysis and direction setting. Generating strategic alternatives is essentially a brainstorming exercise where alternatives are identified and described without being evaluated' (Viljoen 1994: 41).

The SWOT (strengths, weaknesses, opprortunities, threats) analysis (also see Table 13.1) may be revisited at this or any other point. Utilising the outcomes of the brainstorming session, and based on the SWOT analysis, choices may be made about which recreational opportunities will be supplied, in what quantity they will be supplied, and how and when they will be supplied.

Strategy implementation, monitoring and evaluation

Plans are useless by themselves. Once decisions have been reached on the elements to be included in the recreation plan or programme, and the strategies and actions to be pursued, the implementation stage has been reached. Strategy implementation concerns the operational strategies and systems that must be used to put the strategy into practice. It involves developing appropriate management, financial and communication systems; acquiring and utilising relevant human, physical, financial and other resources; developing an appropriate organisational structure; and ensuring that the organisational culture of the organisation is managed in a way that complements the organisation's tasks. These are interrelated activities. For instance, an organisation may acquire new people, whose ideas are innovative to the extent that new recreation supply opportunities, or management techniques or pricing structures are identified. If these ideas are subjected to scrutiny in a strategic manner and then implemented, the nature of recreation demand may change, thereby perhaps bringing further alterations to existing structures, practices and strategies. In implementing plans and programmes, a number of techniques have been employed to keep the plan and its integral components (e.g. budget, resources and timing) on track. These techniques include: programme evaluation review technique (PERT), critical path analysis, milestone scheduling, and Goals Achievement Matrices.

Implementation of a plan is not an end in itself. If a plan or programme is to satisfy the needs of users for whom it was designed, it must be subjected to monitoring and evaluation. Monitoring is concerned with the collection of information about the developing state of a system to which any planning and management process is being addressed (see Haynes 1973). Evaluation is an activity designed to collect, analyse and interpret information, concerning the need for particular policies, plans or programmes, as well as the formulation, implementation, outcomes and impacts of policies, plans or programmes. The days of 'finishing the Master Plan', which remained unaltered over time, have hopefully gone. The plan or programme should not be something set in concrete, but something to be worked and reviewed, and further refined and adapted, as needed, during

Table 13.1 Strategic analysis

Type of analysis	Information to be gathered
Aspirations analysis	*Determine* stakeholders (i.e. those with an interest in the recreational activity or product: staff of the management agency; providers; users), their power bases, and their possible reactions to particular strategies, the effects of their reactions, and the means of shaping strategies to account for their desires.
Environment nature analysis	*Examine* the complexity of the planning environment (e.g. the number and nature of agencies involved in managing a multiple use resource, or competing for use of a resource).
	Examine how the recreational planning environment interacts with other planning environments (e.g. recreation planning in environmentally sensitive areas or in urban environments).
	Examine flexibility in recreational choice or recreational supply (e.g. potentials for substitutability, degrees of recreational specialisation required).
Analysis of the structure of the environment	*Analyse* potential and actual markets (e.g. market segmentation analysis; assessment of latent demand for recreational activity).
	Analyse environmental forces (see chapter 11 and above) that might impinge on recreation supply and demand.
Resource Analysis	*Develop* an inventory of resources (physical; human; systems and intangibles) in current areas of operation (finance; personnel; research and development; marketing; capital).
SWOT Analysis	*Identify* significant opportunities for, and threats to, the supply of outdoor recreation opportunities (e.g. improved water treatment facilities for dams so that a wider range of recreational activities can be developed, or opposition from conservationists to recreation activities in sensitive/protected areas, respectively).
	Identify specific strengths (e.g. large land and water base; strong political support; sound financial backing; strong leadership and community backing) and weaknesses (e.g. very small land base; shrinking financial and other resources; weak community and political support) in the organisation's ability to supply recreational opportunities.

Source: Adapted from Viljoen (1994); Hall and McArthur (1996)

implementation. The analysis of issues and best courses of actions should not stop once the plan is in place, but remain a constant approach. In brief:

> ... the effectiveness of a planning system must be judged by its continuing ability to influence change toward desired ends, and in its responsiveness to pressures to alter those ends in conformity with societal goals.
>
> (Haynes 1973: 5)

The general roles of, and reasons for, monitoring and evaluation in strategic planning include:

- assessing the degree of need for particular plans;
- continuous functioning of the planning process to enlighten, clarify and improve plans;
- conceptual and operational assistance to planners and decision-makers;
- specification of plan outcomes and impacts;
- assessing or measuring the efficiency and effectiveness of recreation plans in terms of resources;
- accountability reporting for resource allocation, distribution and redistribution;
- symbolic reasons – to demonstrate that something is being done;
- political reasons – the evaluation is directed and run in a way that will deliberately dictate the findings (e.g. evaluations may be motivated by self-service as well as public service, or by a desire to use analysis as ammunition for partisan political purposes) (see Hall and Jenkins 1995).

By way of example, Shivers (1967) identified several purposes of evaluation with respect to recreation plans and programmes (see Table 13.2)

Finally, two main forms of evaluation have been noted. *Formative evaluation* is periodic monitoring as the plan or programme is implemented, with a view to making necessary changes and to 'fine tuning' the plan. *Summative evaluation* is the assessment of the plan after it has been completed. Ideally, formative evaluations and, ultimately, summative evaluations, should be used.

The remainder of this chapter examines the principles and practices in setting goals and objectives, difficulties in implementing recreation and tourism plans and policies, and a number of broader recreation planning challenges.

Establishing goals and objectives: principles and practices

Conventionally, the planning process can be set out in terms of goals and objectives, which form the basis for programmes and courses of action to achieve the stated objectives, and move closer to attaining the goal(s) specified. As used in this discussion, the term 'goal' refers to a preferred state or condition towards which action is to be directed. An 'objective' is a specific, positive step, attainable as part of progress towards a particular goal. Goals are long-range targets, which

Table 13.2 The purposes of evaluating a public recreation program

---

1 To ensure that the recreation program meets the stated needs and desires of the people in the community.
2 To promote professional growth and education among staff members of the recreation service system.
3 To ascertain the flexibility of policies within the system.
4 To appraise personnel quality and qualifications in relation to specific functions within the system.
5 To develop firmer grounds of agency philosophy so that a logical frame of reference is developed.
6 To effectively gauge public sector sentiment, attitudes and awareness of the recreation system.
7 To increase knowledge gained through practice and to additionally test current practice as to applicability in the public recreation setting.
8 To appraise existing facilities, physical property and plant as to their adequacy, accessibility, safety, attractiveness, appropriateness and utilisation.
9 To seek out and eliminate any detrimental features within the program or agency.
10 To add any feasible and constructive devices, methods and experiences to the system in order to provide the most efficient and effective service to the community.
11 To promote recognition of the agency on the part of the community.
12 To replace outmoded concepts and invalid ideas which the public may have concerning the recreation agency.
13 As far as possible, to promote the professionalisation of agency personnel and the services provided.
14 To avoid unnecessary expenditure of public monies because of inadequate co-ordination in the provision of recreation services.
15 To ensure the agency and its personnel safeguard against political upheaval and partisan politics.
16 To ensure the adequate provision of spaces, areas and facilities will be safeguarded against any encroachment by establishing protection in perpetuity through dedication of all physical property for public recreation purposes only.

Source: Shivers (1967: 452)

give purpose to the planning process; objectives represent a yardstick by which achievement may be measured. The explicit identification of goals and objectives allows the definition of criteria to identify and evaluate alternative strategies, proposals and policies. It also helps determine the urgency of issues and problems, and the setting of priorities.

In the context of recreation planning, goals may be identified broadly with quantitative and qualitative advances in the availability of a range of opportunities for participation in recreation. The realisation of such goals should be in harmony with the maintenance, enhancement and, where necessary, restoration of desirable features of the physical and social environment. These goals may then be translated into attainable, quantifiable objectives, oriented towards specific aspects of recreation management and linked to designated policies, programmes and courses of action to achieve the stated objectives.

In the planning of urban recreation space, for example, a goal might be:

to provide the community with the widest choice and maximum diversity of recreational opportunities consistent with economic feasibility, the expressed needs of the population, and societal goals.

Examples of objectives derived from this goal include:

- to provide a 'defined quantity' of open space of various types;
- to provide a hierarchy of parks of varying size and distribution from each other;
- to provide for multiple use of specified areas of open space;
- to locate various types of open space in a given proximity to residential areas;
- to link open space into a linear network;
- to allow for conservation of specified areas of natural bushland.

An example of a coastal region's goals, objectives and strategies to address increasing environmental stress from tourism is provided in Table 13.3.

Based on the objectives that are set, policies and programmes can be formulated and courses of action specified, as essential inputs to the implementation phase of the planning process. These could include:

- inventory of existing supply of urban recreation space and its characteristics, distribution and use, relative to identified and predicted needs;
- identification of potential areas of open space to complement existing recreation space;
- determination of priorities for implementation;
- preparation of site development plans and management programmes for existing recreation space and areas to be acquired; and
- monitoring and review of the effectiveness of the planning process in addressing the changing recreation space needs of a dynamic urban system.

Strategic planning provides a framework for a very wide ranging, yet integrated approach to recreation planning. The emphasis on provision of choice and diversity in recreation opportunities, tempered by relevance and realism, should ensure an effective contribution to the long-range goals of human welfare and environmental enhancement, with respect to recreation supply.

At the same time, specifying the goals and objectives for the planned use of resources is a demanding task. By definition, the goals adopted should be broad, comprehensive and long-range. However, care is needed to ensure that they do not become merely 'parenthood-type' statements, of little practical relevance or application. Nor should they be geared to an excessively narrow perception of the future, but be sufficiently flexible to accommodate new concepts and information as they emerge.

Table 13.3 Strategies for Germany's Baltic coast (Rügen)

Tourism has been a major source of income for the north coast of Germany for many decades. The most important tourist resorts on the southern Baltic coast of Germany are, Schleswig-Holstein, Fehmarn Island, the coast of Mecklenburg-Vorpommern, and the islands off the shallow coast, especially Rügen.

Increasing visitor numbers have led to a rise in environmental stress. According to Helfer (1993), tourism and leisure-time infrastructure on the island of Rügen have had a damaging impact, particularly on the forests protecting the coast (e.g. exposure, damage and destruction of tree roots; damage to trunks and branches; destruction of ground vegetation from trampling and driving on it; soil erosion and contamination; moderate-to-strong thinning out of the coastal-protective forest for car parks and paths; clearings for activities).

A development plan at the county and regional level was developed in 1990 and revised in 1991. That plan includes guidelines for tourism and recreation development. In the area of tourism and recreation, the general goal is:

to develop the area for the improvement of social, economic and environmental living conditions for the populace, of and visitors to, the county [of Rügen].

Of particular importance for this goal are the following points, which amount to strategic objectives for Key Result Areas:

- the securing of jobs;
- the preservation and development of the characteristic structure of settlements and the typical landscape;
- the preservation and development of natural values and functions.

In particular, a considerable increase in the quality of the tourism sector is planned, in which the scenic and cultural aspects of Rügen and its environs will serve to develop a characteristic range of touristic and recreational attractions.

For the coastal area the 'zone ordinance' of the structural plan makes the stipulations listed below. Some of these are more specific than strategic objectives, and amount to a combination of operational objectives, performance targets and strategy selection:

- in the entire shore zone, which is at least 100m wide, tourist use must take place only on the basis of the building plans;
- a strip of shoreline 100m wide is to be kept free of construction;
- on the long, drawn-out beaches of the Schaabe and between Gohren and Thiessow, the hinterland is to be kept free of construction and car parks as far as possible, in order to protect the landscape; access is to be secured by means of public transportation;
- the characteristic of the resorts as spas and medicinal baths should be built up again and developed, in connection with improving the environmental situation;
- camping grounds are to be removed from the forests protecting the coasts;
- solid buildings which have been constructed without pertinent regulations, should be examined as to their compatibility with the structure of settlements and landscape, and removed if need be;
- aquatic sports, particularly involving pleasure boats, must be offered suitable facilities, without constructing unprofitable overcapacities, e.g. aquatic sports are to be restricted at moorings, in protected areas and shore zones;
- sports which are not in keeping with nature-compatible recreation in the landscape are to be prohibited.

Source: German Federal Agency for Nature Conservation (1997: 184)

Objectives, as defined above, represent progressive steps towards goal attainment. Although the distinction is not always clear-cut, objectives may generally be expressed in more tangible terms. It is important, too, that the objectives delineated are not restricted only to recognised and familiar strategies. The kind of planning challenges identified in earlier chapters call for a strategic approach, incorporating a range of possible options, so that an appropriate response can be made as and when circumstances change.

It is important to note, also, that recreation planning, despite its importance, is only one component in an array of means of achieving the type of society and environment to which a nation and its people aspire. Clearly, the recreation planning process must be consistent with planning policies in associated areas of public and private sector responsibility. Therefore, goals and objectives related to recreation planning, should take account of, and complement, those perceived to be worth pursuing by society as a whole. Taking those goals and objectives to fruition brings the focus to the implementation phase.

The implementation phase

Any number of examples exist of detailed plans for recreation, prepared by both public agencies and private interests, which have had no practical outcome. Apparently, a gulf can exist between plan formulation and implementation. Even when plans are approved and adopted, formidable barriers can arise which frustrate attempts to translate them into action. Elements of a plan may be discarded or amended, so that progress towards the achievement of objectives and, ultimately, goals, is interrupted and perhaps stalled altogether.

An implementation gap has been identified across a range of planning initiatives (Pigram 1992). The fact that it is more often encountered in planning for recreation and tourism, probably reflects the relative lack of urgency and lower priority given to these activities by decision-makers, compared with those perceived as more fundamental to social and economic development. Impediments that stand in the way of progress towards realisation of planning outcomes determine the extent of the implementation gap. Bridging this gap is essentially a balancing exercise between political and social acceptability, economic and technological feasibility, and administrative reality (Pigram 1992).

Implementation of recreation plans and policies becomes increasingly more difficult when it crosses different tiers of responsibility, especially where both the public and private sector are involved. Studies examining the barriers that inhibit implementation of recreation plans at the local government level reveal the following constraints (Reid 1989):

- Inadequate funding;
- Lack of skilled personnel;
- Need for structural and operational change within city administration to accommodate planning for recreation;
- Environmental conditions – physical, socioeconomic and political;

- Complexity of the plan itself;
- Lack of opportunity for community endorsement and participation in the planning process and the implementation phase.

Overcoming these barriers is central to recreation planning, and to the successful culmination of the implementation phase.

In summary, strategic planning is a continuous process, whereby resource analysis; developing strategic directions; choosing between potential strategies implementing strategies; and monitoring and evaluating the processes, outcomes and impacts of decisions and actions are ongoing. Planners and managers may even move backwards and forwards among these interrelated processes, because of changing circumstances. For example, the acceptance and facilitation of recreation activities in protected areas may be significantly influenced by political forces (e.g. changes in government), while in some urban areas waterfront development, shopping malls, gambling facilities, and festivals and events, appear to be the 'flavours of the month' in terms of the commitment of government resources.

The challenge of recreation planning

Most recreation experiences do not just happen; they have to be provided for in some way. Earlier, it was noted how the availability of recreation opportunities, services and facilities influences choice in outdoor recreation. It is in the expansion of choice, through provision of a diversity of outlets for leisure to meet the many aspirations of people and society, where planning plays an essential role. By providing a wider range of alternative recreation opportunities, the planner is contributing to the potential of leisure to stimulate and satisfy.

In short, planning for leisure environments of the future must progress beyond establishing a series of services or facilities, such as parks and playgrounds. The challenge is to create a physical and social environment in which individuals can satisfy their recreation interests within the economic limitations and resource constraints likely to be encountered. The emphasis is on recognising the multiplicity of individual and societal goals for leisure, and on the need for diversity, substitutability and choice, rather than uniformity, in addressing those goals. The recreation planner's concern, then, is with generating an appropriate array of leisure opportunities, rather than with provision of specific facilities alone. It is the interaction of people's values, needs and wants with those facilities and services, which generates leisure opportunities, and, ultimately, leads to participation and satisfaction – the end-products of the planning process.

Recreation planning is complex, partly because of the unstructured nature of recreation itself and the many conflicting interests and constituencies which have to be catered for, and also because of the rapidly changing context in which planning takes place (Mitchell 1983). In meeting the challenge of recreation planning, a number of aspects need to be considered – some limiting, and others with potential to contribute to positive outcomes. In particular:

1 The apparent inevitability of dwindling public sector support for provision of opportunities for recreation, and the consequent need to build partnerships with private enterprise, and to harness the promise of self-help schemes and voluntarism.
2 The need to plan within the capability of the resource base, the supporting infrastructure, and the thresholds of tolerance of affected communities, while, simultaneously, applying evolving technologies to expand these constraining horizons.
3 The need to recognise the plurality of the 'market' for recreation planning; to build in diversity and flexibility to accommodate change and compensate for equity deficiencies; and to use recreation opportunities, where possible, to offset negative social forces.
4 The adoption of an integrative perspective. Recreation planning would be one important component of overall planning for community welfare and environmental integrity, based on strategic management frameworks, and encompassing appropriate recreation planning frameworks such as the Recreation Opportunity Spectrum, the Limits of Acceptable Change, and the Visitor Impact Management framework and Visitor Activity Management Process.
5 The blending of 'bottom-up' responses from the participating public in the recreation planning process with balanced 'top-down' assessments from business interests and professional advisers and policy makers.
6 Education about leisure to facilitate individual self-determination and freedom of choice.

In a new age, where some predictions suggest that the central focus of society will be on leisure rather than work, the challenge is to use leisure to find fulfilment and satisfaction in the true meaning of the term, 'recreation'. In an increasingly complex world, the task of creating and enhancing meaningful leisure environments and recreation opportunities becomes a most pressing issue. Failure to meet the challenge, means that society must accept and tolerate existing constraints on leisure behaviour, and continue to condone the legacy of *ad hoc* development, out of balance with community needs and inevitably perpetuating the deficiencies of an inadequate system of provision for recreation (Pigram, 1983). Moreover, without planning, there is little prospect of recreation receiving proper priority in resource allocation, against the claims of competing uses.

Guide to further reading

- For detailed discussions of *strategic planning and management*, see Viljoen (1994); Richardson and Richardson (1994); Mintzberg (1994).
- For detailed discussions of *sustainable recreation and tourism planning and development* see Pigram (1993); Hall (1995); Harrison and Husbands (1996); Butler *et al.* (1998).

- For broad overviews of *recreation and plan and programme evaluation* see: Theobald (1979); Hollick (1993); Howe (1980; 1993); Owen (1993); Rossman (1995). For *triangulation in recreation evaluation*, see Fuszek (1987, in Rossman 1995).
- *Leisure policy and planning*: Henry (1993); Veal (1994).
- *Outdoor recreation policy and planning*: Driver (1970b); LaPage (1970); Hutcheson *et al.* (1990); Mercer (1981b); Mercer (1994b).
- *Tourism policy*: Pigram (1992); Hall and Jenkins (1995).

Review questions

1 Why should we plan for outdoor recreation?
2 What are the main responsibilities of the public sector in outdoor recreation planning in your local area? Is the public sector fulfilling its responsibilities in that regard? Justify your answers utilising primary and secondary information sources.
3 Define planning. Define strategic planning. What are the main steps or features in strategic planning? Discuss the benefits and problems associated with strategic planning approaches generally, and with reference to recreation and tourism specifically.
4 What are the critical issues that will affect future demand for outdoor recreation activities?
5 What are the critical issues that will affect future supply for outdoor recreation activities?
6 How can planners best respond to uncertain futures in outdoor recreation demand and supply?

BIBLIOGRAPHY

ABS *see* Australian Bureau of Statistics

Absher, J. and Brake, L. (1996) 'Aboriginal involvement in park management in Australia', *Trends*, 33, 4: 9–15.

Aitken, S. (1991) 'Person–environment theories in contemporary perceptual and behavioural geography I: personality, attitudinal and spatial choice theories', *Progress in Human Geography*, 15, 2: 179–93.

Alderson, W.T. and Low, S.P. (1996) *Interpretation of Historic Sites*, 2nd edition, Walnut Creek: Sage.

Archer, B. (1973) 'The uses and abuses of multipliers', *Tourist Research Paper TUR 1*, Bangor: University College of North Wales.

Ashley, R. (1990) 'The Visitor Activity Management Process and Canadian national historic parks and sites – a new commitment to the visitor', in Graham, R. and Lawrence, R. (eds) *Towards Serving Visitors and Managing Our Resources*, 249–56, Proceedings of a North American Workshop on Visitor Management in Parks and Protected Areas, Waterloo: Tourism Research and Education Centre, University of Waterloo.

Ashworth, G.J. (1989) 'Urban tourism: an imbalance in attention', in Cooper, C.P. (ed.) *Progress in Tourism, Recreation and Hospitality Management*, 1, 33–54, London: Belhaven.

— (1992) 'Planning for sustainable tourism: slogan or reality?', *Town Planning Review* 63, 3, 325–30.

Ashworth, G.J. and Tunbridge, J. (1990) *The Tourist-Historic City*, London: Belhaven.

Atkinson, J. (1991) *Recreation in the Aboriginal Community*, Canberra: Australian Government Publishing Service.

Australian Bureau of Statistics (ABS) (1986).

— (1993) *Disability, Ageing and Carers in Australia: Summary of Findings*, (Catalogue Number 4430.0), Canberra: ABS.

Bammel, G. and Bammel, L. (1992) *Leisure and Human Behaviour*, Dubuque: Brown.

Bannon, J. (1976) *Leisure Resources*, Englewood Cliffs, New Jersey: Prentice Hall.

Barbier (1987) 'The concept of sustainable economic development', *Environmental Conservation*, 14, 2: 101–10.

Barker, R. (1968) *Ecological Psychology*, Stanford: Stanford University Press.

Baron, R.V. (1975) *Seasonality in Tourism*, London: Economist Intelligence Unit.

Barrett, J. (1958) 'The seaside resort towns of England and Wales', unpublished Doctoral thesis, London: University of London.

Barrett, S. and Hough, M. (1989) 'Rethinking city landscapes', *Recreation Research Review*, 14, 3: 7–13.

Bateson, P., Wyman, S. and Sheppard, D. (eds) (1989) *National Parks and Tourism Seminar*, Sydney: NSW National Parks and Wildlife Service.

Beaulieu, J. and Schreyer, R. (1985) 'Choices of wilderness environments – differences between real and hypothetical choice situations', *Proceedings – Symposium on Recreation Choice Behaviour*, USDA Forest Service General Technical Report, INT-184: Ogden, 38–45.

Beckman, E.A. and Russell, R. (comp.) (1995) *Interpretation and the Getting of Wisdom*, Papers from the fourth annual conference of the Interpretation Australia Association, Canberra: Australian Nature Conservation Agency and Australian Heritage Commission.

Bengtsson, A. (ed.) (1972) *Adventure Playgrounds*, New York: Praeger.

Biddlecomb, C. (1981) *Pacific Tourism, Contrasts in Values and Expectations*, Suva: Pacific Conference of Churches.

Black, A. and Breckwoldt, R. (1977) 'Evolution of systems of national park policy-making in Australia', in Mercer, D.C. (ed.) *Leisure and Recreation in Australian*, 190–9, Malvern: Sorrett.

Blenkhorn, A. (1979) 'The attraction of water', *Parks and Recreation*, 44, 2: 17–23.

Blockley, M. (1996) 'Editorial: rights of access', *Interpretation*, 1, 2: 3–4.

Blood, R.O. and Wolfe, D.M. (1960) *Husbands and Wives*, Glencoe, IL: Free Press.

Boden, R. (1977) 'Ecological aspects of outdoor recreational planning', in Mercer, D. (ed.) *Leisure and Recreation in Australia*, 222–31, Melbourne: Sorrett.

Boden, R. and Baines, G. (1981) 'National parks in Australia – origins and future trends', in Mercer, D. (ed.) *Outdoor Recreation: Australian Perspectives*, 148–55, Malvern: Sorrett.

Boniface, B. and Cooper, C. (1987) *The Geography of Travel and Tourism*, London: Heinemann.

Bonn, M.A., Furr, H.L. and Uysal, M. (1992) 'Seasonal variation of coastal resort visitors: Hilton Head Island', *Journal of Travel Research*, 31, 1: 50–6.

Boo, E. (1990) *Ecotourism: The Potentials and Pitfalls*, 2 vols, Washington, DC: World Wildlife Fund.

Borowski, A. (1990) *Australia's Population Trends and Prospects*, Canberra: Bureau of Immigration Research.

Bowler, I.R., Bryant, C.R. and Nellis, M.D. (eds) (1992) *Contemporary Rural Systems in Transition: Economy and Society, Vol 1, Agriculture and Environment?*, Wallingford: CAB International.

Boyle, R. (1983) 'A survey of the use of small parks', *Australian Parks and Recreation*, November: 31–6.

Bradshaw, J. (1972) 'The concept of social need', *New Society*, 496, 30 March: 640–3.

Bramwell, B. and Lane, B. (1993) 'Sustainable tourism: an evolving global approach', *Journal of Sustainable Tourism*, 1, 1: 6–16.

Braverman, H. (1975) *Labor and Monopoly Capital; the Degradation of Work in the Twentieth Century*, New York: Monthly Review Press

Britten, R. (1979) 'Some notes on the geography of tourism', *Canadian Geographer*, 23, 3: 276–82.

Britton, S. (1980) 'A conceptual model of tourism in a peripheral economy', in Pearce, D. (ed.) *Tourism in the Pacific*, Rarotonga: Proceedings of UNESCO Tourism Workshop, 1–12.

— (1982) 'The political economy of tourism in the Third World', *Annals of Tourism Research*, 9: 331–58.

Brotherton, D. (1973) 'The concept of carrying of countryside recreation areas', *Recreation News Supplement*, 9: 6–11.

Brown, P. (1977) 'Information needs for river recreation planning and management', in *Proceedings, River Recreation Management Research Symposium*, 193–201, USDA Forest Service General Technical Report NC-28, Minneapolis: USDA Forest Service.

Buckley, R. and Pannell, J. (1990) 'Environmental impacts of tourism and recreation in national parks and conservation reserves', *Journal of Tourism Studies*, 1, 1: 24–32.

Budowski, G. (1976) 'Tourism and conservation: conflict, coexistence, or symbiosis?', *Environmental Conservation*, 3, 1: 27–31.

Bull, A. (1995) *The Economics of Travel and Tourism*, 2nd Edition, Melbourne: Longman.

Burton, J. (1975) *The Recreational Use of Malpas Reservoir*, Armidale: University of New England.

Bury, R. (1976) 'Recreation carrying capacity – hypothesis or reality', *Parks and Recreation*, 11: 22–5 and 56–8.

Butler, R.W. (1975) 'Tourism as an agent of social change', in *Proceedings of IGU Working Group on Geography of Tourism and Recreation*, Peterborough: University of Trent, 85–90.

— (1980) 'The concept of a tourist area cycle of evolution: implications for management of resources', *Canadian Geographer*, 24,1: 5–12.

— (1984) 'The impact of informal recreation on rural Canada', in Bunce, M.F. and Troughton, M. J. (eds) *The Pressure of Change in Rural Canada*, Downsview, York University, Ontario: Geographical Monograph 14.

— (1990) 'Alternative tourism: pious hope or trojan horse', *Journal of Travel Research*, 28, 3: 40–5.

— (1991) 'Tourism, environment, and sustainable development', *Environmental Conservation*, 18, 3: 201–9.

— (1995) 'Introduction', in Butler, R.W. and Pearce, D.G. (eds) *Change in Tourism: People, Places, Processes*, 1–11, London: Routledge.

Butler, R.W. and Clark, G. (1992) 'Tourism in rural areas: Canada and the United Kingdom', in Bowler, I.R., Bryant, C.R. and Nellis, M.D. (eds) *Contemporary Rural Systems in Transition: Economy and Society*, 161–85, Wallingford: CAB International.

Butler, R.W., Hall, C.M. and Jenkins, J.M. (eds) (1998) *Tourism and Recreation in Rural Areas*, Chichester: John Wiley and Sons.

Butler, R.W. and Mao, B. (1997) 'Seasonality in tourism: problems and measurement', in Murphy, P. (ed.) *Quality Management in Urban Tourism*, 9–24, Chichester: Wiley.

Butler, R.W. and Waldbrook, L.A. (1991) 'A new planning tool: the tourism opportunity spectrum', *Journal of Tourism Studies*, 2,1: 2–14.

Calantone, R.J. and Jotindar, J.S. (1984) 'Seasonal segmentation of the tourism market using a benefit segmentation framework', *Journal of Travel Research*, 23, Fall: 14–24.

Calder, J. (1974) 'Recreational accessibility for the handicapped', in Australian Department of Tourism and Recreation, *Leisure A New Perspective*, 7.1–7.4.1, Canberra: Australian Department of Tourism and Recreation.

Caldwell, L.L., Smith, E.A. and Weissinger, E. (1992) 'Development of a leisure experience battery for adolescents: parsimony, stability and validity', *Journal of Leisure Research*, 24, 4: 361–76.

Canadian Environmental Advisory Council (1991) *A Protected Areas Vision for Canada*, Ottawa: Environment Canada.

Carmichael, B., McBoyle, G. and Wall, G. (1995) 'Responding to environmental change and variability', in Thomson, J.L., Lime, D.W., Gartner, B. and Sames, W.M. (eds), *Proceedings of the Fourth International Outdoor Recreation and Tourism Trends Symposium and the 1995 National Recreation Resources Planning Conference*, 14–17 May, Minneapolis: University of Minnesota, 513–20.

Carroll, J. (1990) 'Foreword', in Hutcheson, J.D., Noe, F.P. and Snow, R.E. (eds) *Outdoor Recreation Policy: Pleasure and Preservation*, xiii–xviii, New York: Greenwood Press.

Cater, E. (1993) 'Ecotourism in the third world: problems for sustainable development', *Tourism Management*, 14, 2: 85–90.

Chadwick, G. (1971) *A Systems View of Planning*, Oxford: Pergamon Press.

Chang, T.C., Milne, S., Fallon, D. and Pohlmann, C. (1996) 'Urban heritage tourism: the global-local nexus', *Annals of Tourism Research*, 23, 2: 284–305.

Charters, T., Gabriel, M. and Prasser, S. (eds) (1996) *National Parks. Private Sector's Role*, Toowoomba: University of Southern Queensland Press.

Cherry, G.E. (1993) 'Changing social attitudes towards leisure and the countryside 1890–1990, in Glyptis, S. (ed.) *Leisure and the Environment*, 22–33, London: Belhaven Press.

Chon, K. (1989) 'Understanding recreational traveler's motivation, attitude and satisfaction', *Tourist Review*, 44, 1: 3–7.

Christaller, W. (1963) 'Some considerations of tourism location in Europe', *Regional Science Association Papers* XII: 95–105, Lund Congress.

Christiansen, M. (1977) *Park Planning Handbook*, New York: Wiley.

Chubb, M. and Chubb, H. (1981) *One Third of Our Time. An Introduction to Recreation Behavior and Resources*, New York: John Wiley & Sons.

Clark, R. (1976) 'Control of vandalism in recreation areas – fact, fiction or folklore', *USDA Forest Service General Technical Report PSW-17*, Oregon: USDA Forest Service.

Clark, R. and Stankey, G. (1979) *The Recreation Opportunity Spectrum: A Framework for Planning, Management and Research*, General Technical Report, PNW-98, Seattle: US Department of Agriculture Forest Service.

Clawson, M. and Knetsch, J. (1966) *Economics of Outdoor Recreation*, Baltimore: John Hopkins Press.

Clawson, M., Held, R. and Stoddard, C. (1960) *Land for the Future*, Baltimore: John Hopkins Press.

Cloke, P. (1993) 'The countryside as commodity: new spaces for rural leisure', in Glyptis, S. (ed.) *Leisure and the Environment*, 53–70, London: Belhaven Press.

Cloke, P. and Goodwin, M. (1992) 'Conceptualising countryside change: from post-Fordism to rural structured coherence', *Transactions of Institute of British Geographers*, 17: 321–36.

— (1993) 'Rural change: Structured coherence or unstructured coherence', *Terra*, 105: 166–74.

Cloke, P. and Park, C. (1985) *Rural Resource Management*, London: Croom Helm.

Cochrane, J. (1996) 'The sustainability of ecotourism in Indonesia', in Parnwell, M and Bryant, R. (eds) *Environmental Change in Southeast Asia: People, Politics and Sustainable Development*, 237–59, London: Routledge.

Cohen, E. (1978) 'The impact of tourism on the physical environment', *Annals of Tourism Research*, 2: 215–37.

Colby, K. (1988) 'Public access to private land: allemansrett in Sweden', *Landscape and Urban Planning*, 15: 253–64.

293

Cole, A. (1977) 'Perception and use of urban parks', in Mercer, D. (ed.) *Leisure and Recreation in Australia*, 89–100, Melbourne: Sorrett.

Coleman, D. (1993) 'Leisure based social support, leisure dispositions and health', *Journal of Leisure Research*, 25, 4: 350–61.

Coleman, D. and Iso-Ahola, S.E. (1993) 'Leisure and health: the role of social support and self determination', *Journal of Leisure Research*, 25, 2: 111–28.

Conway, H. (1991) *People's Parks: the Design and Development of Victorian Parks in Britain*, Cambridge: Cambridge University Press.

Coppock, T., Duffield, B. and Sewell, D. (1974) 'Classification and analysis of recreation resources', in Lavery, P. (ed.) *Recreational Geography*, 231–58, London: David and Charles.

Countryside Commission (1970) *Countryside Recreation Glossary*, London: Countryside Commission.

— (1979) *Leisure and the Countryside*, Cheltenham: Countryside Commission.

— (1988) *Landscape Assessment of Farmland*, Cheltenham: Countryside Commission.

— (1989) *Forests for the Community*, Cheltenham: Countryside Commission.

— (1991) *Countryside Commission News*, July/August, Cheltenham: Countryside Commission.

— (1992) *City Links with National Parks*, Cheltenham: Countryside Commission.

— (1995) *London's Green Corridors*, Cheltenham: Countryside Commission.

— (1998) *Countryside Research*, January, Cheltenham: Countryside Commission.

Countryside Recreation Network (CRN) (1994) '1993 UK day visits survey: Summary', *Countryside Recreation Network News*, 2,1, 7–12.

Cox, P. (1985) 'The architecture and non architecture of tourism developments', in Dean, J. and Judd, B. (eds) *Tourist Developments in Australia*, 46–51, Canberra: Royal Australian Institute of Architects Education Division.

Craig-Smith, S. and Fagence, M. (eds). (1995) *Recreation and Tourism as a Catalyst for Urban Waterfront Redevelopment*, Westport: Praeger.

Crawford, D.W. and Godbey, G. (1987) 'Reconceptualizing barriers to family leisure', *Leisure Sciences*, 9: 119–27.

Crawford, D.W., Jackson, E. and Godbey, G. (1991) 'A hierarchical model of leisure constraints', *Leisure Sciences*, 13: 309–20.

CRN *see* Countryside Recreation Network.

Cullington, J.M. (1981) 'The public use of private land for recreation', unpublished Master of Arts thesis, Department of Geography, University of Waterloo, Waterloo, Ontario.

Cunningham, C. and Jones, M. (1987) 'Play needs of pre-adolescent children', paper presented to National Seminar of the Child Accident Prevention Foundation of Australia, Melbourne.

— (1994) 'The child-friendly neighbourhood', *International Play Journal*, 2: 79–95.

Curry, N. (1996) 'Access: policy directions for the late 1990s', in Watkins, C. (ed.) *Rights of Way: Policy Culture and Management*, 24–34, London: Pinter.

Cushman, G. and Hamilton-Smith, E. (1980) 'Equity issues in urban recreation services', in Mercer, D. and Hamilton-Smith, E. (eds) *Recreation Planning and Social Change in Urban Australia*, 167–79, Melbourne: Sorrett.

Cushman, G. and Laidler, A. (1990) *Recreation, Leisure and Social Policy*, Occasional Paper No. 4, Canterbury, New Zealand: Department of Recreation and Tourism, Lincoln University.

Cushman, G., Veal, A.J. and Zuzanek, J. (eds) (1996a) 'Cross-national leisure participation research: a future', in Cushman, G., Veal, A.J. and Zuzanek, J. (eds) *World Leisure*

Participation: Free Time in the Global Village, 237–58, Wallingford: CAB International.

— (1996b) (eds) *World Leisure Participation: Free Time in the Global Village*, Wallingford: CAB International.

Dales, J. (1972) 'The property interface', in Dorfman, R. and Dorfman, N. (eds) *Economies of the Environment*, 308–22, New York: Norton.

Dann, G. (1976) 'Anomie, ego-enhancement and tourism', *Annals of Tourism Research*, 4, 4: 184–94.

Darling, F.F. and Eichhorn, N.D. (1967) 'The ecological implications of tourism in national parks', in *Ecological Impact of Recreation and Tourism Upon Temperate Environments*, IUCN Proceedings and Papers, New Series 7: 98–101, Morges, Switzerland.

Dasmann, R.F. (1985) 'Achieving the sustainable use of species and ecosystems', *Landscape Planning*, 12: 211–9.

Davidson, J. and Wibberley, G. (1977) *Planning and the Rural Environment*, Oxford: Pergamon.

Davidson, P. (1996) 'The holiday and work experiences of women with young children', *Leisure Studies*, 15: 89–103.

de Kadt, E. (1979) *Tourism. Passport to Development*, New York: Oxford University Press.

Deem, R. (1982) 'Women, leisure and inequality', *Leisure Studies*, 1, 1: 29–46.

DeGrazia, S. (1962) *Of Time, Work and Leisure*, New York: Twentieth Century Fund.

Delin, C.R. and Patrickson, M. (1994) 'An investigation into aspects of leisure among busy people', *Australian Journal of Leisure and Recreation*, 4: 5–12.

Department of Environment, Housing and Community Development (1977) *Leisure Planning Guide for Local Government*, Canberra: Australian Government Publishing Service.

Desbarats, J. (1983) 'Spatial choice and constraints on behavior', *Annals of the Association of American Geographers*, 73, 3: 340–57.

Dibb, J.A. (1980) 'Coastal recreation for the disabled', *Australian Parks and Recreation*, May, 24–33.

Dillon, P. (1993) 'Irish in a stew over battle of the Burren', *National Parks Today*, 36: 5.

Ditwiler, C. (1979) 'Can technology decrease natural resource use conflicts in recreation', *Search*, 10, 12: 439–41.

Don, A. (1997) 'National parks, nature conservation and heritage in Ireland', *Australian Parks and Recreation*, 33, 2: 26–9.

Dower, J. (1965) *The Fourth Wave: The Challenge of Leisure: A Civic Trust Survey*, London: Civic Trust.

Doxey, G.V. (1974) 'A causation theory of visitor–resident irritants: methodology and research inferences', in *Proceedings of the Travel Research Association 6th Annual Conference*, 195–8, San Diego: Travel Research Association.

Dredge, S. and Moore, S. (1992) 'A methodology for the integration of tourism in town planning', *Journal of Tourism Studies*, 3, 1: 8–22.

Driver, B. (1970a) 'Some thoughts on planning, the planning process and related decision processes', in Driver, B. (ed.) *Elements of Outdoor Recreation Planning*, 195–212, Ann Arbor: The University of Michigan Press.

— (ed.) (1970b) *Elements of Outdoor Recreation Planning*, Ann Arbor: The University of Michigan Press.

Driver, B. and Tocher, S.R. (1974) 'Toward a behavioural interpretation of recreational engagements, with implications for planning', in Driver, B. (ed.) *Elements of Outdoor*

Recreation Planning, 9–31, Ann Arbor, The University of Michigan Press.

Driver, B. and Brown, P. (1978) 'A social-physiological definition of recreation demand, with implications for recreation resource planning', Appendix A of *Assessing Demand for Outdoor Recreation*, Washington DC: US Bureau of Outdoor Recreation.

Driver, B., Brown, P. and Peterson, G.L. (1991) *Benefits of Leisure*, State College, PA: Venture.

Driver, B., Brown, P., Stankey, G. and Gregoire, T. (1987) 'The ROS planning system: evolution, basic concepts and research needed', *Leisure Sciences*, 9: 201–12.

Duffield, B. and Owen, M. (1970) *Leisure and Countryside: A Geographical Appraisal of Countryside Recreation in Lanarkshire*, Edinburgh: University of Edinburgh.

Dutton, I. and Hall, C.M. (1989) 'Making tourism sustainable: the policy/practice conundrum', *Proceedings of the Environment Institute of Australia Second National Conference*, Melbourne, 9–11 October.

Dwyer, J.F. and Stewart, S.I. (1995) 'Restoring urban recreation opportunities: an overview with illustrations', in *Proceedings of the Fourth International Outdoor Recreation and Trends Symposium and the 1995 National Recreation Resource Planning Conference*, 606–9, St Paul, Minnesota: University of Minnesota, 606–9.

Eagles, P. F. J. (1996) 'Issues in tourism management in parks: the experience in Australia', *Australian Leisure*, June: 29–37.

Eckbo, G. (1967) 'The landscape of tourism', *Landscape*, 18, 2: 29–31.

Edington, J.M. and Edington, M.A. (1986) *Ecology, Recreation and Tourism*, Cambridge: University of Cambridge.

Egyptian Environmental Affairs Agency (undated) *Ras Mohammed National Park Sector*, Cairo: Department of Natural Protecturates.

Eliot-Hurst, M. (1972) *A Geography of Economic Behavior*, North Scituate, MA: Duxbury.

Elson, J.B. (1978) 'Recreation demand forecasting: a misleading tradition', in *Planning for Leisure*, Seminar Proceedings, 7–8 July, Warwick: University of Warwick.

Environment Canada (1978) *Canada Land Inventory Report No. 14, Land Capability for Recreation: Summary Report*, Ottawa: Environment Canada.

— (1988) *Making a Difference: The Canada Land Inventory*, Fact Sheet 88–5, Ottawa: Environment Canada.

Fagan, R.H. and Webber, M. (1994) *Global Restructuring: The Australian Experience*, Melbourne: Oxford University Press.

Farina, J. (1980) 'Perceptions of time', in Goodale, T. and Witt, P. (eds) *Recreation and Leisure*, 19–29, State College of Pennsylvannia: Venture Publishing.

Fedler, A.J. (1987) 'Introduction: are leisure, recreation and tourism interrelated?', *Annals of Tourism Research*, 14, 3: 311–3.

Ferrario, F. (1988) 'Emerging leisure market among the South African black population', *Tourism Management*, March: 23–38.

Ferris, A.L. (1962) *National Recreation Survey*, Study Report Number 19, Washington DC: Outdoor Recreation Resources Review Commission,

Field, D. (1997) 'Parks and neighbouring communities: a symbiotic relationship', in Pigram, J. and Sundell, R. (eds) *National Parks and Protected Areas: Selection, Delimitation and Management*, 419–28, Armidale: Centre for Water Policy Research.

Firestone, J. and Shelton, B.A. (1994) 'A comparison of women's and men's leisure time: subtle effects of the double day', *Leisure Sciences*, 16: 45–60.

Fitzsimmons, A. (1976) 'National parks: the dilemma of development', *Science*, 191: 440–4.

FNNPE (1993) *Loving Them to Death? Sustainable Tourism in Europe's Nature and National Parks*, Grafenau: Federation of National and Nature Parks of Europe.

Forster, J. (1973) *Planning for Man and Nature in National Parks*, Morges: IUCN.

Forsyth, P. and Dwyer, L. (1996) 'Tourism in the Asia-Pacific region', *Asia-Pacific Literature*, 10, 1: 13–22.

Foster, J. (1979) 'A park system and scenic conservation in Scotland', *Parks*, 4, 2: 1–4.

Fraser, R. and Spencer, G. (1998) 'The value of an ocean view', *Australian Geographical Studies*, 36, 1: 94–8.

Freeman, P. (1992) *The Sydney Morning Herald*, 20 March: 10.

Galbraith, J.K. (1972) *The New Industrial State*, 2nd edition, Harmondsworth: Penguin.

Garling, T. and Golledge, R. (eds) (1993) Behavior and Environment, Amsterdam: North Holland.

Gartner, W. (1996) *Tourism Development: Principles, Processes and Policies*, New York: Van Nostrand Reinhold.

Geering, D. (1989) *Managing the Full Range of River Resources*, Sydney: New South Wales Department of Water Resources (unpublished).

German Federal Agency for Nature Conservation (ed.) (1997) *Biodiversity and Tourism: Conflicts on the World's Seacoasts and Strategies for Their Solution*, Berlin: Springer-Verlag.

Getz, D. (1987) 'Tourism planning and research: traditions, models and futures', paper presented to the Australian Travel Research Workshop, Bunbury, Western Australia, November.

Gittins, J. (1973) 'Conservation and capacity: a case study of Snowdonia National Park', *The Geographical Journal*, 139, 3: 482–6.

Glyptis, S. (1981) 'People at play in the countryside', *Geography*, 66, 4: 277–85.

—— (1991) *Countryside Recreation*, Harlow: Longman.

—— (1992) 'The changing demand for countryside recreation', in: Bowler, I.R., Bryant, C.R. and Nellis, M.D. (eds) *Contemporary Rural Systems in Transition: Economy and Society*. Wallingford: CAB International.

—— (ed.) (1993) *Leisure and the Environment*, London: Belhaven Press.

Godbey, G. (1981) *Leisure in Your Life*, Philadelphia: Saunders.

—— (1985) 'Non-use of public leisure services: a model', *Journal of Recreation and Park Administration*, 3: 1–12.

Godbey, G. and Parker, S. (1976) *Leisure Studies and Services*, Philadelphia: W.B. Saunders.

Godfrey, K.B. (1996) 'Towards sustainability: tourism in the Republic of Cyprus', in Harrison, L.C. and Husbands, W. (eds) *Practicing Responsible Tourism: International Case Studies in Tourism Planning, Policy, and Development*, New York: Wiley & Sons.

Godin, V. and Leonard, R. (1977) 'Design capacity for backcountry recreation management planning', *Journal of Soil and Water Conservation*, 32, 161–4.

Gold, S. (1972) 'Non-use of neighbourhood parks', *Journal of the American Institute of Planners*, XXXVIII, 6: 369–78.

—— (1973) *Urban Recreation Planning*, Philadelphia: Lea and Febiger.

—— (1974) 'Deviant behaviour in urban parks', *Journal of Health, Physical Education, Recreation*, Nov–Dec: 18–20.

—— (1980) *Recreation Planning and Design*, New York: McGraw-Hill.

—— (1988) 'Urban open space preservation: the American experience', *Australian Parks and Recreation*, 25, 1: 21–28.

Goldsmith, F. and Manton, R. (1974) 'The ecological effects of recreation', in Lavery, P. (ed.) *Recreational Geography*, 259–69, London: David and Charles.

Goodall, B. and Whittow, J. (1975) *Recreation Requirements and Forest Opportunities*, Geographical Papers No. 37, Reading: Department of Geography, University of Reading.

Gordon, W.R. and Caltabiano, M.L. (1996) 'Youth leisure experiences in rural and urban North Queensland', *Australian Leisure*, 7, 2: 37–41.

Graefe, A.R. (1990) 'Visitor impact management', in Graham, R. and Lawrence, R. (eds) *Towards Serving Visitors and Managing Our Resources*, Proceedings of a North American Workshop on Visitor Management in Parks and Protected Areas, 213–34, Waterloo: Tourism Research and Education Centre, University of Waterloo.

— (1991) 'Visitor impact management: an integrated approach to assessing the impacts of tourism in national parks and protected areas', in Veal, A.J., Jonson, P. and Cushman, G. (eds) *Leisure and Tourism: Social and Environmental Change,* Papers from the World Leisure and Recreation Association Congress, Sydney, Australia, 74–83, Sydney: University of Technology, Sydney.

Graefe, A.R., Vaske, J.J. and Kuss, F.R. (1984) 'Social carrying capacity: an integration and synthesis of twenty years of research', *Leisure Sciences*, 6, 4: 395–431.

Graham, R. (1990) 'Visitor management in Canada's national parks', in Graham, R. and Lawrence, R. (eds) *Towards Serving Visitors and Managing Our Resources*, Proceedings of a North American Workshop on Visitor Management in Parks and Protected Areas, 271–96, Waterloo: Tourism Research and Education Centre, University of Waterloo.

Graham, R. and Lawrence, R. (eds) (1990) *Towards Serving Visitors and Managing Our Resources*, Proceedings of a North American Workshop on Visitor Management in Parks and Protected Areas, Waterloo: Tourism Research and Education Centre, University of Waterloo.

Graham, R., Payne, R.J. and Nilsen, P. (1988) 'Visitor activity planning and management in Canadian national parks', *Tourism Management*, 9, 1: 44–62. per G 155, AITS

Grandage, J. and Rodd, R. (1981) 'The rationing of recreational land use', in Mercer, D. (ed.) *Outdoor Recreation: Australian Perspectives*, 76–91, Melbourne: Sorrett.

Gray, D. and Pelegrino, D. (1973) *Reflections on the Recreation and Park Movement*, Dubuque: Brown.

Gray, H.P. (1970) *International Travel – International Trade*, Lexington: Heath Lexington.

Graziers' Association of New South Wales (1975) *Submission to the Select Committee of the Legislative Assembly upon the Fishing Industry*, Sydney: Graziers' Centre.

Green, B. (1977) 'Countryside planning: compromise or conflict?', *The Planner*, 63: 67–9.

Green, P. (1992) 'Parks management in New Zealand', *UNEP Industry and Environment*, 15, 3–4: 16–21.

Groome, D. (1993) *Planning and Rural Recreation in Britain*, Aldershot: Avebury.

Gunn, C. (1972) *Vacationscape, Designing Tourist Regions*, Bureau of Business Research, Austin: University of Texas.

— (1979) *Tourism Planning*, New York: Crane Russack.

— (1988) *Vacationscape: Designing Tourist Regions*, 2nd edition, New York: Van Nostrand Reinhold.

— (1994) *Tourism Planning*, 3rd edition, Washington: Taylor and Francis.

Gunn, S.L. and Peterson, C.A. (1978) *Therapeutic Recreation Program Design*, Englewood Cliffs: Prentice Hall.

Gutteridge, Haskins and Davey (1988) *Draft Plan of Management: Wallis Island Crown Reserve*, Pyrmont: Industrial Publishing.

Hall, C.M. (1992) *Wasteland to World Heritage: Preserving Australia's Wilderness*, Melbourne: Melbourne University Press.

— (1994) *Tourism in the Pacific Rim: Development, Impacts and Markets*, Melbourne: Longman.

— (1995) *Introduction to Tourism in Australia: Impacts, Planning and Development*, Melbourne: Longman.

— (1997) *Tourism in the Pacific Rim*, 2nd edition, Melbourne: Longman.

Hall, C.M. and Jenkins, J.M. (1995) *Tourism and Public Policy*, London: Routledge.

Hall, C.M., Jenkins, J.M. and Kearsley, G. (eds) (1997) *Tourism Planning and Policy in Australia and New Zealand: Cases, Issues and Practice,* Sydney: Irwin.

Hall, C.M., Jenkins, J.M. and Kearsley, G. (eds) (1997) 'Tourism planning and policy in urban areas: introductory comments', in Hall *et al.* (eds) *Tourism,Policy and Planning in Australia and New Zealand: Cases, Issues and Practice*, Sydney: Irwin.

Hall, C.M. and McArthur, S. (eds) (1993) *Heritage Management in New Zealand and Australia*, Oxford: Oxford University Press.

— (eds) (1996) *Heritage Management in Australia and New Zealand: The Human Dimension*, Melbourne: Oxford University Press.

Hall, C.M. and Page, S.J. (eds) (1996) *Tourism in the Pacific: Issues and Cases*, London: International Thomson Business Press.

Hamilton-Smith, E. (1975) 'Issues in the measurement of community need', *Australian Journal of Social Issues*, 10, 1.

Hamilton-Smith, E. and Mercer, D.C. (1991) *Urban Parks and their Visitors*, Victoria: Melbourne and Metropolitan Board of Public Works.

Hammitt, W.E. (1990) 'Wildland recreation and resource impacts: A pleasure-policy dilemma', in Hutcheson, J.D., Noe, F.P. and Snow, R.E. (eds) *Outdoor Recreation Policy: Pleasure and Preservation*, 17–30, New York: Greenwood Press.

Hammitt, W. and Cole, D. (1987) *Wildland Recreation Ecology and Management*, New York: Wiley.

— (1991) *Wildland Recreation Ecology and Management*, 2nd edition, New York: Wiley.

Harper, G. and Weiler, B. (eds) (1992) *Ecotourism*, Canberra: Bureau of Tourism Research.

Harrington, M., Dawson, D. and Bolla, P. (1992) 'Objective and subjective constraints on women's enjoyment of leisure', *Loisir et Société, Society and Leisure*, 15, 1: 203–21.

Harrington, R. (1975) 'Liability exposure in the operation of recreation facilities', *Outdoor Recreation Action*, 35: 22–5.

Harrison, A. (1977) 'Getting your story across – interpreting the river resource', in *Proceedings, River Recreation Management Research Symposium*, USDA Forest Service General Technical Report NC-28, Minneapolis, 125–38.

Harrison, J. (1992) 'Protected area management guidelines', *Parks*, 3, 2: 22–5.

Harrison, L.C. and Husbands, W. (eds) (1996) *Practicing Responsible Tourism: International Case Studies in Tourism Planning, Policy, and Development*, New York: Wiley and Sons.

Hart, W. (1966) *A Systems Approach to Park Planning*, Morges: International Union for the Conservation of Nature.

Haulot, A. (1978) 'Cultural protection policy in the field of tourism', *Parks*, 3, 3: 6–8.

Havighurst, R.J. (1961) 'The nature and values of meaningful free-time activity', in Kleemeier, R.W. (ed.) *Aging and Leisure*, New York: Oxford.

Hawkins, D. (1997) 'The virtual tourism environment', in Cooper, C. and Wanhill, S. (eds) *Tourism Development*, 43–58, Chichester: Wiley.

OUTDOOR RECREATION MANAGEMENT

Haynes, P. (1973) 'Towards a concept of monitoring', *Town Planning Review*, 45, 1.

Hayslip, B. and Panek, P.E. (1989) *Adult Development and Aging*, New York: Harper & Row.

HaySmith, L. and Hunt, J.D. (1995) 'Nature tourism: impacts and management', in Knight, R.L. and Gutzwiller, K.J. (eds) *Wildlife and Recreationists: Coexistence through Management and Research*, 203–19, Washington DC: Island Press.

Haywood, K.M. (1989) 'Responsible and responsive approaches to tourism planning in the community', *Tourism Management*, 9, 2: 105–18.

Heit, M. and Malpass, D. (undated) *Do Women Have Equal Play?*, Ottawa: Ministry of Culture and Recreation.

Helman, P., Jones, A., Pigram, J.J. and Smith, J. (1976) *Wilderness in Australia*, Armidale: Department of Geography, University of New England.

Hendee, J., Stankey, G. and Lucas, R. (1978) *Wilderness Management*, USDA Forest Service Miscellaneous Publication 1365, Washington.

Henderson, K.A. (1991) 'The contribution of feminism to an understanding of leisure constraints', *Journal of Leisure Research*, 23: 363–77

— (1994a) 'Broadening an understanding of women, gender and leisure', *Journal of Leisure Research*, 23: 1–7.

— (ed.) (1994b) 'Special issue on women, gender and leisure', *Journal of Leisure Research*, 26, 1.

Henderson, K.A., Bedini, L.A., Hecht, L. and Shuler, R. (1993) 'The negotiation of leisure constraints by women with disabilities', in Fox, K. (ed.) *Proceedings of the 7th Canadian Congress on Leisure Research*, 235–41, Winnipeg, Manitoba: Faculty of Physical Education and Recreation, University of Manitoba.

— (1995) 'Women with physical disabilities and the negotiation of leisure constraints', *Leisure Studies*, 14: 17–31.

Henderson, K.A., Bialeschki, M.D., Shaw, S.M. and Freysinger, V.J. (1989) *A Leisure of One's Own*, State College, PA: Venture Publishing Inc.

Hendry, L.B. (1983) *Growing Up and Growing Out: Adolescents and Leisure*, Aberdeen: Aberdeen University Press.

Henry, I. (1993) *The Politics of Leisure Policy*, Basingstoke: Macmillan.

Hersch, G. (1991) 'Leisure and aging, physical and occupational therapy', *Geriatrics*, 9, 2: 55–72.

Heywood, J. 1989 'Recreation opportunity: the social setting', *Australian Parks and Recreation*, 25, 2: 18–20.

Hogg, D. (1977) 'The evaluation of recreational resources', in Mercer, D. (ed.), *Leisure and Recreation in Australia*, Melbourne: Sorrett.

Hollenhorst, S. (1995) 'Risk, technology driven other new activity trends', in *Proceedings of the Fourth International Outdoor Recreation Trends Symposium*, 65–7, Minneapolis: University of Minnesota.

Hollick, M. (1993) *An Introduction to Project Evaluation*, Melbourne: Longman.

Hollings, C. (1978) *Adaptive Environmental Assessment and Management*, Chichester: Wiley.

Hovinen, G. (1981) 'A tourist cycle in Lancaster County, Pennsylvannia', *Canadian Geographer*, 25, 3: 283–6.

— (1982) 'Visitor cycles outlook for tourism in Lancaster County, Pennsylvannia', *Annals of Tourism Research*, 9: 563–83.

Howard, A. (1997) 'Conservation reserve boundaries and management implications in New South Wales', in Pigram, J. and Sundell, R. (eds) *National Parks and Protected Areas: Selection, Delimitation and Management*, 387–401, Armidale: Centre for Water Policy Research, University of New England.

Howe, C.Z. (1980) 'Models for evaluating public recreation programs: What the literature shows', *Journal of Physical Education and Recreation*, 51, 8: 36–8.

—— (1993) 'The evaluation of leisure programs: applying qualitative methods', *Journal of Physical Education, Recreation, and Dance*, 64, 8: 43–7.

Hultsman, W.Z. (1992) 'Constraints to activity participation in early adolescence', *Journal of Early Adolescence*, 12: 280–99.

—— (1993) 'The influence of others as a barrier to recreation participation among early adolescents', *Journal of Leisure Research*, 25: 150–64.

Hultsman, W.Z. and Kaufman, J.E. (1990) 'The experience of leisure by youth in a therapeutic milieu', *Youth and Society*, 21: 496–510.

Hutcheson, J.D., Noe, F.P. and Snow, R.E. (eds) (1990) *Outdoor Recreation Policy: Pleasure and Preservation*, New York: Greenwood Press.

Hutchison, R. (1994) 'Women and the elderly in Chicago's public parks', *Leisure Sciences*, 16: 229–47.

Ibrahim, I. (1991) *Leisure and Society*, Dubuque: Brown.

Ibrahim, H. and Cordes, K. (1993) *Outdoor Recreation*, Dubuque: Brown and Benchmark.

Ilbery, R. (ed.) (1997) *The Geography of Rural Change*, London: Longman.

Inskeep, E. (1991) *Tourism Planning: An Integrated and Sustainable Development Approach*, New York: Van Nostrand Reinhold.

Iso-Ahola, S.E. (1980) *The Social Psychology of Leisure and Recreation*, Dubuque: William C. Brown.

—— (1988) 'The social psychology of leisure: past, present and future research', in Barnett (ed.) *Research About Leisure. Past, Present and Future*, Champaign, Illinois: Sagamore.

Iso-Ahola, S.E. and Crowley, E.D. (1991) 'Adolescent substance abuse and leisure boredom', *Journal of Leisure Research*, 23, 3: 260–71.

Iso-Ahola, S.E. and Weissinger, E. (1990) 'Perceptions of boredom in leisure: conceptualization, reliability and validity of the Leisure Boredom Scale', *Journal of Leisure Research*, 22, 1: 1–17.

Ittelson, W., Franck, K. and O'Hanlon, T. (1976) 'The nature of environmental experience', in Wapner, S., Cohen, S. and B. Kaplan (eds), *Experiencing the Environment*, New York: Plenum Press.

IUCN (1994) *Guidelines for Protected Area Management Categories*, Switzerland: Gland.

Jackson, E.L. (1988) 'Leisure constraints: a survey of past research', *Leisure Sciences*, 10: 203–15.

—— (1990) 'Variations in the desire to begin a leisure activity: evidence of antecedent constraints?', *Journal of Leisure Research*, 22: 150–64.

—— (1991) 'Leisure constraints/constrained leisure: special issue introduction', *Journal of Leisure Research*, 22: 55–70.

—— (1994) 'Geographical aspects of constraints on leisure and recreation', *The Canadian Geographer*, 38: 110–21.

Jackson, E.L. and Burton, T.L. (eds) (1989) *Understanding Leisure and Recreation: Mapping the Past, Charting the Future*, State College, PA: Venture Publishing Inc.

Jackson, E.L., Crawford, D.W. and Godbey, G. (1993) 'Negotiation of leisure constraints', *Leisure Sciences*, 15,1: 1–11.

301

Jackson, E.L. and Henderson, K.A. (1995) 'Gender-based analysis of leisure constraints', *Leisure Sciences*, 17, 1: 31–51.

Jacob, G. and Schreyer, R. (1980) 'Conflict in outdoor recreation: a theoretical perspective', *Journal of Leisure Research*, 12, 4: 368–80.

Jafari, J. (1977) 'Editor's page', *Annals of Tourism Research*, 5: 6–11.

Jamrozik, A. (1986) 'Leisure as social consumption: some equity considerations for social policy', in Castle, R., Lewis, D. and Mangan, J. (eds) *Work, Leisure and Technology*, 184–209, Melbourne: Longman.

Janiskee, R. (1976) 'On the recreation appeals of extra-urban environments', Mimeograph, 72nd Annual Meeting of Association of American Geographers, New York.

Jansen-Verbeke, M. (1992) 'Urban recreation and tourism: physical planning issues', *Tourism Recreation Research*, 17, 2: 33–45.

Jansen-Verbeke, M. and Van de Wiel, E. (1995) 'Tourism planning in urban revitalization projects: lessons from the Amsterdam waterfront development', in Ashworth, G.J. and Dietvorst, A. (eds) *Tourism and Spatial Transformations: Implications for Policy and Planning*, 129–45, Wallingford: CAB International.

Jefferies, B.E. (1982) 'Sagamartha National Park: the impact of tourism in the Himalayas', *Ambio*, 11, 5: 274–82.

Jenkins, J.M. (1998) *Crown Lands Policy-Making in New South Wales, 1856–1991: The Life and Death of an Organisation, its Culture and a Project*, Canberra: Centre for Public Sector Management.

Jenkins, J.M. and Prin, E. (1998) 'Rural landholder attitudes: the case of public recreational access to private rural lands', in Butler, R.W., Hall, C.M. and Jenkins, J.M. (eds) *Tourism and Recreation in Rural Areas*, Chichester: John Wiley and Sons.

Jubenville, A. (1976) *Outdoor Recreation Planning*, Philadelphia: Saunders.

— (1978) *Outdoor Recreation Management*, Philadelphia: Saunders.

Just, D. (1987) 'Appropriate amounts and design of open spaces', *Australian Parks and Recreation*, 25, 2: 32–9.

Kaiser, C. and Helber, L. (1978) *Tourism Planning and Development*, Boston: CBI Publishing Company.

Kando, T.M. (1975) *Leisure and Popular Culture*, Saint Louis: The C.V. Mosby Company.

Kane, P. (1981) 'Assessing landscape attractiveness', *Applied Geography*, 1: 77–96.

Kaplan, M. (1975) *Leisure: Theory and Practice*, New York: John Wiley.

Kaplan, R. and Kaplan, S. (1989) *The Experience of Nature. A Psychological Perspective*, Cambridge: Cambridge University Press.

Kariel, H.G. (1989) 'Tourism and development: perplexity or panacea?', *Journal of Travel Research*, 28, 1: 2–6.

Kates, Peat, Marwick and Co. (1970) *Tourism and Recreation in Ontario*, Ontario: Minister of Tourism and Information.

Kearsley, G.K. (1997) 'Tourism planning and policy in New Zealand', in Hall, C.M., Jenkins, J.M. and Kearsley, G. (eds) (1997) *Tourism Planning and Policy in Australia and New Zealand: Cases, Issues and Practice*, 49–60, Sydney: Irwin.

Kelleher, G.G. and Kenchington, R.A. (1982) 'Australia's Great Barrier Reef Marine Park: making development compatible with conservation', *Ambio*, 11, 5: 262–7.

Kelly, J.R. (1974) 'Socialization toward leisure: a developmental approach', *Journal of Leisure Research*, 6: 181–93.

— (1990) 'Leisure and aging: A second agenda', *Society and Leisure*, 13, 1: 145–67.

Kelly, J.R., Steinkamp, M.W. and Kelly, J.R. (1987) 'Later life satisfaction: does leisure contribute?', *Leisure Sciences*, 8: 189–200.

Kimble, G. (1951) 'The inadequacy of the regional concept', in Stamp, L. and Wooldridge, S. (eds) *London Essays in Geography*, London.

King, B.E.M. (1997) *Creating Island Resorts*, London: Routledge.

Kirkby, S. (1996) 'The World Wide Web as a provider of multimedia information to the ecotourist', in *Proceedings of National Seminar on the Role of Technology in Parks and Recreation*, Adelaide, April.

Knetsch, J. (1972) 'Interpreting demands for outdoor recreation', *Economic Record*, 48, September: 429–32.

— (1974) *Outdoor Recreation and Water Resources Planning*, Water Resources Monograph No. 3, Washington DC: American Geophysical Union.

Knight, R.L. and Gutzwiller, K.J. (eds) (1995) *Wildlife and Recreationists: Coexistence Through Management and Research*, Washington DC: Island Press.

Knopf, R. (1990) 'The limits of acceptable change (LAC) planning process: potentials and limitations', in Graham, R. and Lawrence, R. (eds) *Towards Serving Visitors and Managing Our Resources*, 201–1, Waterloo: University of Waterloo Press.

Knudson, D.M., Cable, T.T. and Beck, L. (1995) *Interpretation of Cultural and Natural Resources*, State College, PA: Venture.

Komarovsky, M. (1967) *Blue-collar Marriage*, New York: Random House.

König, U. (1998) *Tourism in a Warmer World: Implications of Climate Change due to Enhanced Greenhouse Effect for the Ski Industry in the Australian Alps*, Vol. 28, Zürich: Universität Zürich-Irchel Geographisches Institut Winterthurerstrasse.

Kozlowski, J., Rosier, J. and Hill, G. (1988) 'Ultimate Environmental Threshold (UET) method in a marine environment (Great Barrier Reef Marine Park in Australia)', *Landscape and Urban Planning*, 15: 327–36.

Kraus, R. (1984) *Recreation and Leisure in Modern Society*, 3rd Edition, Glenview, IL: Scott Foresman and Company.

Kreutzwiser, R. (1989) 'Supply', in Wall, G. (ed.) *Outdoor Recreation in Canada*, 21–41, Toronto: Wiley.

Krippendorf, J. (1987) *The Holiday Makers: Understanding the Impact of Leisure and Travel*, Oxford: Heinemann Professional Publishing.

Krumpe, E. (1988) 'The role of information in people's leisure decision making process', in Killin, N., Paradice, W. and Engel, M. (eds) *Information in Planning and Management of Recreation and Tourist Services*, Newcastle: Hunter Valley Research Foundation.

Kruss, F., Graefe, A. and Vaske, J. (1990) *Visitor Impact Management* Vol. I, II, Washington: National Parks and Conservation Association.

Lacey, P. (1996) 'Current recreation trends', *Australian Leisure*, 7, 4: 15–8.

Lambley, D. (1988) 'The economic significance of different types of leisure, recreation and tourism in national parks: the Myall Lakes National Park, unpublished paper, Sydney: National Parks and Tourism Seminar.

Landals, A.G. (1986) 'The tourists are ruining the parks', in *Tourism and the Environment: Conflict or Harmony? Proceedings of a Symposium sponsored by the Canadian Society of Environmental Biologists*, 89–99, Alberta Chapter, Calgary, Canada, 18–19 March 1986, Alberta: CSEB.

Lane, B. (1994) 'What is rural tourism?', *Journal of Sustainable Tourism*, 2, 1–2: 7–21.

LaPage, W. (1967) *Some Observations on Campground Trampling and Ground Cover Response*, Washington: USDA Forest Service Research Paper NE-68.

— (1970) 'The mythology of outdoor recreation planning', *Southern Lumberman*, Dec.: 118–21.

Lavery, P. (1974) *Recreational Geography*, London: David and Charles.

Law, C.M. (1992) 'Urban tourism and its contribution to economic regeneration', *Urban Studies*, 29, 3/4: 599–618.

— (1993) *Urban Tourism: Attracting Visitors to Large Cities*, London: Mansell.

Lawrence, G. (1987) *Capitalism and the Countryside*, Sydney: Pluto Press.

Lea, J. (1988) *Tourism and Development in the Third World*, London: Routledge.

Leatherbury, E. (1979) 'River amenity evaluation', *Water Resources Bulletin*, 15, 5: 1281–92.

Lee-Gosselin, M. and Pas, E.I. (1997) 'The implications of emerging contexts for travel-behaviour research', in Stopher, P. and Lee-Gosselin, M. (eds) *Understanding Travel Behaviour in an Era of Change*, 1–28, Oxford: Elsevier Science.

Leiper, N. (1979) 'The framework of tourism', *Annals of Tourism Research*, 6, 4: 390–407.

— (1981) 'Towards a cohesive curriculum in tourism', *Annals of Tourism Research*, 8, 2: 69–84.

— (1984) 'Tourism and leisure: the significance of tourism in the leisure spectrum', in *Proceedings 12th New Zealand Geography Conference*, Christchurch: New Zealand Geography Society.

— (1995) *Tourism Management*, Collingwood: TAFE publications.

Leopold, A., Cain, S., Cottam, C., Gabrielson, I. and Kimball, T. (1963) *Report of the Advisory Committee to the National Park Service on Research*, Washington DC: National Academy of Sciences.

Levy, J. (1977) 'A paradigm for conceptualising leisure behaviour', *Journal of Leisure Research*, 11, 1: 48–60.

Lew, A. A. (1989) 'Authenticity and sense of place in the tourism development experience of older retail districts', *Journal of Travel Research*, 27, 4: 15–22.

Liddle, M. (1997) *Recreation Ecology*, London: Chapman & Hall.

Lime, D. (1974) 'Locating and designing campgrounds to provide a full range of camping opportunities', in *Outdoor Recreation Research: Applying the Results*, General Technical Report NC-9, 56–66, St Paul: USDA Forest Service.

Lime, D. and Stankey, G. (1971) 'Carrying capacity: Maintaining outdoor recreation quality', in *Proceedings, Forest Recreation Symposium*, 174–84, Syracuse: New York College of Forestry.

Lindberg, K. (1991) *Policies for Maximizing Nature Tourism's Ecological and Economic Benefits*, Washington, DC: World Resources Institute.

Lindberg, K. and Hawkins, D.E. (eds) (1993) *Ecotourism: A Guide for Planners and Managers*, Vermont: The Ecotourism Society.

Lindsay, J. (1980) 'Trends in outdoor recreation activity conflicts', in *Proceedings 1980 National Outdoor Recreation Trends Symposium*, Vol. 1, 215–21, Broomall: USDA Forest Service.

Linz, W. and Linz, M. (1996) 'Letter to the editor', *National Parks Magazine*, 70, 1–2: 10.

Lipscombe, N. (1986) 'Supply and demand in outdoor recreation: which should concern us most?', *Australian Parks and Recreation*, 23, 1: 16–8.

— (1993) 'Recreation planning: where have all the frameworks gone?', in McIntyre, N. (ed.) *Proceedings, Track to the Future: Managing Change in Parks and Recreation*, Queensland: Royal Australian Institute of Parks and Recreation.

Lobo, F. (1995) 'Recreation for all: an inappropriate concept for the unemployed', *Australian Parks and Recreation*, Winter: 1–26.

Lockwood, R. and Lockwood, A. (1993) *Participation in Sport by People with Disabilities: A National Perspective*, A Report to the National Sports Commission.

London, M., Crandall, R. and Fitzgibbons, D. (1977) 'The psychological structure of leisure: activities, needs, people', *Journal of Leisure Research*, 9, 4: 252–63.

Lopata, H.Z. (1972) 'The life cycle of the social role of the housewife', in Bryant, C.D. (ed.) *The Social Dimensions of Work*, 128–44, Englewood Cliffs: Prentice-Hall.

Losier, G.F., Bourque, P.E. and Vallerand, R.J. (1993) 'A motivational model of leisure participation in the elderly', *The Journal of Psychology*, 127: 153–70.

Lucas, B. (1992) 'The Caracas Declaration', *Parks*, 3, 2: 7–8.

Lucas, R. (1964) *The Recreation Capacity of the Quetico – Superior Area*, St Paul: USDA Forest Research Paper, LS-15.

Lundberg, D.E., Stavenga, M.H. and Krishnamoorthy, M. (1995) *Tourism Economics*, New York: John Wiley and Sons.

Lunn, A. (1986) 'Military matters', *Tarn and Tor*, 9: 6.

Lynch, R. and Veal, A.J. (1996) *Australian Leisure*, Melbourne: Longman.

Mak, J. and White, K. (1992) 'Comparative tourism development in Asia and the Pacific', *Journal of Travel Research*, 31, 1: 14–23.

Mannell, R.C. and Zuzanek, J. (1991) 'The nature and variability of leisure constraints in daily life: the case of physically active leisure of older adults', *Leisure Sciences*, 13: 337–51.

Manning, R. (1979) 'Strategies for managing recreational use of national parks', *Parks*, 4, 1: 13–5.

Manning, R., McCool, S. and Graefe, A. (1995) 'Trends in carrying capacity', in Thompson, J., Lime, D., Gartner, B. and James, W. (compilers), *Proceedings of the Fourth International Outdoor Recreation and Tourism Trends Symposium*, 334–41, St Paul: University of Minnesota.

Markwell, K. (1996) 'Towards a gay and lesbian leisure research agenda', *Australian Leisure*, 7, 2: 42–4 & 48.

Martin, L., Bennett, R. and Gregory, D. (1985) 'The thirsty Algarve', *Geographical Magazine*, 57: 321–4.

Martin, W. and Mason, S. (1976) 'Leisure 1980 and beyond', *Long Range*, 9, 2: 58–65.

Mather, A.S. (1986) *Land Use*, Harlow: Longman Scientific and Technical.

Mathieson, A. and Wall, G. (1982) *Tourism: Economic, Physical and Social Impacts*, London: Longman.

Mattyasovsky, E. (1967) 'Recreation area planning: some physical and ecological requirements', *Plan*, 8, 3: 91–109.

Mayo, E.J. and Jarvis, L.P. (eds) (1981) *The Psychology of Leisure and Travel Behaviour*, Massachusetts: CBI.

McCool, S. (1990a) 'Limits of acceptable change: evolution and future', in Graham, R. and Lawrence, R. (eds) *Towards Serving Visitors and Managing Our Resources*, 185–93, Waterloo: University of Waterloo Press.

— (1990b) 'Limits of acceptable change: some principles', in Graham, R. and Lawrence, R. (eds) *Towards Serving Visitors and Managing Our Resources*, 195–9, Waterloo: University of Waterloo Press.

305

McCosh, R. (1973) 'Recreation site selection', in Gray, D. and Pelegrino, D. (eds) *Reflections on the Recreation and Park Movement*, 290–5, Dubuque: Brown.

McDonald, G. and Wilks, L. (1986a) 'Economic and financial benefits of tourism in major protected areas', *Australian Journal of Environmental Management*, 2, 2: 19–39.

— (1986b) 'The regional impact of tourism and recreation in national parks', *Environment and Planning B: Planning and Design*, 13: 349–66.

McIntyre, N. (1990) 'Recreation involvement: the personal meaning of participation', unpublished Ph.D. Thesis, University of New England, Armidale.

— (1993) 'Recreation planning for sustainable use', *Australian Journal of Leisure and Recreation*, 3, 2: 31–7 & 49.

McIntyre, N. and Boag, A. (1995) 'The measurement of crowding in nature-based tourism venues: Uluru National Park', *Tourism Recreation Research*, 20, 1:37–42;

McIntyre, N., Cuskelly, G. and Auld, C. (1991) 'The benefits of urban parks', *Australian Parks and Recreation*, 27, 4: 11–8.

McKercher, B. (1998) *The Business of Nature-Based Tourism*, Melbourne: Hospitality Press.

McLoughlin, L. (1997) 'Sydney and the bush... Sydney or the bush', *Australian Planner*, 34, 3: 165–70.

McMeeking, D. and Purkayastha, B. (1995) 'I can't have my mom running me everywhere: adolescents, leisure and accessibility', *Journal of Leisure Research*, 27, 4: 360–78.

McNeely, J.A. and Thorsell, J.W. (1989) 'Jungles, mountains and islands: how tourism can help conserve the natural heritage', *World Leisure and Recreation*, 31, 4: 29–39.

MacPherson, B.D. (1991) 'Aging and leisure benefits: a life cycle perspective', in Driver, B.L., Brown, P.J. and Peterson, G.L. (eds), *Benefits of Leisure*, Pennsylvannia: Venture Publishing.

Medlik, S. (1993) *Dictionary of Travel, Tourism and Hospitality*, Oxford: Heinemann.

Mercer, D.C. (1970) 'The geography of leisure – contemporary growth point', *Geography*, 55: 261–73.

— (1972) *Planning for Coastal Recreation*, Monograph 1, Melbourne: Universities Recreation Research Group.

— (1975) 'The concept of recreational need', *Journal of Leisure Research*, 5, 1: 37–51.

— (1977) 'The factors affecting recreational demand', in Mercer, D. (ed.) *Leisure and Recreation in Australia*, 59–68, Melbourne, Sorrett.

— (1980a) *In Pursuit of Leisure*, Melbourne: Sorrett.

— (1980b) 'Themes in Australian urban leisure research', in Mercer, D. and Hamilton-Smith, E. (eds) *Recreation Planning and Social Change in Urban Australia*, 1–25, Melbourne: Sorrett.

— (1981a) 'Trends in recreational participation', in Mercer, D. (ed.) *Outdoor Recreation: Australian Perspectives*, 24–44, Melbourne: Sorrett.

— (1981b) *Outdoor Recreation: Australian Perspectives*, Melbourne: Sorrett.

— (1994a) 'Monitoring the spectator society: An overview of research and policy issues', in Mercer, D.C. (ed.) *New Viewpoints in Australian Outdoor Recreation Research and Planning*, 1–28, Melbourne: Hepper Marriott and Associates.

— (ed.) (1994b) *New Viewpoints in Australian Outdoor Recreation Research and Planning*, Melbourne: Hepper Marriott and Associates.

— (1995) *A Question of Balance: Natural Resources Conflict Issues in Australia*, 2nd Edition, Annandale: The Federation Press.

Mercer, D. and Hamilton-Smith, E. (eds) (1980), *Recreation Planning and Social Change in Urban Australia*, Melbourne: Sorrett.

Meyer-Arendt, K.J. (1990) 'Recreational Business Districts in Gulf of Mexico seaside resorts', *Journal of Cultural Geography*, 11: 39–55.

Middleton, V.Y.C. (1982) 'Tourism in rural areas', *Tourism Management*, 3: 52–8.

Mieckzowski, Z. (1990) *World Trends in Tourism and Recreation*, New York: Peter Lang.

Mill, R.C. and Morrison, A.M. (1985) *The Tourism System: An Introductory Text*, Englewood Cliffs: Prentice Hall.

Ministry for Planning and Environment. (1989) *Planning Guide for Urban Open Space*, Melbourne: Ministry for Planning and Environment.

Mintzberg, H. (1994) *The Rise and Fall of Strategic Planning*, New York: Prentice Hall.

Mitchell, L. (1968) 'An evaluation of central place theory in a recreation context', *Southeastern Geographer*, 8, 46–53.

— (1969) 'Toward a theory of public urban recreation', in *Proceedings of Association of American Geographers*, 1: 103–8.

— (1983) 'Future directions of recreation planning', in Lieber, S. and Fesenmaier, D. (eds) *Recreation Planning and Management*, 323–38, State College: Venture Publishing.

Mobily, K.E. and Bedford, R.L. (1993) 'Language, play and work among elderly persons', *Leisure Studies*, 12: 203–19.

Mobily, K.E., Leslie, D.K., Lemkie, J.H., Wallace, R.B. and Kohout, F.J. (1986) 'Leisure patterns and attitudes of the rural elderly', *Journal of Applied Gerontology*, 5: 201–14.

Moir, J. (1995) 'Regional parks in Perth, Western Australia', *Australian Planner*, 32, 2: 88–95.

Morris, S. (1975) 'Owner rights and co-operation', in *Landscape Conservation*, Papers of Australian Conservation Foundation Conference, Canberra.

Morrison, J. (1994) 'Men at leisure: the implications of masculinity and leisure for "househusbands" ', unpublished MA thesis, School of Leisure and Tourism Studies, University of Technology, Sydney.

Moseley, M. (1979) *Accessibility: The Rural Challenge*, London: Methuen.

Mowforth, M. and Munt, I. (1998) *Tourism and Sustainability: New Tourism in the Third World*, London: Routledge.

Mueller, E., Gurin G. and Wood, M. (1962) *Participation in Outdoor Recreation: Factors Affecting Demand Among American Adults*, Outdoor Recreation Resources Review Commission, Study Report Number 20, Washington DC: US Government Printing Service.

Mullins, P. (1991) 'Tourism urbanization', *International Journal of Urban and Regional Research*, 3, September: 326–42.

Murdock, S.H., Backman, K., Nazrul Hoque, M. and Ellis, D. (1991) 'The implications of change in population size and composition on future participation in outdoor recreational activities', *Journal of Leisure Research*, 23, 3: 238–59.

Murphy, P.E. (1985) *Tourism: A Community Approach*, New York: Methuen.

Murphy, P.E. (ed.) (1997) *Quality Management in Urban Tourism*, Chichester: Wiley.

Myers, N. (1972) 'National parks in savannah Africa', *Science*, 178: 1255–63.

— (1973) 'Impending crisis for Tanzanian wildlife', *National Parks and Conservation Magazine*.

National Capital Development Commission (1978) *Planning Concept Papers*, Canberra: Australian Government Publishing Services.

National Parks Review Panel (1991) *Fit for the Future: Report of the National Parks Review Panel* (CCP 335), Cheltenham: Countryside Commission.

Nelson, J. and Butler, R.W. (1974) 'Recreation and the environment', in Manners, I. and Mikesell, M. (eds) *Perspectives on Environment*, 290–310, Washington DC: Association of American Geographers.

Netherlands Ministry for Cultural Affairs (1976) *National Parks and Landscape Parks in Netherlands*, The Hague: Netherlands Government Publishing House.

Neulinger, J. (1982) 'Leisure lack and the quality of life', *Leisure Studies*, 1, 1: 53–64.

New South Wales Department of Lands (1986) *Land Assessment and Planning Process for Crown Lands in NSW*, Sydney: New South Wales Department of Lands.

New South Wales Department of Planning (1989) *Tourism Development Near Natural Areas*, Sydney: New South Wales Department of Planning.

New South Wales National Parks and Wildlife Service (1997a) *Draft Nature Tourism and Recreation Strategy*, Sydney: New South Wales National Parks and Wildlife Service.

— (1997b) *Draft National Parks Public Access Strategy*, Sydney: New South Wales National Parks and Wildlife Service.

New Zealand National Parks Authority (1978) *Guidelines for Interpretative Planning*, Wellington: National Parks Authority.

Newcomb, R. (1979) *Planning the Past*, Dawson: Folkestone.

Nolte, C. (1995) 'Few favor parking lot in Yosemite Valley', *San Francisco Chronicle*, 22 June: A21.

NSW Department of Lands – see New South Wales Department of Lands.

NSW Department of Planning – see New South Wales Department of Planning.

NSW National Parks and Wildlife Service – see New South Wales National Parks and Wildlife Service.

OECD (1990) *Partnerships for Rural Development*, Paris: OECD.

— (1993) *What Future for Our Countryside? A Rural Development Policy*, Paris: OECD.

Ohmann, L. (1974) 'Ecological carrying capacity', in *Outdoor Recreation Research: Applying the Results*, St Paul: USDA Forest Service General Technical Report, NC – 9, 24–8.

Okotai, T. (1980) 'Research requirements of tourism in the Cook Islands', in Pearce, D. (ed.) *Tourism in the South Pacific*, Proceedings of UNESCO Tourism Workshop, 169–76: Rarotonga.

Olokesusi, F. (1990) 'Assessment of the Yankari Game Reserve, Nigeria: problems and prospects', *Tourism Management*, 11, 2: 153–63.

Opperman, M. and Chon, K.S. (1997) *Tourism in Developing Countries*, London: International Thomson Business Press.

Orcutt, J.J. (1984) 'Contrasting effects of two kinds of boredom on alcohol use', *Journal of Drug Issues*, 14: 161–73.

O'Riordan, T. (1971) *Perspectives on Resource Management*, London: Pion.

Ovington, J., Groves, K., Stevens, P. and Tanton, M. (1972) *A Study of the Impact of Tourism at Ayers Rock – Mt Olga National Park*, Canberra: Department of Forestry.

Owen, J.M. (1993) *Program Evaluation: Forms and Approaches*, St. Leonards: Allen and Unwin.

Owen, P. L. (1984) 'Rural leisure and recreation research: a retrospective evaluation', *Progress in Human Geography*, 8, 2: 157–87.

Page, S. (1994) *Transport for Tourism*, London: Routledge.

— (1995) *Urban Tourism*, London: Routledge.

Page, S. and Getz, D. (1997) *The Business of Rural Tourism: International Perspectives*, London: International Thomson Business Press

Parker, S. (1971) *The Future of Work and Leisure*, London: Granada Publishing.

— (1983) *Leisure and Work*, London: Allen & Unwin.

Parker, S. and Paddick, R. (1990) *Leisure in Australia*, Melbourne: Longman.

Patmore, A.J. (1973) *Land and Leisure*, Harmondsworth: Penguin.

— (1983) *Recreation and Resources: Leisure Patterns and Leisure Places*, Oxford: Blackwell.

Patterson, I. and Pegg, S. (1995) 'Leisure, community integration and people with disabilities', *Australian Leisure*, 5, 1: 32–38.

Payne, R. and Nilsen, P. (1997) 'The role of social science research in establishing national parks: a Canadian case study', in Pigram, J. and Sundell, R. (eds) *National Parks and Protected Areas: Selection, Delimitation and Management*, 403–417, Armidale: Centre for Water Policy Research, University of New England.

Pearce, D. (1978) 'A case study of Queenstown', in *Tourism and the Environment*, Information Series No. 6, Wellington: Department of Lands and Survey.

— (1979) 'Towards a geography of tourism', *Annals of Tourism Research*, 6, 3: 245–72.

— (ed.) (1980) Tourism in the South Pacific, Proceedings of UNESCO Tourism Workshop, Rarotonga, Christchurch: University of Canterbury.

— (1987) *Tourism Today: A Geographical Analysis*, Harlow: Longman Scientific and Technical.

— (1989) *Tourist Development*, 2nd edition. Harlow: Longman Scientific and Technical.

Pearce, P.L. (1982) *The Social Psychology of Tourist Behaviour*, Sydney: Pergamon Press.

Pearce, P.L., Morrison, A.M. and Rutledge, J.L. (1998) *Tourism: Bridges Across Continents*, Sydney: McGraw-Hill.

Pearce, P.L., Moscardo, G. and Ross, G.F. (1996) *Tourism Community Relationships*, Oxford: Pergamon.

Pearson, K. (1977), 'Leisure in Australia', in Mercer, D. (ed.) *Leisure and Recreation in Australia*, 25–34, Melbourne: Sorrett.

Penning-Rowsell, E. (1975) 'Constraints on the application of landscape evaluation', *Transactions, Institute of British Geographers*, 66: 49–55.

Perdue, R.R., Long, P.T. and Allen, L. (1987) 'Rural resident tourism perceptions and attitudes', *Annals of Tourism Research*, 14: 420–9.

Perez de Cuellar, J. (1987) 'Statement', *World Leisure and Recreation*, 29, 1: 3.

Peterson, G., Stynes, D., Rosenthal, D. and Dwyer, J. (1985) 'Substitution in recreation choice behaviour', in *Proceedings – Symposium on Recreation Choice Behaviour*, 19–30, General Technical Report, INT-184, St Paul: USDA Forest Service.

Phillips, A. and Roberts, M. (1973) 'The recreation and amenity value of the countryside', *Journal of Agricultural Economics*, 24, 1: 85–102.

Phillips, K. (1977) 'Forests and Recreation', unpublished guest lecture, Armidale: Department of Geography, University of New England.

Pieper, J. (1952) *Leisure the Basis of Culture*, London: Faber.

Pigram, J.J. (1977) 'Beach resort morphology', *Habitat International*, 2, 5–6: 525–41.

— (1980) 'Environmental implications of tourism development', *Annals of Tourism Research*, 7, 4: 554–83.

— (1981) 'Outdoor recreation and access to the countryside: focus on the Australian experience', *Natural Resources Journal*, 21, 1: 107–23.

— (1983) *Outdoor Recreation and Resource Management*, London: Croom Helm.

— (1990) 'Sustainable tourism: policy considerations', *Journal of Tourism Studies*, 1, 2: 2–9.

— (1992) 'Alternative tourism: tourism and sustainable resource management', in Smith, V. and Eadington, W. (eds) *Tourism Alternatives*, 76–87, Philadelphia: Pennsylvannia Press.

— (1993) 'Planning for tourism in rural areas: bridging the policy implementation gap', in Pearce, D.G. and Butler, R.W. (eds) *Tourism Research: Critiques and Challenges*, 156–74, London: Routledge.

— (1995) 'Resource constraints on tourism: water resources and sustainability', in Pearce, D.G. and Butler, R.W. (eds), *Change in Tourism: People, Places, Processes*, 208–28, London: Routledge.

Pigram, J.J. and Hobbs, J. (1975) 'The weather, outdoor recreation, and tourism', *Journal of Physical Education and Recreation*, 46, 9, 12–3.

Pigram, J.J. and Jenkins J.M. (1994) 'The role of the public sector in the supply of rural recreation opportunities', in Mercer, D. (ed.) *New Viewpoints in Outdoor Recreation Research in Australia*, 119–28, Williamstown: Hepper Marriott and Associates.

Pigram, J.J. and Wahab, S. (1997) 'Sustainable tourism in a changing world', in Wahab, S. and Pigram, J.J. (eds) *Tourism, Development and Growth*, 17–32, London: Routledge.

Pigram, J.J., Nguyen, S. and Rugendyke, B. (1997) 'Tourism and national parks in emerging regions of the developing world: Cat Ba Island National Park, Vietnam', Paper presented to Meeting of the International Academy for the Study of Tourism, Malacca, June.

Plog, S. (1972) 'Why destination areas rise and fall in popularity', unpublished paper presented to Southern California Chapter, Travel Research Association, Los Angeles.

Poole, M. (1986) 'Adolescent leisure activities: social class, sex and ethnic differences', *Australian Journal of Social Issues*, 21, 1: 42–56.

Poon, A. (1993) *Tourism Technology and Competitive Strategies*, Wallingford: CAB International.

Prentice, R. (1993) 'Motivations of the heritage consumer in the leisure market: an application of the Manning-Haas demand hierarchy', *Leisure Sciences*, 15: 273–90.

Pressey, R., Bedward, M. and Nicholls, A. (1990) 'Reserve selection in mallee lands', in Noble, J., Joss, P. and Jones, G. (eds) *Proceedings of the National Mallee Conference*, 167–78, Melbourne: CSIRO.

Price, R. (1980) 'Tourism and outdoor recreation – some geographical definitions', paper presented to Annual Meeting of Travel Research Association, Savannah.

— (1981) 'Tourist landscapes and tourist regions' paper presented to 77th Annual Meeting of Association of American Geographers, Los Angeles.

Priddle, G. (1975) 'Identifying scenic rural roads', paper presented to 77th Annual Meeting, Association of American Geographers, Los Angeles.

Prior, T. and Clark, R. (1984) 'Technique for identifying scenic/recreational roads in the Hunter Valley, NSW', *Australian Geographer*, 16, 1: 50–4.

Przeclawski, K. (1986) *Humanistic Foundations of Tourism*, Warsaw: Institute of Tourism.

Pugh, D.A. (1990) 'Decision frameworks and interpretation', in Graham, R. and Lawrence, R. (eds) *Towards Serving Visitors and Managing Our Resources*, Proceedings of a North American Workshop on Visitor Management in Parks and Protected Areas, 355–6, Waterloo: Tourism Research and Education Centre, University of Waterloo.

Pullen, J. (1977) *Greenpeace and the Cities*, Canberra: Australian Institute of Urban Studies.

Rapoport, R. and Rapoport, R.N. (1975) *Leisure and the Family Life Cycle*, London: Routledge and Kegan Paul.

Ravenscroft, N. (1992) *Recreation Planning and Development*, Basingstoke, UK: Macmillan.

— (1996) 'New access initiatives: the extension of recreation opportunities or the diminution of citizen rights?', in Watkins, C. (ed.) *Rights of Way: Policy Culture and Management*, London: Pinter.

Reid, D. (1989) 'Implementing senior government policy at the local level', *Journal of Applied Recreation Research*, 15, 1: 3–13.

Relph, E. (1976) *Place and Placelessness*, London: Pion.

Renard, Y. and Hudson, L. (1992) 'Community-based management of national parks', paper presented to Fourth World Parks Congress, Caracas, February.

Reynolds, R.J. (1990) 'VAMP and its application to camping: the Glacier National Park example', in Graham, R. and Lawrence, R. (eds) *Towards Serving Visitors and Managing Our Resources*, Proceedings of a North American Workshop on Visitor Management in Parks and Protected Areas, 257–70, Waterloo: Tourism Research and Education Centre, University of Waterloo.

Richardson, B. and Richardson, R. (1994) *Business Planning: An Approach to Strategic Management*, 2nd Edition, London: Pitman.

Richins, H., Richardson, J. and Crabtree, A. (eds) (1996) *Ecotourism and Nature-Based Tourism: Taking the Next Steps*, The Ecotourism Association of Australia National Conference Proceedings, Alice Springs, Northern Territory, 18–23 November.

Richter, L.K. (1989) *The Politics of Tourism in Asia*, Honolulu: University of Hawaii Press.

Roberts, K. (1983) *Youth and Leisure*, London: Allen and Unwin.

Robinson, D., Laurie, J., Wagar, J. and Traill, A. (1976) *Landscape Evaluation*, Manchester: University of Manchester.

Robinson, G. (1990) *Conflict and Change in the Countryside*, London: Belhaven.

Roehl, W. (1987) 'An investigation of the perfect information assumption in recreation destination choice models', paper presented to the Annual Conference of the Association of American Geographers, Portland, April.

Ross, G.F. (1994) *The Psychology of Tourism*, Melbourne: Hospitality Press.

Rossman, J.R. (1995) *Recreation Programming: Designing Leisure Experiences*, 2nd edition, Champaign, IL: Sagamore.

Rothman, R. (1978) 'Residents and transients: community reaction to seasonal visitors', *Journal of Travel Research*, 16, 3: 8–13.

Rutledge, A. (1971) *Anatomy of a Park*, New York: McGraw-Hill.

Ryan, C. (1995) *Researching Tourist Satisfaction: Issues, Concepts and Problems*, London: Routledge.

Ryan, C. and Montgomery, D. (1994) 'The attitudes of Bakewell residents to tourism and issues in community responsive tourism', *Tourism Management*, 15, 5: 358–70.

Sax, J. (1980) *Mountains without Handrails: Reflections on the National Parks*, Ann Arbor: University of Michigan Press.

Scharff, R. (1972) *Canada's Mountain National Parks*, Banff: Lebow Books.

Schneider, D.M. and Smith, R.T. (1973) *Class Differences and Sex Roles in American Kinship and Family Structures*, Englewood Cliffs: Prentice Hall.

Schomburgk, C. (1985) 'Urban parkland – how much, where and what types?', *Australian Parks and Recreation*, 22, 1: 21–7.

Schreyer, R., Knopf, R. and Williams, D. (1985) 'Reconceptualizing the motive/environment link in recreation choice behavior', *Proceedings – Symposium on Recreation Choice Behavior*, Ogden: USDA General Technical Report, INT. 184, 9–18.

Seabrooke, W. and Miles, C. (1993) *Recreational Land Management*, 2nd edition, London: Spon.

Searle, M.S. and Jackson, E.L. (1985) 'Socioeconomic variations in perceived barriers to recreation among would-be participants', *Leisure Sciences*, 7: 227–49.

Seedsman, T.A. (1995) 'More to life! The value of leisure, recreation and education for the aged', *Australian Parks and Recreation*, 31, 4: 31–6.

Shackley, M. (1996) *Wildlife Tourism*, London: International Thomson Business Press.

— (ed.) (1998) *Visitor Management: Case Studies from World Heritage Sites*, Oxford: Butterworth-Heinemann.

Sharpley, R. and Sharpley, J. (1997) *Rural Tourism: An Introduction*, London: International Thomson Business Press.

Shaw, S.M. (1985) 'Gender and leisure: inequality in the distribution of leisure time', *Journal of Leisure Research*, 17: 266–82.

— (1992) 'Dereifying family leisure: an examination of women's and men's everyday experiences and perceptions of family time', *Leisure Sciences*, 14: 271–86.

Shaw, S.M., Bonen, A. and McCabe, J.F. (1991) 'Do more constraints mean less leisure? Examining the relationship between constraints and participation', *Journal of Leisure Research*, 23, 4: 286–300.

Shelby, B. and Heberlein, T.A. (1986) *Carrying Capacity in Recreation Settings*, Corvallis, Oregon: Oregon State University Press.

Shivers, J. (1967) *Principles and Practices of Recreational Service*, New York: Macmillan.

Shoard, M. (1996) 'Robbers v. revolutionaries: what the battle for access is really all about', in Watkins, C. (ed) *Rights of Way: Policy, Culture and Management*, 11–23, London: Pinter.

Simmons, B.A. and Dempsey, I. (1996) 'National profile of away from home leisure activities of persons aged 60 and over', *Australian Leisure*, 7, 1: 41–6.

Simmons, D. (1994) 'Community participation in tourism planning', *Tourism Management*, 15, 2: 98–108.

Simmons, D. and Leiper, N. (1993) 'Tourism: a social scientific perspective', in Perkins, H.C. and Cushman, G., *Leisure Recreation and Tourism*, Auckland: Longman Paul.

Simmons, I.G. (1975) *Rural Recreation in the Industrial World*, London: Edward Arnold.

Simmons, R., Davis, B.W., Chapman, R.J.K. and Sager, D.D. (1974) 'Policy flow analysis: a conceptual model for comparative policy research', *Western Political Quarterly*, 27, 3: 457–68.

Simpson, R. (1995) 'Channel vision of the future', *National Parks Today*, 39: 5.

Singleton, J.F. (1985) 'Activity patterns of the elderly', *Society and Leisure*, 8, 2: 805–19.

Slatter, R. (1978) 'Ecological effects of trampling on sand dune vegetation', *Journal of Biological Education*, 12, 2: 89–96.

Smale, B. (1990) 'Spatial equity in the provision of urban recreation opportunities', *Proceedings of Sixth Canadian Congress on Leisure Research*, Waterloo: University of Waterloo.

Smale, B.J.A. and Dupuis, S.L. (1993) 'The relationship between leisure activity participation and psychological well being across the lifespan', *Journal of Applied Recreation Research*, 18, 4: 281–300.

Smith, R.A. (1991) 'Beach resorts: a model of development evolution', *Landscape and Urban Planning*, 21: 189–210.

— (1992) 'Review of integrated beach resort development in Southeast Asia', *Land Use Policy*, 9: 209–17.

Smith, V. (ed.) (1978) *Hosts and Guests: The Anthropology of Tourism*, Oxford: Basil Blackwell.

— (1994) 'Privatisation in the Third World: small scale tourism enterprises', in Theobald, W.F. (ed.) *Global Tourism: The Next Decade*, 163–73, Oxford: Butterworth-Heinemann.

Snepenger, D.J. (1987) 'Segmenting the vacation market by novelty-seeking role', *Journal of Travel Research*, 26, 2: 8–14.

Snyder, E. and Spreitzner, E. (1973) 'Correlates of sport participation among adolescent girls', *Research Quarterly*, 47: 804–809.

Sorensen, A.D. and Epps, R. (eds) (1993) *Prospects and Policies for Rural Australia*, Melbourne: Longman.

Spinew, K., Tucker, D. and Arnold, M. (1996) 'Free time as a workplace incentive. A comparison between women and men', *Journal of Applied Recreation Research*, 21, 3: 195–212.

Sports Council (1991) *A Countryside for Sport: Towards a Policy for Sport and Recreation in the Countryside*, London: Sports Council.

Stafford, F.P. (1980) 'Women's use of leisure time converging with men's', *Monthly Labor Review*, 103: 57–9.

Stanfield, C. (1969) 'Recreation land use patterns within an American seaside resort', *The Tourist Review*, 24: 128–36.

Stanfield, C. and Rickert, J. (1970) 'The recreational business district', *Journal of Leisure Research*, 4: 213–25.

Stankey, G. (1977) 'Some social concepts for outdoor recreation planning', in *Proceedings of Symposium on Outdoor Advances in the Application of Economics*, Washington DC: US Department of Agriculture, Forest Service, General Technical Report WO-2.

— (1982) 'Carrying capacity, impact management and the recreation opportunity spectrum', *Australian Parks and Recreation*, May: 24–30.

Stankey, G., Cole, D., Lucas, R., Peterson, M. and Frissell, S. (1985) *The Limits of Acceptable Change (LAC) System for Wilderness Planning*, Ogden: USDA Forest Service.

Stankey, G., McCool, S. and Stokes, G. (1984) 'Limits of acceptable change: a new framework for managing the Bob Marshall Wilderness Complex', *Western Wildlands*, Fall.

Stephens, W.N. (1983) *Explanations for failures of youth organizations*, ERIC Document No. ED 228 440.

Stokowski, P. (1990) 'The social networks of spouses', in Smale, B. (ed.) *Proceedings of the Sixth Canadian Congress on Leisure Research*, Waterloo: University of Waterloo.

Stopher, P. and Lee-Gosselin, M. (eds) (1997) *Understanding Travel Behaviour in an Era of Change*, Oxford: Elsevier Science.

Stott, D. (1998) 'Korea crisis slashes tourist numbers', *The Sydney Morning Herald*, January 7: 2.

Sullivan, R. (1993) *Recreation. A Healthy Alternative to Crime*, Woodville: Royal Australian Institute of Parks and Recreation.

Sundell, R. (1991) 'The use of spatial modelling to evaluate park boundaries and delineate critical resource areas', unpublished Ph.D. thesis, Northwestern University, Evanston.

Swinglehurst, E. (1994) 'Face to face: the socio-cultural impacts of tourism', in Theobald, W.F. (ed.) *Global Tourism: The Next Decade*, 92–102, Oxford: Butterworth-Heinemann.

Swinnerton, G. (1982) *Recreation on Agricultural Land in Alberta*, Edmonton: Environment Council of Alberta.

313

Tangi, M. (1977) 'Tourism and the environment', *Ambio*, 6: 336–41.

Taylor, P.W. (1959) ' "Need" statements', *Analysis* 19: 106–11.

The Australian (1993) 10 November.

Theberge, J. (1989) 'Guidelines to drawing ecologically sound boundaries for national parks and nature reserves', *Environmental Management*, 13, 6: 695–702.

— (1992) 'Concepts of conservation biology and boundary delineation in parks', in *Proceedings of a Seminar on Size and Integrity Standards for Natural Heritage Areas in Ontario*, 16–24, Toronto: Ministry of Natural Resources.

Theobald, W.F. (1979) *Evaluation of Recreation and Park Programs*, New York: John Wiley and Sons.

Thompson, P. (1992) ' "I don't feel old": subjective ageing and the search for meaning in later life', *Ageing and Society*, 12: 23–47.

Thomson, J., Lime, D., Gartner, W. and Sames, W. (1995) *Proceedings Fourth International Outdoor Recreation and Tourism Trends Symposium*, St Paul: University of Minnesota.

Thomson, K. and Whitby, M. (1976) 'The economics of public access in the countryside', *Journal of Agricultural Economics*, 27, 3: 307–19.

Tilden, F. (1977) *Interpreting Our Heritage*, 3rd edition, Chapel Hill: North Carolina University Press.

Tinsley, H.E.A., Colbs, S.L., Teaff, J.D. and Kauffman, N. (1987) 'The relationship of age, gender, health and economic status to the psychological benefits older adults report from participation in leisure activities', *Leisure Sciences*, 9: 53–65.

Towner, J. (1996) *An Historical Geography of Recreation and Tourism in the Western World 1540–1940*, Chichester: John Wiley.

Toyne, P. (1974) *Recreation and Environment*, London: Macmillan.

Trapp, S., Gross, M. and Zimmerman, R. (1994) *Signs, Trails, and Wayside Exhibits: Connecting People and Places*, University of Wisconsin, Stevens Point: UW-SP Foundation Press Inc.

Tribe, J. (1995) *The Economics of Leisure and Tourism: Environments and Markets*, Oxford: Butterworth-Heinemann.

Turner, A. (1987) 'The management of impacts in recreational use of nature areas', paper presented to 22nd Conference of the Institute of Australian Geographers, Canberra, August.

— (1994) 'Managing impacts: measurement and judgement', in Mercer, D. (ed.) *New Viewpoints in Outdoor Recreation Research and Planning*, 129–40, Melbourne: Hepper Marriott and Associates.

Turner, L. (1976) 'The international division of leisure: tourism and the Third World', *Annals of Tourism Research*, 4, 1: 12–24.

UNESCO (1976) 'The effects of tourism on socio-cultural values', *Annals of Tourism Research*, 4, 2: 74–105.

United States Bureau of Outdoor Recreation (1973) *Outdoor Recreation. A Legacy for America*, Washington DC: Department of the Interior.

— (1975) *Assessing Demand for Outdoor Recreation*, Washington, DC: US Bureau of Outdoor Recreation.

United States Department of Agriculture (USDA) (1981) *National Agricultural Lands Study*, Washington DC: USDA.

United States Department of the Interior (1978) *National Urban Recreation Study*, Washington DC: United States Department of the Interior.

Unwin, K. (1975) 'The relationship of observer and landscape in landscape evaluation', *Transactions, Institute of British Geographers*, 66: 130–4.

USA Today, 11 April 1994: 7.

US Bureau of Outdoor Recreation – see United States Bureau of Outdoor Recreation.

US Department of Agriculture – see United States Department of Agriculture.

US Department of the Interior – see United States Department of the Interior.

Uysal, M. and Hagan, L.A.R. (1994) 'Motivation of pleasure travel and tourism', in Khan, M.A., Olsen, M.D. and Var, T. (eds) *VNR's Encyclopedia of Hospitality and Tourism*, New York: Van Nostrand Reinhold.

Uysal, M. Fesenmaier, D.R. and O'Leary, J.T. (1994) 'Geographic and seasonal variation in the concentration of travel in the United States', *Journal of Travel Research*, 32, 3: 61–4.

Valentine, P.S. (1991) 'Ecotourism and nature conservation: a definition with some recent development in Micronesia', in Weiler, B. (ed.) *Ecotourism: Incorporating the Global Classroom*, 4–10, Canberra: Bureau of Tourism Research.

— (1992) 'Review: Nature-based Tourism', in Weiler, B. and Hall, C.M. (eds) *Special Interest Tourism*, 105–28, London: Belhaven.

Van Lier, H.N. and Taylor, P.D. (eds) (1993) *New Challenges in Recreation and Tourism Planning*, Amsterdam: Elsevier Science Publishers B.V.

Vanhove, N. (1997) 'Mass tourism: benefits and costs', in Wahab, S. and Pigram, J.J. (eds), *Tourism, Development and Growth*, 50–77, London: Routledge.

Vaske, J.J., Decker, D.J. and Manfredo, M.J. (1995) 'Human dimensions of wildlife management: an integrated framework for coexistence', in Knight, R.L. and Gutzwiller, K.J. (eds) *Wildlife and Recreationists: Coexistence through Management and Research*, 33–47, Washington DC: Island Press.

Veal, A.J. (1987) *Leisure and the Future*, London: Allen and Unwin.

— (1994) *Leisure Policy and Planning*, Essex: Longman.

Veverka, J.A. (1994) *Interpretive Master Planning*, Montana: Falcon Press.

Viljoen, J. (1994) *Strategic Management: Planning and Implementing Successful Corporate Strategies*, Melbourne: Longman.

Vining, J. and Fishwick, L. (1991) 'An exploratory study of outdoor recreation site choices', *Journal of Leisure Research*, 23, 2: 114–32.

Wade, M.G. (ed.) (1985) *Constraints on Leisure*, Springield, IL: Charles C Thomas.

Wahab, S. and Pigram, J.J. (eds) (1997) *Tourism, Development and Growth*, London: Routledge.

Walker, B. and Nix, H. (1993) 'Managing Australia's biological diversity', *Search*, 24, 5: 173–78.

Wall, G. (1989) *Outdoor recreation in Canada*, New York: Wiley.

Wall, G. and Wright, C. (1977) *The Environmental Impact of Outdoor Recreation*, Department of Geography Publications Series No. 11, Waterloo: University of Waterloo.

Walmsley, D.J., Boskovic, R. and Pigram, J.J. (1981) *Tourism and Crime, Report to the Criminology Research Council*, Armidale: University of New England.

Walmsley, D.J. and Jenkins, J.M. (1994) 'Evaluations of recreation opportunities: tourist images of the New South Wales North Coast', in Mercer, D.C. (ed.) *New Viewpoints in Australian Outdoor Recreation Research and Planning*, 89–98, Melbourne: Hepper Marriott and Associates.

Walters, C. (1986) *Adaptive Management of Renewable Resources*, New York: Macmillan.

Washburne, R.F. (1982) 'Wilderness recreation carrying capacity: are numbers necessary?', *Journal of Forestry*, 80: 726–8.

Watkins, C. (ed) (1996) *Rights of Way: Policy, Culture and Management*. London: Pinter.

WCED – See World Commission on Environment and Development.

Weinmayer, M. (1973) 'Vandalism by design: a critique', in Gray, D. and Pelegrino, D. (eds) *Reflections on the Recreation and Park Movement*, 246–8, Dubuque: Brown.

Whelan, T. (1991) 'Ecotourism and its role in sustainable development', in Whelan, T. (ed.) *Nature Tourism*, Washington: Island Press.

WHO – See World Health Organisation.

Wight, P.A. (1993) 'Sustainable ecotourism: balancing economic, environmental and social goals within an ethical framework', *Journal of Tourism Studies*, 4, 2: 54–66.

Wilcox, A.T. (1969) 'Professional preparation for interpretive services', *Rocky Mountain-High Plains Parks and Recreation Journal*, 4, 1: 11–4.

Wilensky, H. (1961) 'The uneven distribution of leisure', *Social Problems*, 9, 1: 107–45.

Williams, S. (1995) *Outdoor Recreation and the Environment*, London: Routledge.

Williamson, P. (1995) 'Occupational therapy and "serious" leisure: promoting productive occupations through leisure', *Australian Journal of Leisure and Recreation*, 5, 1: 61–4.

Willits, W. and Willits, F. (1986) 'Adolescent participation in leisure activities: "the less the more" or "the more the more"?', *Leisure Sciences*, 8: 189–205.

Wilson, P. and Biberbach, P. (1994) 'Community forests – northern experience', *Countryside Campaigner*, Spring: 23.

Wilson, W. (1941) 'The study of administration', *Political Science Quarterly*, 55: 481–506.

Wingo, L. (1964) 'Recreation and urban development: a policy perspective', *Annals of the American Academy of Political Science*, 35: 129–40.

Wolfe, R. (1964) 'Perspectives on outdoor recreation', *Geographical Review*, 54: 203–38.

World Commission on Environment and Development (WCED) (1987) *Our Common Future*, New York: Oxford University Press.

World Health Organisation (WHO) (1980) *International Classification of Impairments, Disabilities and Handicaps*, Geneva: World Health Organisation.

World Tourism Organisation (WTO) (1977) *World Travel Statistics*, Madrid: WTO.

—— (1979) *World Travel Statistics*, Madrid: WTO.

—— (1991) *Resolutions of International Conference on Travel and Tourism (Recommendation No. 29)*, Ottawa, Canada.

—— (1992) *Guidelines: Protection of National Parks and Protected Areas for Tourism*, Madrid: WTO.

—— (1993a) *Recommendations on Tourism Statistics*, Madrid: WTO.

—— (1993b) *Global Tourism Forecasts to the Year 2000 and Beyond*, Madrid: WTO.

—— (1994) *Compendium of Tourism Statistics*, Madrid: WTO.

—— (1997a) *Tourism Highlights 1996*, Madrid: WTO.

—— (1997b) *WTO News*, March, Madrid: WTO.

World Travel and Tourism Council (WTTC) (1995) *Travel and Tourism: A New Economic Perspective*, London: Pergamon.

WTO – see World Tourism Organisation.

Wunderlich, G. (1979) 'Land ownership: a status of facts', *Natural Resources Journal*, 19, 1: 97–118.

Young, S. (1973) *Tourism. Blessing or Blight?*, London: Penguin.

Youniss, J. (1980) *Parents and Peers in Social Development*, Chicago: The University of Chicago Press.

Zimmerman, E. (1951) *World Resources and Industries*, New York: Harper.

INDEX

Havinghurst, R.J. 52
Hawaii 239
Hawkins, D. 269, 274
Haynes, P. 280, 282
Hayslip, B. 51–2
HaySmith, L. 128, 264
Haywood, K.M. 82
hazards 275; and fire 109; and risk 86; survey of 115
Heberlein, T.A. 120
Heit, M. 56
Helber, L. 240
Helman, P. 199
Hendee, J. 120
Henderson, K.A. 41, 49, 55
Hendry, L.B. 15, 56
Henry, I. 11, 45, 289
heritage 77, 145, 164, 170, 230, 251, 264
Heritage Conservation and Recreation Service 132
Hersch, G. 51–2
Heywood, J. 34
historic interest 235, preservation 235; sites 192
Hobbs, J. 255
Hochhanz National Park 191
Hogg, D. 79
Hollenhorst, S. 274
Hollick, M. 289
Hollings, C. 103
Honolulu 139
host communities 234, 253, 263; and culture 262; and guests 231
Hough, M. 136
Hovinen, G. 240
Howard, A. 204–5
Howe, C.Z. 289
Hudson, L. 223
Hultsman, W.Z. 53
human rights 41, 43–4; 232
Hunt, J.D. 128, 264
Hunter Valley 81
Husbands, W. 251, 288
Hutcheson, J.D. 289
Hutchison, R. 50
Hyde Park 139

Ibrahim, H. 62, 177
Ibrahim, I. 38
Ilbery, R. 177
immigration 13
impacts on wildlife 265
Indonesia 196

industrialisation 229
inequity 154
inner city 137–40; and decay 137
Inskeep, E. 266
institutional arrangements 46
integration 48; vertical 262; horizontal 262
international tourism 260
International Union for the Conservation of Nature (IUCN) 178, 181, 186, 190–1, 200, 222
Internet 9, 274
interpretation 19, 82, 219–22; methods and personal services 221; methods and self-directed services 221
Ireland 189, 234
Iso-Ahola, S.E. 18–9, 51, 53, 56, 160
Ittelson, W. 32–3

Jackson, E.L. 17, 40–2, 50, 56
Jacob, G. 36–7
Jafari, J. 227
Jamrozik, A. 3, 8–9
Janiskee, R. 142
Jansen-Verbeke, M. 134, 159, 246
Japan 228–9
Jarvis, L.P. 38
Jefferies, B.E. 225
Jenkins, J.M. 2, 163–4, 166, 170, 172–3, 177, 200, 289
Jerusalem 234
Jones, M. 146
Jotindar, J.S. 269
Jubenville, A. 113, 115–6, 118, 220
Just, D. 154
Kaiser, C. 240
Kakadu 192, 223
Kando, T.M. 3–4
Kane, P. 82
Kaplan, M. and S. 2, 5, 17, 31–2
Kariel, H.G. 245
Kates, Peat, Marwick and Co. 20
Kaufman, J.E. 53
Kearsley, G.W. 78
Kelleher, G.G. 225
Kelly, J.R. 51, 53
Kenchington, R.A. 225
Kenya 196
Khmer Rouge 232
Kilimanjaro 196
Kimble, G. 203
King, B. 239, 243
Kirkby, S. 274
Knetsch, J. 2, 20–1, 59–60, 63

Tribe, J. 251
trip generation 29
Tsavo 196
Tucker, D. 56
Tunbridge, J. 159, 251
Tunbridge Wells 234
Turner, A. 96, 99–100
Turner, L. 260
Tweed Heads 243

Ultimate Environmental Threshold (UET) 99, 110
Uluru 87, 192, 223
unemployment 54, 271
UNESCO 262
United Kingdom (UK) 12, 15, 17, 43, 47, 64, 161–2, 169–71, 175–6, 187, 191, 193, 198, 204, 224, 234–5, 240
United Nations Year of the Disabled 47
United States *see* US
Unwin, K. 76
Upper Yarra Valley 194
urban: areas 53; change 130; decay 135; development 134, 136–7; fringe 137, 142–3; heritage 159; open space 143–6; 156, 159; redevelopment 130, 135; regeneration 136; restoration projects 132; sprawl 9
urban recreation: opportunities 132, 148, 157; planning 152–9; space 146–52, 157
urban tourism 159
urbanisation 14, 130, 175
US 12, 15, 17, 43, 61, 64, 77, 117, 122, 130, 132–3, 145, 150, 161, 179–89, 184, 195, 202, 204, 221, 224, 228, 234–5, 239
US Bureau of Outdoor Recreation 19, 139
US Department of the Interior 133
US National Parks Service 183
US National Reserves 194
USA Today 179
Uysal, M. 269

Valentine, P.S. 265, 269
Vallerand, R.J. 51
values 45–6, 47, 287
Van de Wiel, E. 246
Van Lier, H.N. 1
Vancouver 134
vandalism 117–8, 168, 252
Vanhove, N. 250
Vaske, J.J. 110, 120–1, 128, 215

Vatican 234
Veal, A.J. 1–3, 14, 17, 40, 44–5, 47–8, 52–6, 113, 289
vegetation 86–7, 89, 95
Veverka, J.A. 220
Victoria 192, 194
Vietnam 196, 200, 232
Viljoen, J. 277–81, 288
Vining, J. 21
Visitor Activity Management Process (VAMP) 120, 122, 125, 127–8, 185, 288
Visitor Impact Management (VIM) framework 120–1, 125, 127–8, 288
visitor management 111–2, 118, 202, 221
visitor pressure 86–7
Voluntary Conservation Agreement 193

Wade, M.G. 56
Wahab, S. 230, 232,249
Waldbrook, L.A. 226, 245–7
Wales 161, 168, 186, 188, 240
Walker, B. 204
walking tracks and trails 60, 114
Wall, G. 82, 85, 87, 90, 110, 177, 225, 245, 248, 252, 257
Wallis Island Crown Reserve 103, 109
Walmsley, D.J. 2, 262
Walters, C. 103
wanderlust 233, 238
Washburne, R.F. 121
Washington DC 139, 182
water 115; acquisition of 19; quality 104, 145; for recreation and tourism 63, 172–6
water resources: management of 61; ownership of 61
water storage 61
water-based recreation 103, 172–6
waterbodies 120
waterfowl 69
waterfront development 159
Watkins, C. 171, 177
weather 255
Weber, F.H. 229
Weiler, B. 269
Weinmayer, M. 118
Weissinger, E. 56
Western Australia 193
Western Europe 130, 198, 229
Westminster 138
Westminster City Council 138
wetlands 86, 103